FDA and Intellectual Property Strategies for Medical Device Technologies

Gerald B. Halt, Jr. · John C. Donch, Jr. ·
Amber R. Stiles · Brandon R. Theiss ·
Michelle Byrne · William F. Hodnick ·
Joshua Eldridge · Randy Goodman ·
Lisa Jenkins VanLuvanee · Dana L. Blue

FDA and Intellectual Property Strategies for Medical Device Technologies

Including Artificial Intelligence, Software and Applications

Second Edition

Springer

Gerald B. Halt, Jr.
United Plaza
Volpe Koenig
Philadelphia, PA, USA

Amber R. Stiles
Erwin, TN, USA

Michelle Byrne
Facet Life Sciences, Inc.
Wayne, PA, USA

Joshua Eldridge
Facet Life Sciences, LLC
Lawrenceburg, KY, USA

Lisa Jenkins VanLuvanee
Facet Life Sciences, Inc.
Phoenixville, PA, USA

John C. Donch, Jr.
United Plaza
Volpe Koenig
Philadelphia, PA, USA

Brandon R. Theiss
United Plaza
Volpe Koenig
Philadelphia, PA, USA

William F. Hodnick
Facet Life Sciences, Inc.
Sparks, NV, USA

Randy Goodman
Facet Life Sciences, Inc.
Avon Lake, OH, USA

Dana L. Blue
Facet Life Sciences, Inc.
Brownsburg, IN, USA

ISBN 978-3-032-12648-1 ISBN 978-3-032-12649-8 (eBook)
https://doi.org/10.1007/978-3-032-12649-8

This Springer imprint is published by the registered company Springer Nature Switzerland AG
The registered company address is: Gewerbestrasse 11, 6330 Cham, Switzerland

Preface

When it comes to inventing, developing and commercializing medical devices, companies will invariably come to an intersection between business, regulatory compliance, intellectual property and the law. These areas are diverse, but an understanding of each is absolutely critical to the successful development of a new medical device product. With years of experience as consultants to entrepreneurs, doctors and medical device companies, the authors of this book have recognized the guidance that is needed and have sought to furnish a guide to issues that are important to medical device companies as they work to bring new medical device technologies to market. Specifically, this book identifies and explains FDA regulatory pathways that are available to medical companies, provides a primer on intellectual property rights, and explores implementation strategies for medical device innovators.

Philadelphia, USA	Gerald B. Halt, Jr.
Philadelphia, USA	John C. Donch, Jr.
Erwin, USA	Amber R. Stiles
Philadelphia, USA	Brandon R. Theiss
Wayne, USA	Michelle Byrne
Sparks, USA	William F. Hodnick
Lawrenceburg, USA	Joshua Eldridge
Avon Lake, USA	Randy Goodman
Phoenixville, USA	Lisa Jenkins VanLuvanee
Brownsburg, USA	Dana L. Blue

Acknowledgments The authors would like to thank everyone at the law firm of Volpe Koenig and Facet Life Sciences, LLC for providing support and encouragement in preparing this book. The authors would also like to individually thank Anthony S. Volpe for his contribution to the intellectual property enforcement and litigation chapter, Frederick Koenig III for his contribution to the copyright chapter, and Dr. Tom Hallam, for his review of the Chapters that the Facet team members authored.

Competing Interests The authors have no competing interests to declare that are relevant to the content of this manuscript.

About This Book

FDA and Intellectual Property Strategies for Medical Device Technologies: Including Artificial Intelligence, Software and Applications is intended to help guide readers through the U.S. Food and Drug Administration (FDA) regulatory compliance process that many medical devices must go through and is meant to shed light on some of the unique challenges medical device technology faces when it comes to intellectual property and regulatory strategy.

Part I of this book serves to provide the reader with a foundation of medical device development in the United States. It begins with a summary of the evolution of drugs and medical devices and how legislation evolved in the United States to protect the public. Examples of each type of medical device, how they are developed, and how the FDA regulates them are presented. Part I also provides recommendations for when and how to meet with the FDA, as well as tips and tricks for successful and productive interactions at each stage in development. It includes a description of the content of the various medical device regulatory applications and summarizes Sponsor requirements once a device is actively being marketed.

Part II of this book provides a comprehensive overview of the most common forms of intellectual property rights. Part II also provides guidelines for how medical device companies can properly secure and implement their intellectual property rights. In Part II of this book, readers will learn how intellectual property rights can apply to their business by presenting examples from real world companies, along with fictitious examples. Case studies are used throughout the book to demonstrate how intellectual property rights, management, and business all work together in industry. The examples throughout the book will relate to well-known companies and products that many readers will be familiar with.

Part III discusses how medical device companies can think about intellectual property strategy, while exploring ways to avoid potential pitfalls and mitigate potential risks. Part III also describes commonly used tools that medical device companies typically use to successfully implement their regulatory strategy and delves into potential complications while providing several strategies for avoiding or preventing regulatory compliance problems.

Examples and Illustrations Throughout This Book

Examples and illustrations will be used throughout this book. Some are real-world scenarios featuring well-known companies. Others are hypothetical scenarios. Each example is intended to enhance the reader's understanding of the subject matter contained in the chapters.

Contents

About the Authors

Gerald B. Halt, Jr., JD is a shareholder of Volpe Koenig, where he counsels a global roster of clients ranging from individual inventors to Fortune 500 companies. With more than 25 years of experience, he focuses his practice on securing, enforcing and licensing intellectual property rights in a wide array of technologies, including significant work with medical devices.

John C. Donch, Jr., JD is the president and a shareholder at Volpe Koenig, with over 20 years of experience. He regularly counsels domestic and multinational clients on strategies for securing, realizing and defending the value of their intellectual property. He has extensive experience with a wide variety of IP issues, including portfolio development and strategy, acquisitions, licensing, domestic and foreign patent prosecution, and infringement/validity analysis.

Amber R. Stiles, JD, MBA is a registered patent attorney with a background in biomedical engineering. Ms. Stiles is a published book author, a former medical devices patent examiner, and is currently employed as a supervisory patent examiner for design patents at the United States Patent and Trademark Office.

Brandon R. Theiss, JD is a licensed professional engineer and a registered patent attorney. An inventor himself, Mr. Theiss previously worked for a leading medical device manufacturer designing FDA compliant manufacturing systems. Mr. Theiss is currently a shareholder at Volpe Koenig.

Michelle Byrne, RAC is an accomplished regulatory strategist with Facet Life Sciences with over 23 years' professional experience in the medical device and pharmaceutical industries. Ms. Byrne is a biomedical engineer with extensive experience in clinical and regulatory affairs in many different therapeutic areas. Notably, Ms. Byrne was formerly a lead reviewer at the United States Food and Drug Administration (Center for Devices and Radiological Health).

William F. Hodnick, Ph.D., DABT is a pharmaceutical scientist with 25 years of pharmaceutical experience in the areas of analytical method development and validation, drug stability, and lot release testing, drug target identification, pharmacological activity screening, mechanism of action studies, drug metabolism and pharmacokinetics, toxicology, and regulatory affairs. Dr. Hodnick is currently employed as a strategic regulatory innovator at Facet Life Sciences.

Joshua Eldridge, Ph.D. is a pharmaceutical scientist with over 13 years of pharmaceutical industry experience covering drug from design, pharmaceutics, scaleup, GMP manufacturing, regulatory CMC consulting, and quality assurance. Dr. Eldridge has been named as an inventor on two transdermal patch (medical device) patents as well as patent applications in the areas of buccal peptide delivery and non-crystallizing forms of cannabidiolic acid. Dr. Eldridge is currently employed as a regulatory strategist with Facet Life Sciences.

Randy Goodman, Ph.D. is a health economist with a background in clinical outcomes research, economic model development and asset valuation with over 25 years of experience in pharmaceutical and medical device product commercialization. Dr. Goodman is currently employed as a commercial and HEOR strategy consultant at Facet Life Sciences.

Lisa Jenkins VanLuvanee, Ph.D. is a research psychologist with a background in cognition and neuroimaging. Dr. VanLuvanee is currently employed as the Chief Operating Officer and Vice President of Research and Development at Facet Life Sciences a regulatory and development services organization devoted to helping small companies get their products to and through FDA.

Dana L. Blue, M.S. is an organic chemist with a background in pharmaceutical development and evaluation with over 25 years' professional experience in clinical and CMC for pharmaceutical product development. Ms. Blue is currently employed as a Strategic Program Champion and Team Manager at Facet Life Sciences.

List of Figures

Chapter 1
Introduction

The medical device industry is a booming area of technological development that is advancing with breakneck speed. Since the original publication of this book, the FDA and the medical device industry have both undergone dramatic changes. This is often the case with a regulated industry affected by any number of factors, including legal and scientific precedent, funding, an ever-changing political landscape, scientific evidence, legal mandates, amongst many other factors. Flux and change are also to be expected in an industry with continuously evolving technology and solutions. The advent of artificial intelligence (AI) and the various ways that AI has been used with medical devices are contributing to revolutionizing the industry. Creative thought, in conjunction with ingenuity, can produce innovative medical devices that have immense value for the medical device company bringing the product to market. Innovative medical devices can also enable healthcare providers to administer care more effectively and can provide patients with better options for prevention, mitigation, treatment, and cure of disease. Innovators are designing and developing new medical devices daily. However, developing new, life-saving medical device technologies is not without challenges. Medical device innovators in the United States must navigate hurdles in both the intellectual property space, as well as the regulatory space by ensuring compliance with federal regulations set out by the Food and Drug Administration (FDA). FDA compliance is often crucial to successfully bringing a medical device product to market and securing intellectual property rights is key to gaining and maintaining a competitive advantage.

FDA and Intellectual Property Strategies for Medical Device Technologies, Including Artificial Intelligence, Software and Applications is intended to serve as a guide for readers who are involved in medical device development and commercialization. The book is broken down into three main parts:

I. FDA Strategies for Medical Device Technology.
II. Overview of Intellectual Property Rights for Medical Device Technology.
III. Implementation Strategies for Medical Device Innovators.

G. B. Halt Jr. et al., *FDA and Intellectual Property Strategies for Medical Device Technologies*, https://doi.org/10.1007/978-3-032-12649-8_1

1.1 Part I: FDA Strategies for Medical Device Technology

To the naïve observer, federal regulation of medical devices may seem confusing and complex. For instance, medical devices are categorized by a class designation (Class I, II or III), but only some classes of medical devices are subject to FDA regulation. Furthermore, there are several different regulatory pathways to choose from (i.e., application processes, such as PMA, 510(k), De Novo, etc.), and determining which is applicable for your medical device can be challenging at best. Each regulatory pathway has various requirements that must be satisfied in order to obtain FDA approval, clearance, or grant. To the uninitiated, FDA regulation of medical devices can feel like a quagmire. The good news is that you've taken your first step toward understanding and simplifying the process—reading this book!

Part I: FDA Strategies for Medical Device Technology will help you understand FDA regulation in the medical device realm. Each type of medical device class is explained, as is how the FDA regulates each type of device. You will learn how strategy can play a role in obtaining FDA approval, clearance, or grant as well as a few helpful tips and tricks for streamlining the FDA review process. You will discover the differences between the various medical device regulatory applications. Finally, this book will help you learn the medical device Sponsor's obligations during the application review process as well as once the product is actively being marketed to the U.S. public.

1.2 Part II: Overview of Intellectual Property Rights for Medical Device Technology

Through the proper use of intellectual property law, one has a much better chance of transforming creativity and innovation into economic value. Intellectual property law recognizes a creator's rights in ideas, innovations and goodwill. Being intangible, intellectual property differs from real property (land) or personal property (physical possessions) that are secured, controlled, and protected using physical means, such as fences, locks, alarms, and guards. Because intellectual property is a product of the mind, there is often no easy way to build a "fence" around it.

There are many intellectual property pitfalls that await the unwary. Different rules apply to different types of intellectual property ("IP"), and you may forfeit your rights if you do not take the appropriate measures to secure and protect them. It is important to understand the types of IP protection that exist and the respective rules that govern each type of IP.

(1) **Patent**. Patents may be granted for the invention of any new and useful process, machine, manufacture, or composition of matter or any new useful improvement thereof. A patent is a property right that grants the inventor or owner the right to exclude others from making, using, selling, or offering to sell the invention as defined by the patent's claims in the United States for a limited period of time.

(2) **Trademark**. A trademark is a word, phrase, symbol, or design, or combination of words, phrases, symbols, or designs which identifies and distinguishes the source of the goods or services of one party from those of others. Trademarks promote competition by giving products corporate identity and marketing leverage.
(3) **Copyright**. Copyrights protect original works of authorship fixed in a tangible medium of expression. Copyrighted works include literary, dramatic, and musical compositions, movies, pictures, paintings, sculptures, computer programs, etc. Copyright protects the expression of an idea, but not the idea itself.
(4) **Trade Secre**t. Generally, a trade secret is any formula, manufacturing process, method of business, technical know-how, etc. that gives its holder a competitive advantage and is not generally known. The legal definition of a trade secret and the protection afforded to a trade secret owner varies from state to state.

Patents, trademarks, trade secrets, and copyrights all have a strong presence in the field of medical device innovation. Part II of this book provides an introduction to intellectual property law as applied to the medical device industry.

1.3 Part III: Implementation Strategies for Medical Device Innovators

Having an understanding of FDA regulation and IP rights is only half the puzzle—medical device innovators can also benefit from guidance on how to avoid potential pitfalls and how to mitigate or eliminate risk in the areas of FDA regulatory compliance and IP procurement, policing, and enforcement.

Obtaining approval, clearance, or grant from the FDA is a good start, but medical device companies will have to demonstrate diligence in order to remain compliant with the rapidly evolving federal regulations once their product has been put on the market. Part III of this book will describe tools (such as implementation of a robust Quality Management System) and techniques (for instance, meticulous record keeping) typically used by medical device companies to successfully keep up with all the changes.

Similarly, obtaining IP rights and registrations is only the first step in a successful IP strategy. Once IP rights are secured, it is important to monitor, police, and exercise those rights. IP rights can be used offensively to initiate infringement litigation or encourage licensing activities; defensively to scare away potential competitors or to fend off litigation; or as a means of generating revenue streams through monetization. Part III of this book will also discuss how medical device companies can think about intellectual property strategy, while exploring ways to avoid potential pitfalls and mitigate potential risks.

Part I
FDA Strategies for Medical Device Technology

Chapter 2
General Overview of Development Process

2.1 History of Medical Device Regulation in the United States

In order to understand medical device development and regulation today, one must look back in history at the evolution of both drugs and medical devices, and how legislation evolved to protect the citizens of the United States. Device regulation is not as "new" as one might think. As with everything, guidance updates are required due to advancements in technology, unmet medical needs, and improved safety requirements.

During the 1800s, there were significant medical advances, including the discovery and effective use of anesthetic for surgery [1], the development of a rabies vaccine [2], and the use of X-rays in medical imaging [3]. However, during this same time period, medical quackery and "snake oil salesmen" gained prominence, and the use of products that were neither safe nor effective was rampant. Until 1906, drugs, medical devices, and foods went unregulated.

On June 30, 1906, the original Food and Drugs Act was passed by Congress and signed into law by President Theodore Roosevelt. This legislation focused on prohibiting interstate commerce of misbranded and adulterated foods, drinks, and drugs. Although the Food and Drugs Act did not include provisions for medical devices, this groundbreaking legislation started the U.S. federal government's mission to protect Americans against threats from harmful medical substances and deceptive practices.

In 1938, the Federal Food, Drug, and Cosmetic (FD&C) Act contained provisions to extend regulation to therapeutic devices. This act enabled the Food and Drug Administration (FDA) to evaluate the efficacy and safety of both drugs and medical devices. If such products were deemed ineffective and/or unsafe, the court system could be used to bring charges against the company marketing the product. However, because premarket testing, review, and approval were not required, products had to

G. B. Halt Jr. et al., *FDA and Intellectual Property Strategies for Medical Device Technologies*, https://doi.org/10.1007/978-3-032-12649-8_2

cause significant harm in order for the FDA to be able to act. The environment was ripe for disaster.

In 1962, such a disaster struck outside the United States. Thalidomide, a sedative and sleeping pill popularly prescribed at the time, was reported to cause birth defects in thousands of babies in western Europe [4]. Dr. Frances Kelsey, an FDA medical officer, was proactive in keeping thalidomide off the U.S. market. Importantly, the thalidomide disaster and the FDA's proactive approach to protecting Americans engendered significant public support for stronger regulations. In 1962, President Kennedy also proposed that medical devices should be regulated like drugs (but under different laws specific to device products), but because of the thalidomide disaster, it would be nearly 10 years before this proposal would materialize into law.

In 1971, the U.S. Public Health Service Bureau of Radiological Health was incorporated into the FDA. Their mission was to protect U.S. citizens against unnecessary exposure to radiation from electronic products in the home, industry, and the "healing arts." Five (5) short years later, laws were passed to ensure safety and effectiveness of medical devices, including diagnostic products, but not before the FDA was made aware of deaths associated with faulty pacemakers and Dalkon Shield intrauterine device injuries.

The 1976 Medical Device Amendments formally differentiated between drugs and devices (drugs exert their effect via a chemical reaction in the body, whereas devices do not). This amendment also required that devices be categorized into one of three classes with increasing controls based on their intended use and possible risk to patients:

- Class I devices pose the lowest risk to patients and include the use of general controls.
- Class II devices pose intermediate risk to patients and include the use of general controls with special controls.
- Class III devices pose the greatest potential risk to patients and include the use of general controls with special controls but require premarket approval.

The amendments also required manufacturers to register with the FDA and follow quality control procedures. This basic device classification is still used today and will be discussed in more detail later.

The FDA expanded their regulatory reach into long-term medical device safety with the Safe Medical Devices Act passed in 1990. This act required nursing homes, hospitals, and other facilities to report to the FDA incidents of death, serious illness, or serious injury of a patient that were likely caused by a medical device. Importantly, this legislation also required device manufacturers to conduct formal postmarketing surveillance on devices implanted in the body that had the risk of causing serious harm or death upon failure. The Safe Medical Devices Act of 1990 empowered the FDA to order device manufacturers to recall faulty devices in the marketplace.

In 2002, major advances in FDA device regulation were codified. The Medical Device User Fee and Modernization Act (MDUFMA I), the foundational law of present day device regulation, enabled the FDA to assess fees from Sponsors of medical device applications. This same year, the FDA announced an initiative focused on FDA enforcement of current good manufacturing practices (cGMP). The goal of the cGMP initiative was to have risk-based manufacturing procedures along with process and product quality standards that were robust but did not stifle innovation.

The success of MDUFMA I led to reauthorization in 2007 (the Medical Device User Fee Amendment [MDUFA II]), and again in both 2012 (MDUFA III), and 2017 (MDUFA IV). In each of these reauthorizations, the FDA published the user fee rates for Sponsor applications and updated their commitments to process improvement, enhanced guidance and support through formal and informal meetings, and decision-making performance goals. More information on each of these topics is provided in Chaps. 3–5. MDUFA is eligible for reauthorization every five years and will be up for renewal in 2028.

Importantly, the current state of device regulation is in significant flux. Although regulation tends to change slowly, the FDA has been particularly active in issuing revised and new medical device guidance documents in the last decade. FDA aims to provide updated guidance, and where applicable, regulations, to keep pace with advancing and rapidly developing technology. Thus, throughout this book, we will refer the reader to specific sections of U.S. Code (U.S.C.) and the U.S. Code of Federal Regulation (C.F.R.) and issued draft and final guidance documents (also referred to as guidances throughout this book) current as of the writing of this book. Guidance documents that are denoted as "draft" are FDA's current thinking on the topic and are intended for public comment whereas "final" guidance documents are in full effect and provide FDA's conclusive recommendations on the topic. The reader should conduct their own search of the device guidance documents on the FDA's website (https://www.fda.gov/regulatory-information/search-fda-guidance-documents) to ensure that you have the most current FDA thinking at your fingertips as you develop your medical device. You may also subscribe to the FDA's electronic notification program which will automatically send an email when the FDA issues a new guidance or updates an existing guidance.

2.2 Regulatory Definitions

2.2.1 What is a Medical Device?

Section 201(h) of the FD&C Act [5] defines the term "device" as an instrument, apparatus, implement, machine, contrivance, implant, in vitro reagent, or other similar or related article, including any component, part, or accessory, which is:

(a) Recognized in the official National Formulary, the United States Pharmacopeia, or any supplement to them.
(b) Intended for use in the diagnosis of disease or other conditions, or in the cure, mitigation, treatment, or prevention of disease, in man or other animals.

(c) Or intended to affect the structure or any function of the body of man or other animals, and which does not achieve its primary intended purposes through chemical action within or on the body of man or other animals.
(d) Not dependent upon being metabolized for the achievement of its primary intended purposes.

Example Medical device examples include items as simple as elastic bandages, disposable gloves, stethoscopes, electric powered toothbrushes, wheelchairs, pregnancy test kits, and condoms, to more complex items such as implantable pacemakers and artificial joints.

The term "device" does not include software functions excluded pursuant to section 520(o) which was added to the FD&C Act as part of the twenty-first Century Cures Act (Cures Act) amendment, see section 3060(a) (Pub. L. 114-255). This exclusion is not for all software functions but for certain software functions. Per the Cures Act, a device that has both software with a device and non-device function, the impact that the software non-device function has on the device function, may be determined by the FDA as part evaluating the safety and effectiveness of the medical device function.

Medical device products that have at least one device function and at least one "other function" are referred to as "multiple function device products." The term "function" is defined as a distinct purpose of the product, which could be the intended use or a subset of the intended use of the product.

Example Medical device product examples include a device that analyzes data and therefore has one function, which is analysis. A medical device product that stores, transfers, and analyzes data has three functions: (1) storage, (2) transfer, and (3) analysis.

The term "other function" is a function that:

(a) Does not meet the definition of device.
(b) Meets the definition of device, but is not subject to premarket review, for example is 510(k)-exempt.
(c) Meets the definition of device, but the FDA has expressed its intention not to enforce compliance with applicable regulatory controls.

An example of an "other function" is a Medical Device Data System (MDDS), which is described in Sect. 2.2.7.

This topic is covered in a 2020 FDA guidance (*Multiple Function Device Products: Policy and Considerations Guidance for Industry and Food and Drug Administration* [6]).

2.2.1.1 Components and Accessories

The FDA defines a component as any raw material, substance, piece, part, software, firmware, labeling, or assembly which is intended to be included as part of the finished, packaged, and labeled device [7]. Examples of components can include wheels provided by a supplier to a wheelchair manufacturer, bedrails for hospital beds (to be assembled by the manufacturer and provided as a finished device prior to use by a hospital), and plastic housing for a glucose monitoring system. In all these cases, each of these components serves no direct medical purpose and the component is used by the device manufacturer in the assembly of the finished device.

One point of caution what may be regulated as a component for one device may be regulated as a device by the FDA for another product. Context is important when considering component classification. FDA defines an accessory as a finished device that is intended to support, supplement, and/or augment the performance of one or more parent devices. A parent device is a finished device whose performance is supported, supplemented, and/or augmented by one or more accessories [8]. Examples of medical device accessories include a tool (accessory) used to accurately place a cranial nerve stimulator (parent device); bone cutting guides (accessories) to assist in the proper placement of artificial joints (parent device); and a stand (accessory) for an infusion pump (parent device).

2.2.2 *What is a Drug?*

The FDA's definition of a drug as provided under section 201(g) of the FD&C Act [9]:

(a) Articles recognized in the official United States Pharmacopoeia, official Homeopathic Pharmacopoeia of the United States, or official National Formulary, or any supplement to any of them.
(b) Articles intended for use in the diagnosis, cure, mitigation, treatment, or prevention of disease in man or other animals.
(c) Articles (other than food) intended to affect the structure or any function of the body of man or other animals.
(d) Articles intended for use as a component of any articles specified in clause (a), (b), or (c).

Example There are thousands of drugs currently approved by the FDA and many more under investigational development. Examples of FDA-approved drugs include acetylsalicylic acid (pain reliever), duloxetine (depression), oxycodone (pain reliever), sildenafil (erectile dysfunction), insulin (diabetes), ezetimibe (cholesterol), and fluticasone (asthma). Each one exerts their primary intended effect through chemical reaction within or on the body.

2.2.3 What is a Biologic?

A biologic is defined by the FDA as a virus, therapeutic serum, toxin, antitoxin, vaccine, blood, blood component or derivative, allergenic product, protein (except any chemically synthesized polypeptide), or analogous product, or arsphenamine or derivative of arsphenamine (or any other trivalent organic arsenic compound), applicable to the prevention, treatment, or cure of a disease or condition of human beings. Products that meet the drug definition and that also meet the definition of biological product are classified as biological products (see section 351(i) of the Public Health Service Act [10]).

Example FDA-approved biological product examples include: monoclonal antibody products such as adalimumab (rheumatoid arthritis), natalizumab (multiple sclerosis), and rituximab (non-Hodgkin's lymphoma; vaccines such as rabies, smallpox, and influenza; and cellular protein products such as talimogene laherparepvec (melanoma) and autologous cellular immunotherapy (refractory prostate cancer).

2.2.4 What is a Combination Product?

Importantly, combination products are defined in 21 C.F.R. § 3.2(e). The term combination product includes:

(a) A product comprised of two or more regulated components, i.e., drug-device, biologic-device, drug-biologic, or drug-device-biologic, that are physically, chemically, or otherwise combined or mixed and produced as a single entity ("single-entity combination products," see 21 C.F.R. § 3.2(e)(1)). An example would be a prefilled insulin injector pen.

(b) Two or more separate products packaged together in a single package or as a unit and comprised of drug and device products, device and biological products, or biological and drug products ("co-packaged combination product," see 21 C.F.R. § 3.2(e)(2)). An example would be a first aid kit that includes bandages as well as an antibacterial ointment.

(c) A drug, device, or biological product packaged separately that according to its investigational plan or proposed labeling is intended for use only with an approved individually specified drug, device, or biological product where both are required to achieve the intended use, indication, or effect and where upon approval of the proposed product the labeling of the approved product would need to be changed, e.g., to reflect a change in intended use, dosage form, strength, route of administration, or significant change in dose ("cross-labeled combination products," see 21 C.F.R. § 3.2(e)(3)). An example would be a light-activated biologic product that is packaged separately from the specific light source device that it must be used with; both the biologic product and device are labeled for use with the other.

(d) Any investigational drug, device, or biological product packaged separately that according to its proposed labeling is for use only with another individually specified investigational drug, device, or biological product where both are required to achieve the intended use, indication, or effect (also "cross-labeled combination products," see 21 C.F.R. § 3.2(e)(4)). An example would be a drug that is under development for assessing its antibacterial properties to be packaged separately but labeled for use with a surgical sealant device that is also under development.

Example FDA-cleared or approved combination product examples include condoms with spermicide, glycopyrrolate inhalation solution (COPD), drug-eluting stents, recombinant human bone morphogenic protein-2 sponge (spine, oral-maxillofacial and orthopedic trauma surgeries), and sumatriptan nasal powder administered via an intranasal device (migraine). Of note, condoms with spermicide and drug-eluting stents are both regulated by the FDA as devices (i.e., Center for Devices and Radiological Health (CDRH) as lead Center and Center for Drug Evaluation and Research (CDER) providing review consultants), whereas the other listed products were regulated as drugs or biologics (i.e., CDER or Center for Biologics Evaluation and Research (CBER), respectively as lead Center and CDRH providing review consults).

2.2.5 *What is an In Vitro Diagnostic (IVD)?*

The FDA defines an in vitro diagnostic (IVD) as a reagent, instrument, and system that is intended for use in the diagnosis of disease or other conditions, including a determination of the state of health, in order to cure, mitigate, treat, or prevent disease or its downstream effects [11]. IVDs are intended for use in the collection, preparation, and examination of specimens taken from the human body. Some IVDs are used exclusively in professional laboratories or other health professional settings (e.g., hospitals), although a greater number of IVDs are being developed for patients to use at home. Most FDA-cleared, granted, or approved IVDs to date are focused on helping determine if patients with an existing disease might benefit from FDA-cleared, granted, or approved therapies that have been shown to be effective in only a subset of patients with that disease. Information regarding the development and approval of IVDs can be found in the guidance document *In Vitro Diagnostic (IVD) Device Studies—Frequently Asked Questions* [12].

Example FDA-approved IVD examples include imatinib mesylate used in the detection of a specific gene rearrangement to aid in the selection of Myelodysplastic/Myeloproliferative Neoplasms (MDS/MPD) patients who would benefit from imatinib mesylate treatment, and pembrolizumab used in the detection of PD-L1 protein in formalin-fixed, paraffin-embedded (FFPE) non-small cell lung cancer (NSCLC) tissue to aid in the selection of NSCLC patients who would benefit from prmbrolizumab treatment.

Laboratory developed tests (LDTs) are a subset of IVDs. Examples include SARS-CoV2 (COVID-19) and flu testing, vitamin D assays, and emerging pathogen assays. Historically, LDTs have been designed, manufactured, and used under Clinical Laboratory Improvement Amendments (CLIA) requirements [13]. For many years now, the FDA has exercised what is called "enforcement discretion" over LDTs which means the FDA does not intend to enforce regulatory requirements for this type of product. However, the regulatory landscape for LDTs is complex and evolving. In May of 2024, the FDA issued a final rule that LDTs are to be regulated as medical devices and are subject to regulation under the FD&C Act [14]. In 2025, a federal court vacated the FDA's final rule, and a possible FDA appeal of the decision is pending at the time of this writing.

2.2.6 What is Software as a Medical Device (SaMD)?

Software as a medical device (SaMD) is defined by the International Medical Device Regulators Forum (IMDRF, a global group of medical device regulators that includes FDA, who strive to harmonize medical product regulatory requirements) as "software intended to be used for one or more medical purposes that perform these purposes without being part of a hardware medical device." Necessarily then, SaMD is not software that is used to operate a medical device's therapeutic functionality (such as an infusion pump's flow and timing parameters) nor is SaMD involved in the manufacturing process of any hardware medical device. Specific examples of SaMD include, but are not limited to:

- Diagnostic Imaging Analysis Software that provides a summary and interpretation of the imaging results (not the software that runs, for instance, a Magnetic Resonance Imaging (MRI) instrument, a Computer Tomography (CT) scanner, or an X-ray machine).
- Clinical Decision Support Systems that provide recommendations to healthcare providers based on patient data, such as electronic health record monitoring applications.
- Remote patient monitoring applications that analyze data from wearable monitoring devices.
- Software applications for cognitive behavioral therapy (i.e., patient interactive stress management or dietary applications).
- Other SaMD applications that incorporate artificial intelligence or machine learning.

As with any regulatory objective across the whole of pharmaceutical and medical device sciences, the FDA utilizes a risk-based approach when considering classification for any SaMD under regulatory review. The classification possibilities are:

- Class I: Low-risk devices (e.g., general wellness apps) with minimal regulatory controls.
- Class II: Moderate-risk devices (e.g., software for analyzing data to aid diagnosis) that often require premarket notification (510(k)).
- Class III: High-risk devices (e.g., software that directly drives treatment decisions) that typically require premarket approval (PMA).

Whatever the case, any software that is meant to diagnose, treat, mitigate, or prevent a disease will qualify as SaMD under FDA jurisdiction. Those familiar with medical device classifications will find the Class I-III system to be intuitive and familiar. Importantly, exemptions from FDA regulation exist for certain software applications, such as general wellness apps (e.g., fitness trackers) or software used solely for administrative purposes (e.g., scheduling), as outlined in the twenty-first Century Cures Act (2016).

The following guidance documents have been issued by the FDA as of this writing, and they should be read and understood by Sponsors who are medical device developers of SaMDs. It is noteworthy that the SaMD field is burgeoning, and new guidance documents are likely to be published often. Also, guidance document updates can be anticipated as the field grows and continues to emerge.

- Software as a Medical Device (SAMD): Clinical Evaluation Guidance for Industry and Food and Drug Administration Staff [15].
- Medical Device Data Systems, Medical Image Storage Devices, and Medical Image Communications Devices Guidance for Industry and Food and Drug Administration Staff [16].
- Policy for Device Software Functions and Mobile Medical Applications Guidance for Industry and Food and Drug Administration Staff [17].

- Marketing Submission Recommendations for a Predetermined Change Control Plan for Artificial Intelligence-Enabled Device Software Functions Guidance for Industry and Food and Drug Administration Staff [18].
- Content of Premarket Submissions for Device Software Functions Guidance for Industry and Food and Drug Administration Staff [19].
- Off-The-Shelf Software Use in Medical Devices Guidance for Industry and Food and Drug Administration Staff [20].
- Cybersecurity in Medical Devices: Quality System Considerations and Content of Premarket Submissions Guidance for Industry and Food and Drug Administration Staff [21].
- Predetermined Change Control Plans for Medical Devices Draft Guidance for Industry and Food and Drug Administration Staff [22].
- Considerations for the Use of Artificial Intelligence to Support Regulatory Decision-Making for Drug and Biological Products; Draft Guidance for Industry and Other Interested Parties [23].
- Artificial Intelligence-Enabled Device 1 Software Functions: Lifecycle 2 Management and Marketing 3 Submission Recommendations Draft Guidance for Industry and Food and Drug Administration Staff [24].

2.2.7 *What is a Medical Device Data System (MDDS)?*

Over the last decade, the twenty-first Century Cures Act modified how the FDA is to regulate MDDS. This modification is nicely summarized in FDA's guidance, *Medical Device Data Systems, Medical Image Storage Devices, and Medical Image Communications Devices Guidance for Industry and Food and Drug Administration Staff*. Medical Device Data Systems (MDDS), or hardware and software that transfer, store, convert formats, or display medical device data and results, have been split into Non-Device-MDDS and Device-MDDS.

Per the guidance, Non-Device-MDDS are *software functions* that are *solely intended* to transfer, store, convert formats, or display medical device data, without controlling or altering the parameters or functions of any connected medical devices that may or may not be intended for active patient monitoring. Thus, these software functions are not subject to applicable device FDA regulatory requirements as they are not devices under section 201(h) of the FD&C Act. Note that software functions that interpret and analyze medical device data (in addition to transferring, storing, converting formats, or displaying data) *are* still subject to FDA's regulatory oversight.

Device-MDDS are *hardware functions* that are *solely intended* to transfer, store, convert formats, or display medical device data and results, without altering data or controlling the parameters or functions of any connected medical devices. FDA does *not* intend to enforce compliance with regulatory controls (e.g., premarket review, Quality System Regulation, registration and listing, and postmarket reporting), for the following hardware functions: medical image communications devices subject to C.F.R. 892.2020, medical image storage devices subject to C.F.R. 892.2010, and

MDDS subject to C.F.R. 880.6310. Note there are exceptions identified in C.F.R. 880.9 and C.F.R. 892.9 that *do* require premarket notification.

Example Non-Device-MDDS examples include software that collects a patient's carbon dioxide levels from a ventilator and transfers the data to a central patient data repository, software that stores patient data such as blood pressure readings for a healthcare provider's review at a later time, software that convers pulse oximeter's digital data into a printable digital format, and software that displays a patient's previously stored electrocardiogram to be viewed by a healthcare provider.

2.3 Inventing a Medical Device

Invention of a medical device has historically been borne out of necessity, unmet medical need, or quite simply, creative innovation. For example, it was commonly believed 2000–3000 years ago that disease was related to issues of the blood. Bloodletting became a common practice to attempt to rid the patient of the diseased blood. Medical devices, including lancets (standard and spring-loaded), fleam sticks, scarificators, and flint cups, were used to extract blood from sick patients [25]. Although bloodletting has, by in large, been abandoned as a legitimate medical practice in the treatment of disease, devices to extract blood from the body are still used in medicine today.

There have been great advances in medical device technology, and medical device regulation has had to evolve to encompass these complexities. Medical devices range from the simplest of adhesive bandages to the most complicated intracranial imaging systems, bionic limbs, artificial skeletons, or web-enabled software applications. The sky has become the limit to innovation in the medical device space.

Throughout the remainder of this book, we will employ three examples of medical devices designed for cardiovascular health to illustrate basic concepts:

1. **Sphygmomanometer (blood pressure monitor)**—A device used to measure blood pressure that is composed of an inflatable cuff to collapse and then release the artery under the cuff in a controlled manner. The cuff is placed around an upper arm at roughly the same vertical height as the heart while the subject is seated with the arm supported. With a manual blood pressure monitor, a stethoscope is used to listen to the blood flow while the healthcare provider slowly releases the pressure in the cuff. As the pressure in the cuff falls, a "whooshing" or pounding sound is heard when blood flow first starts again in the artery. The pressure at which this sound begins is recorded as the systolic blood pressure. The cuff pressure is further released until the sound can no longer be heard. This is recorded as the diastolic blood pressure. Blood pressure monitoring is used to

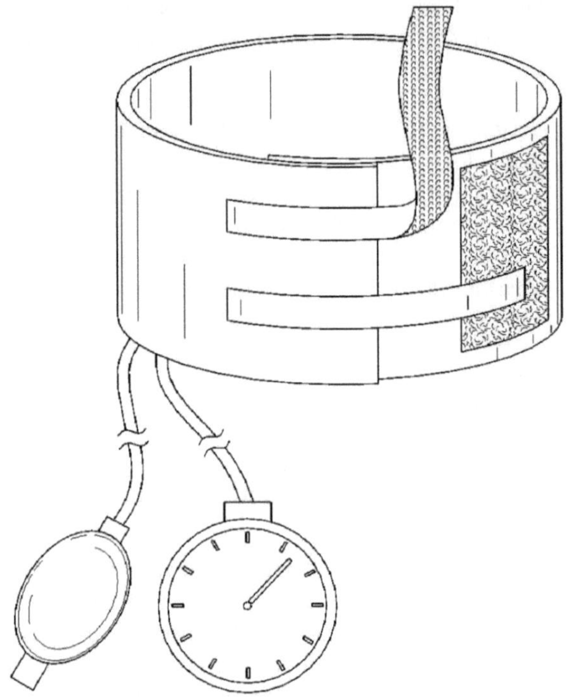

Fig. 2.1 Typical sphygmomanometer/blood pressure monitor

assess general health and evaluate ongoing health conditions (e.g., high or low blood pressure, heart disease, or other conditions) (Fig. 2.1).

2. **Coronary stent**—A device placed in an artery to keep the artery open for blood to flow freely. Typically, the expandable tube (stent) is threaded in its collapsed state through an artery to a site of occlusion (blockage), where it is enlarged and left as a permanent implant to hold the artery open. Coronary stents are used to restore blood flow to and from the heart and have been shown to reduce chest pain and improve survivability in the event of a heart attack (Fig. 2.2).
3. **Concept device: The CardioColander**—A device that contains a high-tech mesh filter that is placed in an artery to filter low-density lipoprotein (LDL) out of the blood. To allow filter changes, access to the device will be from a percutaneous port implanted in the patient's skin. High levels of LDL have been associated with a buildup of fatty deposits in the arteries (atherosclerosis) and raise the risk of heart attack, stroke, and peripheral artery disease (Figs. 2.3 and 2.4).

These three examples will be carried through the remainder of this book to help illustrate a number of concepts.

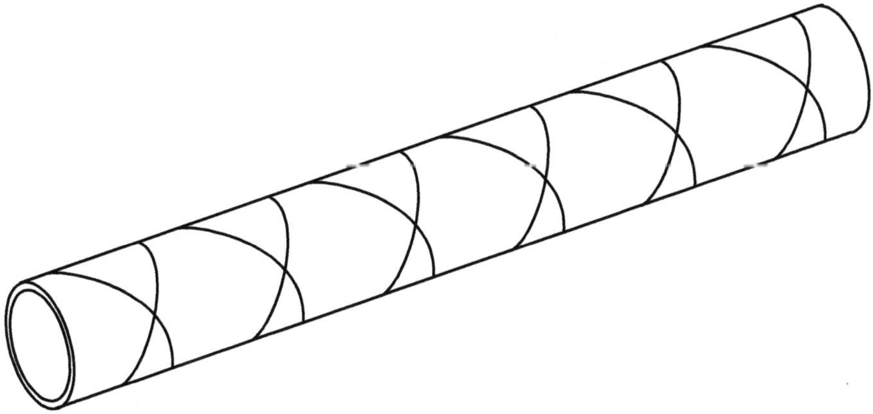

Fig. 2.2 Coronary stent

Fig. 2.3 Side view of the CardioColander device

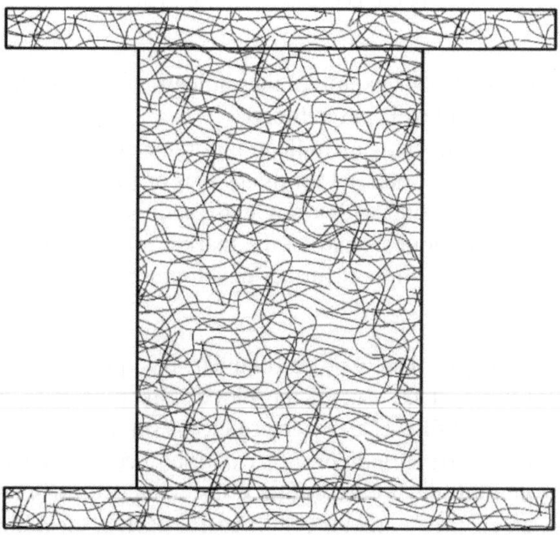

2.3.1 *Concept Development and Indications for Use*

Medical devices can have humble beginnings. Sketches on a piece of paper in a lab notebook, drawings on a white board in a conference room, or conceptual ideas jotted down in an email. Concept development typically identifies the medical need and establishes the indications for use, or, in other words, what we would like the device to do to fulfill that medical need. In the case of the well-known sphygmomanometer, the indication for use is to accurately measure blood pressure. For the typical stent, the indication for use is to improve coronary diameter. Finally, for the fictional

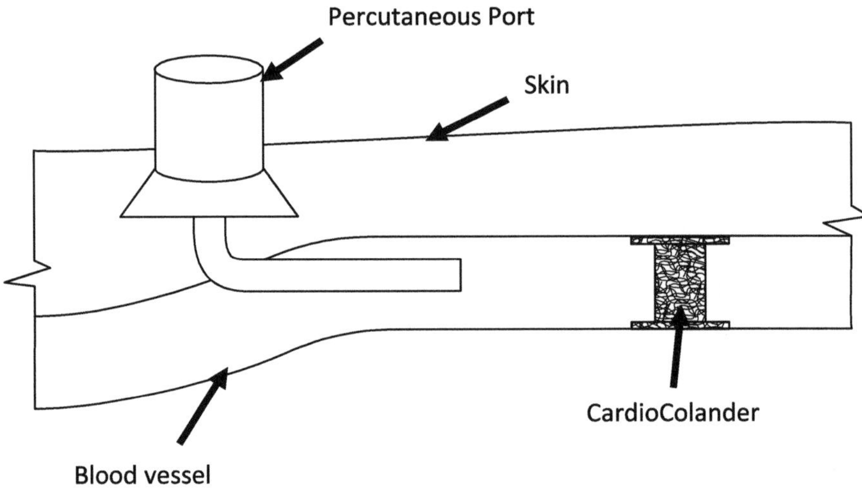

Fig. 2.4 View of CardioColander implanted in a blood vessel of a patient

CardioColander example, the indications for use for our device are to lower LDL cholesterol levels in the blood.

When choosing an indication for use, remember to keep it simple to start. Several devices have potential for many uses or for use in a different setting. As with everything in life, start with a single use and then expand as your budget allows. It is often best to get your device on the market for a specific use and then allow the revenue from sales to support further development for new indications for use.

Example The CardioColander has potential for additional uses beyond filtering LDL out of the blood. Our plan is to first get approval for use as a blood filter. Once our product is on the market, revenue will pay for future design development. For example, because device access is easy, additional possibilities for expanded IFUs include allowing direct drug administration in emergency situations (i.e., heart attacks), the addition of monitoring capabilities to provide direct feedback to both the patient and physician, and eventually even accessories (such as designer plates (color, jewels, emblems, etc.) for the percutaneous port portion of the device).

2.3.2 Design Development

Device design should follow basic design engineering principles and practical use parameters based on the indications for use of the device.

How do design engineering principles play a role? The design of a medical device should proceed in logical stages. The first step is to identify the requirements for the medical device. The FDA terms this "**design input**" and defines it as the physical and performance requirements of a device that are used as a basis for device design [7, 26]. In 1997, the FDA put forth a seminal guidance on this topic (*Design Control Guidance For Medical Device Manufacturers Guidance for Industry* [27]) which is still in effect today and provides detailed descriptions of each stage of device design and development.

Most medical device manufacturers know that design input is the single most important step in development. Poor requirements specifications can lead to costly late-stage refinements, quality issues, and delays to market. Design input involves a detailed evaluation and description of:

- What the device does, what the inputs are, and what the outputs are.
- How the device performs in terms of speed, accuracy, limits of operation, etc.
- The situations of use and device reliability across varied environments where the device will be used. This includes ensuring the device will be used as intended and will reproducibly provide the correct/same outcome.
- The requirements for compatibility with external systems, including patient interactions with the device.

Diligent work on design input leads to evaluation of the design input and output through the generation, testing, and refinement of a prototype. Adequate evaluation of design input requirements involves evaluation of:

- Design output procedures which contain or make reference to acceptance criteria and ensure that design outputs are essential for the proper functioning of the device.
- Design outputs are documented, reviewed and approved before release of the device.

In order to bring this to life, let's return to our example of the manual sphygmomanometer. Our indications for use help us to define the requirements for our device: it must reliably measure the force of circulating blood. In order to do this, we need an arm cuff to modify blood flow in order for a healthcare provider to listen to changes in the patient's pulse using a stethoscope (another medical device).

At a high level, these are our requirements, but in device development, the devil is in the details. Let's just talk about the cuff—how wide should it be, should it differ for adults and children, men and women, etc.? From what material should it be made? Where does it need to be placed on the body to get an informative reading? How is the cuff secured? How is the cuff inflated and to what pressure? How is the cuff deflated? These are just a few questions that must be answered as part of our device design development. Once we answer these questions, we must establish and evaluate output procedures to ensure that our device is operating as intended within set parameters.

Device development is incomplete until the intended user is able to use the medical device as intended without making errors that jeopardize safety or impede the efficacy of the device. Therefore, device design also must consider the needs, preferences, and behaviors of the user [28]. If we want our manual sphygmomanometer to be used properly, we need to ensure that the healthcare provider has the ability to readily secure the cuff to the patient's arm. If the cuff is too loose, a false reading may result, thus compromising the efficacy of the device. This should inform our device design. Today, many blood pressure monitoring systems use hook and loop fasteners to secure the cuff because it provides fast and flexible application with the needed flexibility for patients with different arm circumferences.

Recently, the FDA has been particularly outspoken about including device design elements based on patient needs and uses, with requirements to conduct human factors testing for medical devices and drug-device combination products (see 21 C.F.R. § 820.30, Design Controls where human factors considerations are implied in regulation, but also see the FDA's website which has detailed information about human factors considerations). The requirement for human factor testing is universal across all devices and intended users, even if the "only user" will be a healthcare provider. More on that later.

Finally, simplicity is a basic engineering principle that has dominated systems development for decades. This is not different in the medical device arena. To fully understand this concept, consider how very complicated systems are made up of a series of subtasks that are linked together. If any one of the subtasks fails, the entire system fails. In medical device design, failure of any aspect of the device can result in a reduction or loss in efficacy and/or patient harm. The simpler the device, the fewer design features there are that can possibly fail. Thus, all aspects of the device should serve a necessary function. With "multiple function device products," to the extent possible, the device function should be separated from "other functions" in design and implementation, to prevent the "other functions" from negatively impacting the device function. It is important to note, however, that some manufacturers build in additional device features solely for marketing or competitive advantage. More on this concept in Chap. 4.

2.3.3 *Prototype Development*

When the design requirements of the device have been identified, reviewed, and finalized (determined to be complete and acceptable for what the device must be able to do), a basic design is conceptualized. Drawings and schematics with technical specifications are often generated. The design requirements are then converted into a tangible and working representation of a prototype of the device. The initial prototype may include one, or only a few, of the necessary and basic feature functions of the device. Later prototypes incorporate all necessary device functions to satisfy the indications for use.

With respect to our manual sphygmomanometer example, constriction of the arteries in the arm is a design requirement. This requirement may be accomplished via a cuff that restricts blood flow in the arm by tightening through the use of an air pump. The design of the cuff itself needs to be converted into a series of specifications, including but not limited to the material of the cuff, the width/height of the cuff, a mechanism to secure the cuff to the arm, and a description of the process and procedures around the inflation/deflation of the cuff. There should be documentation to specify what the design of the cuff is intended to do while also describing why specific design solutions have been eliminated. For example, the design specifications may require the cuff to have a hook and loop fastener strip to secure the cuff to the patient's arm and may specifically restrict the use of a sticky gel adhesive because such a method would not be able to flex appropriately when the cuff is inflated/deflated.

The initial prototype serves as the foundation for further development and refinement to ensure that the device satisfies all the desired design requirements for the indications for use. Just like the design input stage of development, prototype development is critical to the overall success of a medical device and should not be rushed.

It would be quite rare indeed if an initial prototype became the final marketed device. The process of identifying the critical and desired device design features often leads to product enhancements. The characteristics, availability, and cost of material goods can lead to design changes as well. As an example, if a hook and loop fasteners material is not available for use in the sphygmomanometer, it must be replaced by a suitable alternative with similar design characteristics.

With respect to our manual sphygmomanometer example, we discussed several aspects of the cuff design that must be specified. With each specification, there must be verification to establish that the design conforms to the set specifications. Because several specifications directly affect the cuff functionality, verification of these design specifications works to establish that the cuff will meet the necessary constriction requirements. Verification testing can take the form of collecting data on patients under a variety of settings (e.g., sitting, standing, bare arm or with clothing). Verification testing provides concrete and objective evidence that the requirements for our cuff for the purpose of constricting blood in the arteries of the arm can be consistently established.

Regulation 21 C.F.R. § 820.3(y) outlines the production specification requirements, which include drawings and documents used to procure components, fabricate, test, inspect, install, maintain, and service the device, such as:

- Assembly drawings.
- Component and material specifications.
- Production and process specifications.
- Software machine code.
- Work instructions.
- Quality assurance specifications and procedures.
- Installation and servicing procedures.
- Packaging and labeling specifications, including methods and processes used.

Once iterative testing is complete, a final device design can be established. If you have not met with the FDA yet, now is an excellent time to open discussion with the

FDA about your device. Any suggestions that they may have could be invaluable to know at this stage before spending time and money on further development. More information about when and how to meet with the FDA is provided in Chap. 3.

It is worth noting here that even after a device is finalized, approved/cleared/granted and marketed, important information about the design and use of that device is gathered for continued improvements. As with our sphygmomanometer example, continued improvements eventually lead to smaller sphygmomanometers, sphygmomanometers for home use, electronic versions, large readouts for vision impaired users, accessories, etc. More information related to postmarketing requirements and product updates is provided in Chap. 5.

2.3.4 *Software as a Medical Device*

When designing SaMD or attempting to innovate any pharmaceutical product for human use, a company should look at the market of similar medical device products that have achieved regulatory clearance, grant, or approval by FDA. With respect to SaMD in particular, these products are regulated under Class I, II, or III, depending on risk, with most falling under Class II (moderate risk) requiring 510(k) clearance. Table 2.1 provides a summary of some SaMD products that have come on the scene in the years recent to the writing of this book.

However, the SaMD market story is far from complete with the short list provided in Table 2.1. Proper due diligence when examining the SaMD market space must be done by searching the relevant databases provided by the FDA for public information. Artificial intelligence/machine learning (AI/ML)-enabled devices are burgeoning in the industry at this writing, with more than 300 AI/ML-enabled medical devices, many of which are SaMD, used in radiology, cardiology, and oncology. Search tips to find similar devices include using the:

- **510(k) Database** and search for "software" or specific SaMD terms. This is the most logical start point for a query, as the majority of SaMDs are submitted to FDA via the 510(k) pathway.
- **De Novo Database** provides a list of novel SaMDs, like IDx-DR and Viz LVO.
- **PMA Database** can be searched for high-risk (Class III) SaMDs.
- **The FDA's AI/ML-enabled device list** includes many SaMDs and is updated regularly.

Needed updates to the list evolve rapidly, especially with AI/ML integration, so it is prudent to visit the list often to understand new market players.

2.3.4.1 Regulatory Considerations for SaMD

When developing your product, it is important to keep in mind that the FDA uses a risk-based approach for SaMDs inspired by the International Medical Device

Table 2.1 Recent FDA decisions on SaMD products

Product	Purpose	Regulatory route	Regulatory decision year
IDx-DR (Digital Diagnostics)	Detects diabetic retinopathy in retinal images	De novo	2018
Viz LVO (Viz.ai)	Analyzes computed tomography scans to detect large vessel occlusions (stroke-related) and alert clinicians	De novo	2018
Caption Guidance (Caption Health)	Guides non-experts in capturing diagnostic-quality ultrasound images of the heart	510(k)	2020
LuminenticsCore (formerly DreaMed Advisor)	Analyzes glucose data to recommend insulin dosing for diabetes management	510(k)	2018
HeartFlow Analysis (HeartFlow)	Creates 3D models of coronary arteries from computed tomography scans to assess blood flow and diagnose coronary artery disease	510(k)	2014
Transpara (2.1.0)	To be used by healthcare providers to improve interpretation of full-field digital mammography and digital breast tomosynthesis	510(k)	2024
InferRead Ct Stroke.ai	For use in the analysis of non-enhanced head computed tomography images	510(k)	2021

Regulators Forum (IMDRF) and considers the health and safety significance of the information provided by the SaMD. To that end, the most critical questions to ask about a SaMD concept in design are relevant to what information the software will provide, and to whom. For instance, does the proposed SaMD inform the diagnosis or treatment of a human disease? If so, is that disease serious or non-serious? These questions are important since the FDA has introduced initiatives like the Digital Health Innovation Action Plan and the Pre-Cert Program to streamline regulation for low-risk SaMDs. Clearly FDA intends to continue developing pathways for the marketing of SaMD products that encourage innovation. Companies with a strong case for these programs may qualify for expedited review. FDA is currently (as of this writing) pursuing faster ways to manage changes to SaMD without the torpor of longer marketing authorization changes.

Predetermined Change Control Guidance

The FDA's regulatory framework adapts to iterative software updates via Predetermined Change Control Plans (PCCP). The relevant guidance document is *Predetermined Change Control Plans for Medical Devices Draft Guidance for Industry and*

Food and Drug Administration Staff Draft Guidance for Industry and FDA Staff. This framework avoids the need to prepare new marketing submissions for every change to the device that could impact safety or effectiveness. Using the PCCP process requires a detailed Description of Modifications, a Modification Protocol, and an Impact Assessment, all of which are described in the draft guidance document in more detail.

For the sake of brevity, AI-enabled devices are also referred to as AI-enabled device software functions or AI-DSFs. AI/ML components exist in these devices to guide the function of the total device (hardware included). Changes to the software can be AI-driven (automatic continuous learning), manually modified by a programmer, or a combination of both approaches. Most devices in this category use ML. In the following sentences of this paragraph, quotations indicate language taken straight from the referenced guidance about PCCP. Specifically, ML is "a set of techniques that can be used to train AI algorithms to improve performance at a task based on data." The training data sets involved are "training data" (disease state and population demographics), "tuning data" (small numbers of AI models with varied architectures or hyperparameters), and "test data" (characterization of performance data that were not previously shared with the AI models).

Cybersecurity in Medical Devices

Obviously, any software-based system must survive the rigors of regulatory scrutiny relevant to cybersecurity. For example, let's assume that the CardioColander is controlled by software that turns on and off the device remotely depending on a patient's blood pressure. A cybersecurity attack could cause an alteration or disruption in the use of the CardioColander which could jeopardize the patient's health.

Currently as of this writing, a SaMD must meet cybersecurity requirements to protect patient data and ensure device functionality, especially for connected software. The FDA has provided final guidance on managing cybersecurity risks as of September 2023, *Cybersecurity in Medical Devices: Quality System Considerations and Content of Premarket Submissions Guidance for Industry and Food and Drug Administration Staff* [29]. Essentially, the final guidance supplants previous guidance drafts and nominates critical pieces for Sponsor consideration when integrating cybersecurity into the design, development, and premarket submission of medical devices. Manufacturers must:

- Establish a cybersecurity risk management process as part of their quality system design controls (21 C.F.R. 820.30).
- Identify, analyze, and mitigate cybersecurity risks throughout the device lifecycle (design, development, and postmarket).

There are a few usual quality system (QS) considerations as well. In keeping with quality by design (QbD), design controls are expected to be in place, and they must incorporate cybersecurity into design inputs, verification, and validation. All cybersecurity requirements must be traceable, and software must be validated to ensure cybersecurity controls are effective (e.g., secure data transfer, authentication, etc.). Vendors of cybersecurity practices for components or software must be qualified and managed with traceability, and postmarket surveillance is expected to monitor cybersecurity threats and vulnerabilities so that updates or patches can be provided as needed.

At the premarket submission stage, cybersecurity documentation should include a cybersecurity risk management report, threat modeling, and a Software Bill of Materials (SBOM). Diagrams and descriptions of the device's connectivity and data flow are also expected. In general, the FDA expects manufacturers to demonstrate that cybersecurity risks are adequately addressed in premarket submissions to ensure safe and effective use.

The reader is encouraged to consider the key points delivered in Sects. 2.2.6 and 2.3.4 when tackling the guidance landscape for a field as rapidly changing as SaMD. It is almost certain to be beneficial to enlist the help of seasoned medical device regulatory professionals when considering SaMD design and launch.

2.3.5 Commercial Considerations During Device Development

Given the current complexities of the U.S. healthcare market (as of this writing), it is essential that device manufacturers consider their commercial, pricing, and reimbursement strategies in the development phase of a product's lifecycle to ensure proper market placement and broad, widespread adoption. These strategies must also be aligned with a company's regulatory strategy to ensure broad market success.

2.3.5.1 Market Landscape Assessment

When you are developing a medical device, it is crucial to evaluate the clinical and market landscapes at the beginning of your journey. Such an evaluation aims to identify gaps in care, suboptimal outcomes, and unmet practitioner and patient needs that a new device could potentially address.

Research on disease prevalence, costs, and current standards of care are an integral part of the market analysis. By leveraging epidemiological databases, health economics projections, electronic medical records data, and other data sets, companies can quantify the clinical burden and market opportunity of a health condition. This information then forms the basis for the value proposition and informs market pricing and reimbursement strategies. Additionally, engaging key opinion leaders, specialty healthcare providers, payers, and patient groups can provide qualitative insights into the limitations of current solutions and desired future functionality.

Overall, understanding key "pain points" in disease diagnosis, monitoring, and treatment is imperative for designing, refining, and positioning new medical technologies for clinical and commercial (financial) success.

In the case of our current example, the CardioColander, which is placed in the artery to filter low-density lipoprotein (LDL) out of the blood, the landscape assessment would need to:

- Provide a comprehensive overview of the cardiac device (and drug/biologic).
- Identify existing gaps in current patient care.
- Address unmet patient need based on the use of existing devices and other treatment options.
- Show clinical outcomes improvements based on the use of this device.

Any suggestion of a change to an accepted "standard of care" must meet an acceptable standard of proof that can be supported by healthcare providers, payers, and key opinion leaders. We would want the landscape assessment to also demonstrate beneficial clinical gains in comparison to stents and other devices currently used in treating this patient population. The results of this assessment would then be used to help define the true value of CardioColander and provide avenues for enhanced value.

In developing a landscape assessment for our CardioColander, we reviewed one specific device, drug eluding stents which are small, mesh like tubes used to open blocked or narrowed arteries, particularly in coronary artery disease. These stents are coated with a medication that is released into the artery to help prevent the artery from narrowing again. The medication helps prevent scar tissue from forming and restricting blood flow. It is not known exactly how many people have received these stents, but there are some estimates showing that they have been used in over one million coronary angioplasty procedures in 2023–2024 [30]. Based on conversations with the Centers for Medicare and Medicaid (CMS) and several commercial health plans, drug eluding stents can be reimbursed at levels ranging from $5400.00 to over $8500, depending on the payer group and where the procedure is performed. It is also believed that these stents have a market growth rate of 8.0% year over year. This data helps inform our landscape assessment of the CardioColander, providing additional perspective into the cardiac device market. Expanding the landscape assessment further, a device manufacturer should consider a review of additional devices and biologics to have a more complete understanding of this market, as well as pricing and reimbursement options.

2.3.5.2 Adding Value in Early Development

Medical device companies must understand the value of their product and add as much value as possible during development to optimize revenue from their technologies. This necessitates proactive payer engagement and evidence generation catered to purchaser priorities surrounding the clinical value and economic impacts. Manufacturers must educate themselves or engage experts on the coding, coverage, and payment procedures that govern how different insurance plans approach assessments of product eligibility for reimbursement. Public and government payers operate differently than private insurers, which creates a complex web of considerations when applying for favorable status. Although that information will be most useful at the time of product launch, it is important to understand that data that is required by these constituencies in order to secure optimal coverage.

For CardioColander, robust health economics and outcomes research (HEOR) can substantiate product value claims by demonstrating cost savings, care efficiencies, and patient benefits relative to standards of care or a newly suggested standard of care for managing LDL cholesterol levels. Packaging clinical trial findings and real-world data into compelling dossiers convey value to payers. Therefore, we would want to design the development program with the end goal in mind. In a crowded cardiac market, it is essential to prove cost benefit and clinical efficiencies (asset value) to gain market acceptance. This would be supported by interviews conducted with several national and regional U.S. health plan organizations. These organizations would require thorough value dossiers on CardioColander that support the clinical and economic claims before considering a path to eventual coverage and reimbursement options for the device. However, for the purpose of establishing coding, a dossier alone would not be sufficient for American Medical Association (AMA) consideration, as validated clinical product value claims would need to be included as part of any application.

2.3.5.3 Late-Stage Development Commercial Activities

Later in development, an intricately coordinated payer strategy paralleling regulatory submissions allows device developers to simultaneously achieve both market realization and coverage crucial for market success.

As your medical device progresses toward clearance, grant, or approval, we recommend you formulate a comprehensive commercialization plan that includes marketing, sales and launch logistics. As part of this activity, it will be important to establish pricing models based on competitive analysis, market considerations and value-based strategies which capture the clinical, operational, and financial upsides enabled through the adoption of your medical device.

Given the pivotal role of reimbursement in the widespread adoption of a medical device, you must understand who determines reimbursement policies and the framework that guides those decisions, including potential reimbursement mechanisms.

A sustainable commercial and reimbursement model must meet the following criteria as also outlined by the Centers for Medicare and Medicaid Services (CMS): ensure trust and patient benefit by meeting foundational criteria using an ethical framework; ensure appropriate utilization and avoidance of overutilization; ensure financial sustainability; ensure a relatively short path length; and allow informed deployment decisions by medical professionals, health systems, and payers through transparency in estimating overall financial impact. Commercial payers in the U.S. generally follow the same criteria and guidelines used by CMS.

With respect to CardioColander, the manufacturer would need to develop a comprehensive, sustainable commercial market plan that incorporates:

- The value points and understanding from a landscape assessment of the cardiac market.
- The clinical and economic value demonstrated in a value dossier.
- A payer strategy that supports a positive coverage position for the device.
- Pricing models based on a competitive market analysis.
- An overview of the financial impact of this device.
- A product launch plan which considers potential risk and market barriers.
- A business and sales plan that details a path to financial success.

The commercial market plan is not a "one-and-done" document. It must be updated throughout development, design changes, and any changes to the market landscape, including potential new competitors that enter into the cardiac space.

2.4 Device Classifications and Regulatory Pathways

So now you have a working prototype, and you need to know if and how your device is regulated. In the United States, device classification depends on the intended use of the device, the indications for use, the perceived risk, and the amount of evidence needed to ensure both the efficacy and safety of the device. As was mentioned previously, there are three device classes defined by increasing risk [31]:

- Class I devices pose the lowest risk to patients and include the use of general controls.
- Class II devices pose intermediate risk to patients and include the use of general controls with special controls.
- Class III devices pose the greatest potential risk to patients and include the use of general controls with special controls.

Because the risk to patients is different for each class, the regulatory pathways, device controls, and requirements to market each type of device differ. Most Class I devices are exempt from Premarket Notification 510(k); most Class II devices require Premarket Notification 510(k); and most Class III devices require Premarket Approval. More information related to the specific requirements for each of these pathways is provided in Chap. 4.

2.4.1 What Class is My Device?

Over 1700 distinct types of devices have been classified by FDA and are organized in the C.F.R. into 16 medical specialties (found in Parts 862–892 of 21 C.F.R.), ranging from anesthesiology to obstetrical and gynecological to neurology. For each type of device classified by FDA, the C.F.R. contains the class to which the device belongs (i.e., Class I, II, or III), the intended use, a general description and information for marketing requirements.

The FDA has a device classification database to help Sponsors, i.e., the company or person who takes responsibility for the direction and oversight of medical device investigations [32], know which class their device might fit into. Classification is done by regulation (C.F.R.) number and the classification database is searchable. Classification numbers are 7-digits and are organized according to the aforementioned 16 medical specialties. As an example, cardiovascular diagnostic devices begin with the medical specialty code of 870. Within this specialty, there are specific devices listed. For example, 21 C.F.R. § 870.1120 is for a blood pressure cuff. Within this C.F.R. citation, the identification of this specific device and classification are listed:

(a) Identification. A blood pressure cuff is a device that has an inflatable bladder in an inelastic sleeve (cuff) with a mechanism for inflating and deflating the bladder. The cuff is used in conjunction with another device to determine a subject's blood pressure.
(b) Classification. Class II (performance standards).

As technology advances and new products are developed, devices that you seek to bring to the U.S. market may not be an exact fit with the classification description that most closely relates to your device. In these scenarios, it may be ideal to request a meeting with FDA, as further described in Sect. 3.1.5 of this book, or to submit a 513(g) Request for Information, as described in FDA's guidance document, *FDA and Industry Procedures for section 513(g) Requests for Information under the Federal Food, Drug, and Cosmetic Act Guidance for Industry and Food and Drug Administration Staff* [33].

2.4.1.1 Components and Accessories

Medical device <u>components</u> that are being sold to an end user as replacement parts are regulated as a medical device. A premarket application, specifically a 510(k), must be submitted to and cleared by the FDA before the component can be sold. This is because the component part must be demonstrated to perform reliably within specifications set by the component manufacturer. However, Sponsors who manufacture device component(s) under another device manufacturer's specifications are not required to submit an application for that component part [34].

Example Wheelchairs need tires. If a site is manufacturing tires for a specific wheelchair manufacturer using that manufacturer's specifications, the site does not need to submit an application for the tires. However, if the site is making tires for use on several different wheelchairs (i.e., knobby tires for use on the beach), the site will need to submit their own application.

The FDA also classifies medical device accessories and regulates them as medical devices of their own accord according to risk. In order to classify an accessory, the FDA first confirms that the article is an accessory and then evaluates the risk of the accessory when used as intended with the parent device(s) (i.e., a finished device whose performance is supported, supplemented, and/or augmented by one or more accessories). The risk profile of the accessory can be different than the risk profile for the parent device. Once the risk is established, the FDA determines what regulatory controls are necessary to provide a reasonable assurance of its safety and efficacy, which then informs the submission pathway required.

An accessory is categorized as a device under two conditions:

(1) If the labeling, promotional materials, or other evidence of intended use demonstrates that a device is intended to be used with one or more parent devices, and it supports, supplements, and/or augments that device.
(2) If the device supports, supplements, or augments the performance of one or more parent devices by enabling or facilitating that device to perform according to its intended use.

The FDA's guidance document, *Medical Device Accessories—Describing Accessories and Classification Pathways Guidance for Industry and FDA Staff* [8] provides an excellent description of the categorization of medical devices as accessories. Most accessories are classified and regulated based on the same classification as the parent device, but there are exceptions. If you are in doubt about how your accessory would be classified by the FDA or if you have a novel accessory, you can undergo an accessory classification request process [35]. FDA's webpage, Medical Device Accessories [36], provides additional information regarding the request process.

2.4.2 Exempt from Premarket Submission

A premarket notification application and formal FDA review is not required for most Class I devices before they are marketed. Devices are considered low risk when the FDA believes that general controls alone are adequate to assure the safety and efficacy of the device. A stethoscope is a great example of a Class I device. Its classification code is 870.1875.

Class I devices are subject to general controls, which include establishment registration of the device manufacturer, medical device listing, cGMP manufacturing [37], and adherence to medical device labeling requirements [38, 39]. More information for each of these topics is provided in Chap. 5. Details regarding the general control requirements are provided in subsequent sections of this chapter.

Note that certain unclassified medical devices are exempt from premarket notification requirements. The FDA believes such devices are sufficiently well understood and therefore do not require premarket notification to assure the safety and efficacy of these device products. In the June 14, 2019 update of the guidance (*Intent to Exempt Certain Unclassified Medical Devices from Premarket Notification Requirements Guidance for Industry and Food and Drug Administration Staff* [40]), the FDA identifies a number of such devices that they intend to exempt from premarket notification requirements.

2.4.3 Premarket Applications

Class II devices are higher-risk devices than Class I devices because the FDA believes that general controls alone cannot provide reasonable assurance of the efficacy and/or safety of the device. Class II devices are devices for which there is sufficient information to establish special controls to provide such assurance. In general, Class II and Class III devices require FDA clearance of a Premarket Notification (510(k)) or an approval of a Premarket Approval (PMA) application, or sometimes FDA grant of a De Novo Classification Request, before the device can be legally marketed in the U.S.

2.4.3.1 Premarket Notification 510(k)

The 510(k) route (named after section 510(k) of the FD&C Act) is designed specifically for Sponsors who have made modifications or enhancements to a legally marketed device, known as the predicate device.

A legally marketed device is a device that was:

- Legally marketed in the U.S. prior to May 28, 1976 ("preamendments device"), which has not been significantly modified or changed since then or for which a PMA was not required.
- Reclassified from Class III to Class II or Class I.
- Found to be substantially equivalent through the 510(k) process.
- Granted marketing authorization via a De Novo that is not exempt from premarket notification requirements.

In the 510(k) application, the Sponsor must demonstrate that their new device is as safe and effective, or "substantially equivalent" (SE), to the predicate device [41]. FDA considers a device "substantially equivalent" if it has:

- The same intended use *and* the same technological characteristics as the predicate device.
- The same intended use as the predicate *and* has different technological characteristics that do not raise different questions of safety and effectiveness *and* the information submitted to FDA demonstrates that the device is as safe and effective as the predicate.

Note that when FDA makes an SE determination, they "clear" the 510(k), they do not "approve" the 510(k). This is commonly misunderstood and incorrectly stated. Further, "FDA-approval" is often used as a blanket statement (in conversations, press releases, marketing materials, etc.) but is only technically correct when referring to FDA's approval of a PMA (or FDA's approval of an NDA for drugs or drug-device combination products or a BLA for biologics or biologic-device combination products).

A digital blood pressure monitor is an example of a Class II device (870.1130) falling under the label of a "noninvasive blood pressure measurement system":

(a) Identification. A noninvasive blood pressure measurement system is a device that provides a signal from which systolic, diastolic, mean, or any combination of the three pressures can be derived through the use of transducers placed on the surface of the body.
(b) Classification. Class II (performance standards).

If the FDA determines that the device is substantially equivalent to a legally marketed medical device, it is placed in the same class. Class II devices are subject to general controls just like Class I devices are, but because of their increased risk to patients, they are also subject to special controls. Special controls include special labeling requirements, mandatory performance standards, human factors testing, and postmarketing surveillance. More information on each of these topics is provided in Chaps. 3–5. Referring back to the information in Sect. 2.4.1.1 of this chapter, note that most medical device components that are being sold directly for use in other devices are regulated via the 510(k) route.

If the FDA determines that the device is not substantially equivalent, known as "NSE," the FDA will regulate that device as a Class III device. The Sponsor will then need to submit a PMA or a De Novo.

Similarly, if a Sponsor wants to market a Class II device for which there is no predicate device, the Sponsor must submit a PMA or De Novo for the device, and the application must be approved or granted, respectively, prior to marketing.

SaMD 510(k) Applications

As stated in Sect. 2.2.6, the majority of SaMD products are submitted to FDA via the 510(k) pathway. A few draft and final guidance documents speak to the 510(k) process for obtaining FDA clearance for an SaMD, and those are also listed in Sect. 2.2.6. Risk-based premarket application considerations for SaMD are also covered in Sects. 2.2.6 and 2.3.4. It is noteworthy that software rarely stays the same as time goes by, and it can be the case that changes to an existing SaMD require a new 510(k) submission. *Deciding When to Submit a 510(k) for a Software Change to an Existing Device Guidance for Industry and Food and Drug Administration Staff* [42] applies to software changes in SaMD or devices with software components whose changes impact safety, effectiveness, or the intended use of the device. Sponsors are encouraged to consider key questions:

- Does the change introduce a new or modified intended use?
- Does the change create new risks or significantly modify existing risks?
- Does the change affect critical device performance or safety specifications?

Generally, a new 510(k) will be required for SaMD when changes to algorithms, the user interface, or software functionality modify clinical decision-making or patient safety. Modifications introducing new indications, new user populations, or new risks (e.g., cybersecurity vulnerabilities) are also fit for new 510(k) submissions. However, minor software updates that don't impact safety, effectiveness, or intended use will not require a new 510(k) submission, nor will changes validated through existing verification and validation processes.

Manufacturers must document the changes, the rationale for the change, and a risk assessment of the changes per Quality System Regulation (21 C.F.R. § 820). Please refer to Sect. 2.3.4 for additional information about 510(k) cleared devices that should be reviewed at the FDA 510(k) database for predicate considerations. It is always advised to review this database for summary documentation, because most SaMDs are cleared by the 510(k) pathway, and a predicate is required to be cleared by this mechanism.

2.4.3.2 Premarket Approval (PMA)

Class III devices are devices that have the greatest risk for patients. These devices either support or sustain life, are critical to preventing debilitating disease or illness, or present a high risk of illness or injury to the patient. For Class III devices, the FDA does not believe that general and special controls are enough to provide reasonable assurance that the device is efficacious and/or safe [43]. Most Class III devices are regulated via the PMA pathway, which must include clinical data demonstrating both the safety and efficacy of the novel device.

The development program for a novel Class III device typically involves conducting several nonclinical studies (i.e., biocompatibility, in vivo safety, proof of concept, performance, wear, stress, shelf life, etc.) and one or more clinical experiments, including a pivotal, multicenter, prospective, randomized and controlled clinical trial. The complexity and extent of the clinical testing program is dictated by the nature of the device, the possible risks of the device, and the proposed use(s) of the device. The FDA must approve the PMA application prior to marketing a Class III device. More information on the different types of testing (nonclinical and clinical as well as human factors testing) is provided in Chap. 3.

Table 2.2 De novo examples

Product	Purpose	Approval year	Summary available?
DEN230035 Dental Monitoring	Uses image processing algorithms to analyze pictures of the oral cavity to aid diagnosis of oral health conditions and/or to monitor treatment progress	2024	Yes
DEN230043 Notal Vision Home Optical Coherence Tomography (OCT) System	For visualization of intraretinal and subretinal hyporeflective spaces in a 10 by 10-degrees area centered on the point of fixation of eyes diagnosed with neovascular age-related macular degeneration (NVAMD). In addition, it provides segmentation and an estimation of the volume of hyporeflective spaces	2024	Yes
DEN210035 Genius TM Digital Diagnostics System with the Genius TM Cervical AI Algorithm	Intended for the creation and viewing of digital images of scanned ThinPrep Pap Test glass slides. Objects of interest selected by the Genius™ Cervical AI algorithm from the scanned digital image are presented in a gallery format next to the image of the whole cell spot on the Genius™ Review Station for review and interpretation	2024	Yes (and a very useful summary indeed)

Class III devices are subject to both general and special controls. A cardiovascular stent is an example of a Class III device that is subject to a PMA application. It is an implanted device that is intended to support/sustain life.

SaMD PMA Applications

Any SaMD that could be considered a PMA candidate should be presented to FDA in a formal meeting with well-defined questions to obtain helpful and actionable feedback.

FDA's Artificial Intelligence and Machine Learning (AI/ML)-Enabled Medical Devices List contains a few examples of PMA-approved SaMDs. For example, PAPNET® Testing System (P940029) was PMA-approved in 1995, and M1000 IMAGECHECKER (P970058) was approved in 1998. Unfortunately, there is no publicly available summary documentation for these products. Reader, Cervical Cytology Slide, Automated is the generic name for PAPNET® Testing system, and that speaks to the probable reason for the PMA pathway for this SaMD back in the 1990s when AI/ML was in its infancy. Pap smears are utilized in diagnosis of cervical cancer, and cervical cancer is a high-risk diagnosis and treatment category. False readings in this space could indeed be life-threatening. However, with changes in time and technology, the reader is encouraged to see a De Novo example of another pap test diagnostic aid SaMD that was granted in 2024 (Table 2.2, DEN210035).

Imagio Breast Imaging System (P200003) was approved much more recently in 2021, and it is indicated *"for use by a trained and qualified healthcare provider for evaluation of palpable and non-palpable breast abnormalities in adult patients who are referred for a diagnostic imaging breast work-up, following clinical presentation or either screening or diagnostic mammography or other imaging examinations."* In this case, a useful summary is publicly available at the FDA's Artificial Intelligence and Machine Learning (AI/ML)-Enabled Medical Devices List site. We recommend that Sponsors who are developing new SaMD review this PMA summary for helpful information.

De Novo Classification Request

There are device "types" that have never been marketed in the U.S., but whose safety profile and technology are reasonably well understood. The De Novo Classification Request (De Novo) provides a pathway to classify the device (as Class I or Class II) for which general controls alone, or general and special controls, provide a reasonable assurance of the device's safety and efficacy for the intended use, but for which there is no legally marketed predicate device. Classification through the De Novo pathway is a risk-based classification process.

When FDA grants a De Novo for a given device, not only can the device be legally marketed in the US, but the grant also triggers the FDA to create a classification regulation for devices of this type and permits the device to serve as a future predicate device, when applicable. In 2021, the FDA issued a final rule under the FD&C Act to establish new regulations for the implantation of the De Novo classification process. This final rule includes processes and criteria for De Novos and their review. At this time, FDA also issued a final guidance, *De Novo Classification Process (Evaluation of Automatic Class III Designation)* [44], that outlines the details behind this regulatory process.

Our CardioColander, the novel device that contains a high-tech mesh filter that is placed in an artery to filter low-density lipoprotein (LDL) out of the blood, would likely be regulated via the De Novo pathway. This device is unlike any other device cleared, granted, or approved by FDA previously; thus, the 510(k) pathway would not be appropriate, and there is no classification in the C.F.R. for it. General and special controls should be adequate to provide reasonable assurance of safety and effectiveness for its intended use; thus, the De Novo pathway would likely be the appropriate route, rather than the PMA pathway.

De Novo SaMD Applications

Table 2.2 summarizes several granted De Novo products on the FDA's Artificial Intelligence and Machine Learning (AI/ML)-Enabled Medical Devices List. This listing is not exhaustive. Developers are encouraged to review every De Novo entry in the FDA's list at the time they are considering the necessary activities and information for their own products, as some of them provide unusual strategic information in their Decision Summaries that speak to specific guidance documents that were utilized in the De Novo submission and decision processes.

For example, Table V in the Decision Summary for Genius™ Digital Diagnostics System with the Genius™ Cervical Al Algorithm represents a rare opportunity to see the ISO, FDA, and other standards or guidance documents that were used to see the product through FDA's grant of the De Novo. Please refer to the Agency's Artificial Intelligence and Machine Learning (AI/ML)-Enabled Medical Devices List for multiple other examples of SaMDs submitted via the De Novo pathway and granted by FDA.

Humanitarian Device Exemption (HDE)

One final regulatory pathway is devoted specifically to Class III devices that treat rare disorders. This pathway is known as a Humanitarian Device Exemption (HDE) (formerly the Humanitarian Use Device (HUD)). Approval of a HDE requires demonstration that the device is safe, and the probable benefits outweigh the probable risks. Although the process may require smaller clinical trials, an HDE device must continue to operate under local Institutional Review Board (IRB) approval at each participating institution, and the Sponsor must continue to collect case report forms akin to an ongoing clinical trial. The FDA issued two guidances on this pathway in 2019 entitled, *Humanitarian Use Device (HUD) Designations Guidance for Industry and Food and Drug Administration Staff* [45], and *Humanitarian Device Exemption (HDE) Program Guidance for Industry and Food and Drug Administration Staff* [46]. Of note, as of late 2016, HUD defines rare disorders as those affecting not more than 8000 patients in the U.S. in a calendar year.

One example of a HUD device is an implantable replacement heart. This FDA-approved device is indicated for use in patients with hearts where neither ventricle works; are not heart transplant candidates; are less than 75 years old; require medications to change their heart rate; are not treatable with mechanical blood circulation support (another device); and cannot be weaned from the mechanical blood circulation support, if they are on such support.

2.4.4 Combination Products

The FDA regulates combination products based on the product's primary mode of action or its primary function. Additionally, the primary mode of action is what dictates which FDA Center (Center for Drug Evaluation and Research [CDER], Center for Biologics Evaluation and Research [CBER], or Center for Devices and Radiological Health [CDRH]) will be the "lead" review center. If a product contains a device and a drug, for example, and the primary purpose of the combination product is based on the action of the drug, the product would be regulated as a drug. This means that CDER would be considered the lead FDA Center for all meetings and regulatory applications, but all applicable FDA Centers (such as CDRH in this example) would be involved in the review of the product. Note that combination product submissions must still contain all the necessary data and evidence to establish the efficacy and safety of all components of the product, regardless of which FDA Center is deemed the lead.

Historically, most combination products that included a drug or biologic (e.g., drug-device, biologic-device, drug-device-biologic) have been regulated by FDA as drugs because the products involved a novel device that delivered a drug intended to exert chemical action in/on the body (i.e., the drug was the primary mode of action of the combination product). However, technology has advanced significantly since the early days of combination products, and some combination products are regulated as medical devices.

Example A condom with spermicide is regulated as a medical device. The FDA's Center for Devices and Radiological Health (CDRH) would serve as the lead Division in any regulatory communications with the Sponsor, and would provide all regulatory review of any submissions because the primary mode of action of the combination product is a mechanistic barrier. CDRH would receive consultation from the Office of New Drugs because the combination product includes a drug (spermicide) as a second line of defense as part of the mechanism of action. Combination product submissions obviously must contain all the necessary data and evidence to establish the efficacy and safety of both the device and drug components of the product.

In fact, because of the complexity of emerging combination products, the FDA has developed a website entirely devoted to combination product assistance and has issued numerous guidances on the topic. If you are developing a device-led combination product, we recommend starting with two specific guidance documents that will give you an overview on combination products, potential regulatory pathways, and interacting with FDA: *Requesting FDA Feedback on Combination Products Guidance for Industry and FDA Staff* [47] and *Principles of Premarket Pathways for Combination Products Guidance for Industry and FDA Staff* [48]. Additional help on combination product development can be obtained from the FDA's Office of Combination Products and through the FDA Combination Products public meetings, conferences, industry presentations, and workshops, all of which are available on the FDA's website.

References

1. Ellis, Harold, *A History of Surgery*, p. 47, Cambridge University Press (2001).

2. Geison G.L., *Pasteur's work on rabies: Reexamining the ethical issues*, 8 (2): 26–33, Hastings Center Report (1978).
3. Nitske, RW., *The Life of W. C. Röntgen, Discoverer of the X-Ray*, University of Arizona Press, (1971).
4. Ridings J.E., *The thalidomide disaster, lessons from the past*, 947:575–586, Methods Mol Biol., (2013).
5. 21 U.S.C. § 321(h).
6. FDA's guidance for Industry and FDA Staff: *Multiple Function Device Products: Policy and Considerations Guidance for Industry and Food and Drug Administration* – FINAL GUIDANCE (July 2020). Available online at https://www.fda.gov/regulatory-information/search-fda-guidance-documents/multiple-function-device-products-policy-and-considerations. Accessed 6 July 2025.
7. 21 C.F.R. § 820.3(c).
8. FDA's guidance for Industry and FDA Staff: *Medical Device Accessories—Describing Accessories and Classification Pathways Guidance for Industry and FDA Staff*—FINAL GUIDANCE (December 2017). Available online at https://www.fda.gov/regulatory-information/search-fda-guidance-documents/medical-device-accessories-describing-accessories-and-classification-pathways. Accessed 6 July 2025.
9. 21 U.S.C. § 321(g).
10. 42 U.S.C. § 262(i).
11. 21 C.F.R. § 809.3.
12. FDA's guidance for Industry and FDA Staff: *In Vitro Diagnostic (IVD) Device Studies—Frequently Asked Questions*—FINAL GUIDANCE (June 2010). Available online at https://www.fda.gov/regulatory-information/search-fda-guidance-documents/in-vitro-diagnostic-ivd-device-studies-frequently-asked-questions. Accessed 6 July 2025.
13. FDA Webpage, Laboratory Developed Tests: FAQs. Available online at https://www.fda.gov/medical-devices/laboratory-developed-tests/laboratory-developed-tests-faqs. Accessed 6 July 2025.
14. FDA Webpage, Laboratory Developed Tests. Available online at https://www.fda.gov/medical-devices/in-vitro-diagnostics/laboratory-developed-tests. Accessed 6 July 2025.
15. FDA's guidance for Industry and FDA Staff: *Software as a Medical Device (SAMD): Clinical Evaluation Guidance for Industry and Food and Drug Administration Staff* – FINAL GUIDANCE (December 2017). Available online at https://www.fda.gov/regulatory-information/search-fda-guidance-documents/software-medical-device-samd-clinical-evaluation. Accessed 6 July 2025.
16. FDA's guidance for Industry and FDA Staff: *Medical Device Data Systems, Medical Image Storage Devices, and Medical Image Communications Devices Guidance for Industry and Food and Drug Administration Staff*—FINAL GUIDANCE (September 2022). Available online at https://www.fda.gov/regulatory-information/search-fda-guidance-documents/medical-device-data-systems-medical-image-storage-devices-and-medical-image-communications-devices. Accessed 6 July 2025.
17. FDA's guidance for Industry and FDA Staff: *Policy for Device Software Functions and Mobile Medical Applications Guidance for Industry and Food and Drug Administration Staff*—FINAL GUIDANCE (September 2022). Available online at https://www.fda.gov/regulatory-information/search-fda-guidance-documents/policy-device-software-functions-and-mobile-medical-applications. Accessed 6 July 2025.
18. FDA's guidance for Industry and FDA Staff: *Marketing Submission Recommendations for a Predetermined Change Control Plan for Artificial Intelligence-Enabled Device Software Functions Guidance for Industry and Food and Drug Administration Staff*—FINAL GUIDANCE (April 2023). Available online at https://www.fda.gov/regulatory-information/search-fda-guidance-documents/marketing-submission-recommendations-predetermined-change-control-plan-artificial-intelligence. Accessed 6 July 2025.
19. FDA's guidance for Industry and FDA Staff: *Content of Premarket Submissions for Device Software Functions Guidance for Industry and Food and Drug Administration Staff*—FINAL

GUIDANCE (June 2023). Available online at https://www.fda.gov/regulatory-information/search-fda-guidance-documents/content-premarket-submissions-device-software-functions. Accessed 6 July 2025.
20. FDA's guidance for Industry and FDA Staff: *Off-The-Shelf Software Use in Medical Devices Guidance for Industry and Food and Drug Administration Staff*—FINAL GUIDANCE (August 2023). Available online at https://www.fda.gov/regulatory-information/search-fda-guidance-documents/shelf-software-use-medical-devices. Accessed 6 July 2025.
21. FDA's guidance for Industry and FDA Staff: *Cybersecurity in Medical Devices: Quality System Considerations and Content of Premarket Submissions Guidance for Industry and Food and Drug Administration Staff*—FINAL GUIDANCE (June 2025). Available online at https://www.fda.gov/regulatory-information/search-fda-guidance-documents/cybersecurity-medical-devices-quality-system-considerations-and-content-premarket-submissions. Accessed 6 July 2025.
22. FDA's draft guidance for Industry and FDA Staff: *Predetermined Change Control Plans for Medical Devices Draft Guidance for Industry and Food and Drug Administration Staff*—DRAFT GUIDANCE (August 2024). Available online at https://www.fda.gov/regulatory-information/search-fda-guidance-documents/predetermined-change-control-plans-medical-devices. Accessed 6 July 2025.
23. FDA's draft guidance for Industry and FDA Staff: *Considerations for the Use of Artificial Intelligence to Support Regulatory Decision-Making for Drug and Biological Products; Draft Guidance for Industry and Other Interested Parties*—DRAFT GUIDANCE (January 2025). Available online at https://www.fda.gov/regulatory-information/search-fda-guidance-documents/considerations-use-artificial-intelligence-support-regulatory-decision-making-drug-and-biological. Accessed 6 July 2025.
24. FDA's draft guidance for Industry and FDA Staff: *Artificial Intelligence-Enabled Device 1 Software Functions: Lifecycle 2 Management and Marketing 3 Submission Recommendations Draft Guidance for Industry and Food and Drug Administration Staff*—DRAFT GUIDANCE (January 2025). Available online at https://www.fda.gov/regulatory-information/search-fda-guidance-documents/artificial-intelligence-enabled-device-software-functions-lifecycle-management-and-marketing. Accessed 6 July 2025.
25. Greenstone G., *The history of bloodletting*, page(s) 12–14 BCMJ, Vol. 52, No. 1, (2010).
26. 21 C.F.R. § 820.3(f).
27. FDA's guidance for Industry and FDA Staff: *Design Control Guidance For Medical Device Manufacturers Guidance for Industry*—FINAL GUIDANCE (March 1997). Available online at https://www.fda.gov/media/116573/download. Accessed 6 July 2025.
28. Money AG, Barnett J, Kuljis J, Craven MP, Martin JL, Young T., *The role of the user within the medical device design and development process: medical device manufacturers' perspectives*, 2011, 11:15, BMC Medical Informatics and Decision Making (2011).
29. FDA's guidance for Industry and FDA Staff: *Cybersecurity in Medical Devices: Quality System Considerations and Content of Premarket Submissions Guidance for Industry and Food and Drug Administration Staff*—FINAL GUIDANCE (September 2023). Available online at https://www.fda.gov/regulatory-information/search-fda-guidance-documents/cybersecurity-medical-devices-quality-system-considerations-and-content-premarket-submissions. Accessed 6 July 2025.
30. Dhruva et al (2020) Attribution of Adverse Events Following Coronary Stent Placement Identified Using Administrative Claims Data. J Amer Heart Assoc 9(4). https://doi.org/10.1161/JAHA.119.013606
31. FD&C Act, section 513.
32. 21 C.F.R. § 812.3.
33. FDA's guidance for Industry and FDA Staff: *FDA and Industry Procedures for Section 513(g) Requests for Information under the Federal Food, Drug, and Cosmetic Act Guidance for Industry and Food and Drug Administration Staff*—FINAL GUIDANCE (August 2024). Available online at https://www.fda.gov/regulatory-information/search-fda-guidance-documents/fda-and-industry-procedures-section-513g-requests-information-under-federal-food-drug-and-cosmetic. Accessed 6 July 2025.

34. FDA Webpage, Premarket Notification 510(k). Available online at https://www.fda.gov/medical-devices/premarket-submissions-selecting-and-preparing-correct-submission/premarket-notification-510k. Updated August 22, 2024. Accessed 6 July 2025.
35. FD&C Act, section 513(f)(6).
36. FDA's Webpage, Medical Device Accessories. Available online at https://www.fda.gov/medical-devices/classify-your-medical-device/medical-device-accessories. Updated 14 August 2020. Accessed 6 July 2025.
37. 21 C.F.R. § 820.
38. 21 C.F.R. § 801.
39. 21 C.F.R. § 809.
40. FDA's guidance for Industry and FDA Staff: *Intent to Exempt Certain Unclassified Medical Devices from Premarket Notification Requirements Guidance for Industry and Food and Drug Administration Staff*—FINAL GUIDANCE (February 2019). Available online at https://www.fda.gov/regulatory-information/search-fda-guidance-documents/intent-exempt-certain-unclassified-medical-devices-premarket-notification-requirements. Accessed 6 July 2025.
41. 21 C.F.R. § 807, Subpart E.
42. FDA's guidance for Industry and FDA Staff: *Deciding When to Submit a 510(k) for a Software Change to an Existing Device Guidance for Industry and Food and Drug Administration Staff*—FINAL GUIDANCE (October 2017). Available online at https://www.fda.gov/regulatory-information/search-fda-guidance-documents/deciding-when-submit-510k-change-existing-device. Accessed 6 July 2025.
43. C.F.R. § 814.
44. FDA's guidance for Industry and FDA Staff: *De Novo Classification Process (Evaluation of Automatic Class III Designation) Guidance for Industry and Food and Drug Administration Staff*—FINAL GUIDANCE (October 2021). Available online at https://www.fda.gov/regulatory-information/search-fda-guidance-documents/de-novo-classification-process-evaluation-automatic-class-iii-designation. Accessed 6 July 2025.
45. FDA's guidance for Industry and FDA Staff: *Humanitarian Use Device (HUD) Designations Guidance for Industry and Food and Drug Administration Staff*—FINAL GUIDANCE (September 2019). Available online at https://www.fda.gov/regulatory-information/search-fda-guidance-documents/humanitarian-use-device-hud-designations. Accessed 6 July 2025.
46. FDA's guidance for Industry and FDA Staff: *Humanitarian Device Exemption (HDE) Program Guidance for Industry and Food and Drug Administration Staff*—FINAL GUIDANCE (September 2019). Available online at https://www.fda.gov/regulatory-information/search-fda-guidance-documents/humanitarian-device-exemption-hde-program. Accessed 6 July 2025.
47. FDA's guidance for Industry and FDA Staff: *Requesting FDA Feedback on Combination Products Guidance for Industry and FDA Staff*—FINAL GUIDANCE (December 2020). Available online at https://www.fda.gov/regulatory-information/search-fda-guidance-documents/requesting-fda-feedback-combination-products. Accessed 6 July 2025.
48. FDA's guidance for Industry and FDA Staff: *Principles of Premarket Pathways for Combination Products Guidance for Industry and FDA Staff*—FINAL GUIDANCE (January 2022). Available online at https://www.fda.gov/regulatory-information/search-fda-guidance-documents/principles-premarket-pathways-combination-products. Accessed 6 July 2025.

Chapter 3
FDA Meetings and Device Testing

3.1 Interactions with the FDA

3.1.1 Publicly Available Information

Along the way, device Sponsors are likely to have a myriad of questions. However, there are many tools that can help. Class I, Class II, and Class III medical device Sponsors all have an obligation to read and understand the regulations and guidances that apply to their product. The regulations and guidances are readily available on the internet and on the FDA's website. The advantage of accessing these regulations and guidances online is that updates are posted dynamically (versus reading them in a book where updates are not possible).

The FDA has also created the Division of Industry and Consumer Education (otherwise known as DICE), whose purpose is to provide general answers to phone and email questions from the medical device industry and consumers regarding medical devices and radiation-emitting electronic products. DICE is also responsible for the educational resources on the FDA's website that help those in the industry better understand and interpret U.S. regulations and FDA device policies. DICE is the medical device Sponsor's single best source for general information.

FDA's webpage, Medical Device Databases [1], contains links to all the different databases that can be searched for information pertaining to medical devices. For devices cleared via a 510(k), you can review the FDA database for 510(k) summaries that contain information about the device's indications for use, product code, technological characteristics, testing conducted, and predicate used to demonstrate substantial equivalence. For devices granted through the De Novo pathway, you can review the FDA database for the Classification Order and Decision Summary, which contains information regarding the regulation name, number, and class; the product code; special controls; testing conducted; and risks and mitigations. For devices approved via a PMA, you can review the FDA database for the Approval Order and the

G. B. Halt Jr. et al., *FDA and Intellectual Property Strategies for Medical Device Technologies*, https://doi.org/10.1007/978-3-032-12649-8_3

Summary of Safety and Effectiveness Data (SSED). These documents provide important information about the development history of a given device, including a description of the nonclinical and clinical studies that served as the basis for approval for that device. Although the information provided in the aforementioned documents is device- and application-specific, it can give the device Sponsor a good feel for the FDA's thinking in a particular area, and the level of evidence needed more broadly. More information about SSED as well as device Sponsor responsibilities related to the SSED is provided in Chap. 5.

Finally, medical device Sponsors can view videos of teaching modules, seminars, and recorded webinars that cover a variety of FDA topics, policies, and guidances under a program entitled CDRH-Learn. Information under this program is updated routinely and often provides up-to-the-minute guidance on the FDA's thinking on specific topics helpful to device Sponsors.

The FDA's expectation is that Sponsors will exhaust all publicly available resources prior to contacting them for additional information. However, despite all the information and resources available today, Sponsors may still have questions, especially those specifically related to their device under development. That's where talking directly with the FDA comes in. The FDA encourages dialog with Sponsors, but there are necessary procedures that companies must follow. There are also some clear "do's" and "don'ts" when trying to talk to the FDA. Throughout this chapter, we provide best practice recommendations for the device Sponsor.

3.1.2 Talking with FDA—U.S. Agent

Medical device Sponsors can contact FDA general help lines (i.e., DICE) directly for general regulatory, educational, and process questions. However, when a Sponsor wants to interact with CDRH, the medical device Center at the FDA, to ask specific questions about an investigational or marketed device, that Sponsor must follow basic FDA rules of engagement.

Prior to marketing a device in the U.S., a company that resides outside the U.S. is not required to have a U.S. Agent in order to submit a Q-Submission (i.e., a meeting request, discussed more in Sect. 3.1.4) nor is a U.S. Agent required to submit a HDE, 510(k), or De Novo. However, a company that does not reside in the U.S. may find it helpful to establish a U.S. Agent or Primary U.S. Correspondent, to be the Sponsor's primary point of contact with the FDA to communicate with the FDA on their behalf during typical U.S. business hours. For an IDE, the Sponsor of the clinical investigation *must* reside in the U.S. [2]. For a PMA, if the applicant does not reside in the U.S. or have a place of business within the U.S., the PMA must be countersigned by an authorized representative residing or maintain a place of business in the U.S [3].

When marketing your device, if you are a company that does not reside in the U.S. or maintain a place of business in the U.S. (a P.O. Box is not acceptable), you are required by the FDA to appoint a U.S. Agent and must submit formal documentation to the FDA (see FDA's webpage, U.S. Agents [4], regarding the FDA Unified Registration and Listing System (FURLS) for additional details). Specifically, any foreign device company engaged in the manufacture, preparation, propagation, compounding, or processing of a device imported into the U.S. must identify a U.S. Agent for that company. The FDA requires a U.S. Agent so that they can communicate directly with a representative of the company during typical U.S. business hours by email or phone. This requirement also applies to drug companies as well as companies developing, manufacturing or importing combination products.

When and if a U.S. Agent or Primary Correspondent is established, it should be noted that the FDA will only interact directly with the U.S. Agent/ Primary Correspondent when contact with the Sponsor is needed. FDA will not speak with individuals who are not authorized to interact with the FDA on the company's behalf.

Example The job of a U.S. Agent/Primary Correspondent is to:

- Assist the FDA in communications with the company.
- Respond to questions concerning the company's device(s) that are imported or offered for import into the U.S.
- Asist the FDA in scheduling inspections of the company. If FDA is unable to contact the company directly or in a timely manner, the FDA may provide information directly to the U.S. Agent. This direct communication is considered by the FDA as legally equivalent to providing that information directly to the company.

For companies that do have one or more offices in the U.S., it is a best practice to assign one consistent contact person to interact with the FDA on the company's behalf. The contact person must be someone who can legally serve as a company's representative. Alternatively, the contact may be a consultant authorized by the company to serve in this role. The latter commonly occurs with small device companies, where the members of the company wish to delegate this responsibility to an experienced regulatory professional. This contact person has all the same job responsibilities as the U.S. Agent does for a foreign company.

As a general guideline, if a company has only a few products, one person may be able to handle all interactions with the FDA. However, if a company has many products, it may be more beneficial to assign a dedicated contact person to champion a particular product or set of related products. The salient point here is that the interaction with the FDA is about building a strong working relationship with the FDA lead reviewer, who is assigned to serve as the sole source of contact with the company for a given submission and often times for a given product (depending on the length of time between submissions for the same product and FDA resources). Over the course of development and marketing, the Sponsor contact person and the FDA review team can have countless interactions. Companies who change their primary regulatory contact person may be less likely to foster collaborative and collegial working relationships with the FDA.

3.1.3 Talking with FDA—Guidance Specific to Sponsor's Device

During the development program for a device, it is common for Sponsors to have questions. These questions can range from the most basic, "How will FDA classify my device?" to more complicated questions surrounding whether clinical trial evidence is required in your submission package, or whether a given proposed clinical study design (patient population, endpoints, sample size) is adequate to definitively demonstrate safety and efficacy. After combing the regulations, guidances, and competitive regulatory landscape, questions may linger, and this is the time to consider meeting with the FDA.

A common mantra in the regulatory industry is that one should meet with the FDA "early and often." As you are developing your medical device, this can seem like overkill. Some products may need only a few interactions with the FDA because there is clear regulatory guidance or regulatory precedent that can be followed. However, the mantra is sound advice, particularly if your device is novel. For the Sponsor of a novel, never-seen-before, innovative device, it's prudent to introduce FDA to the device and discuss key development activities and regulatory milestones to ensure rapid, cost-effective development. Board members, key opinion leaders, and investors see increased value in programs where FDA input has been sought and agreement has been achieved. In fact, FDA encourages Sponsors to come to them to get answers to their device questions and has strongly suggested that interactions can save device manufacturers significant time and financial resources over the course of development by getting timely FDA feedback.

One of the most common questions device manufacturers have is how the new device will be classified. We provided some answers to this question in Chap. 2. Other common topics for which Sponsors seek guidance from the FDA are the choice of the predicate and how to establish equivalence for products submitted via the 510(k) pathway, test data needed to support a premarket application and future product labeling, design of nonclinical and clinical studies, and what type of human factor (HF)/usability data is required in the submission package. All of these topics are critically important to the regulation of your device and FDA's review of your submission package to obtain FDA's authorization to legally market your device in the U.S. We will discuss these topics in a little more detail later in this chapter.

3.1.4 History of FDA Meeting Types

Before we discuss the types of meetings available to device Sponsors today, it is helpful to have a brief history lesson on meeting opportunities with CDRH. In 1997, the Food and Drug Administration Modernization Act (FDAMA) established two types of formal, early collaboration meetings: Determination Meetings (under section 513(a)(3)(D) of the FD&C Act) and Agreement Meetings (under section 520(g)(7) of the FD&C Act), which allowed Sponsors to obtain clear feedback on adequate clinical testing for certain devices to support marketing. Since 1995, a pre-IDE (pre-Investigational Device Exemption) meeting was made available to device Sponsors. The purpose of the pre-IDE meeting was to obtain feedback on questions related to a future IDE application. Over time, however, Sponsors sought feedback on other types of device applications in pre-IDE meetings. The laws needed to change!

In 2012, the FDA issued a draft guidance entitled, *Requests for Feedback on Medical Device Submissions: The Pre-Submission Program and Meetings with Food and Drug Administration Staff*, that expanded the pre-IDE meeting offering and made meetings with the FDA more transparent to Sponsors. Sponsors now had the ability to talk with the FDA through several different types of meetings:

- Pre-Submissions.
- Informational meetings.
- Study risk determinations.
- Formal early collaboration meetings (i.e., agreement and Determination Meetings).
- Submission issue meetings.
- PMA Day 100 meetings.

The 2012 guidance defined the meeting request process, differentiated between the types of meetings, and provided clear timeframes for meeting scheduling, thereby enhancing predictability and transparency.

The FDA updated the guidance again and in May 2025 released the final guidance entitled *Requests for Feedback and Meetings for Medical Device Submissions: The Q-Submission Program* [5], which supersedes the previous versions of the guidance. The May 2025 guidance, summarized herein, provides information regarding the scheduling of meetings and the timing of FDA feedback.

Requests for formal agency feedback are generically referred to as "Q-Submissions" or "Q-Subs." The name Q-Sub comes about because the agency assigns a Q-number to each original Q-Sub (i.e., Q18003). A supplement, or a new request for feedback, is denoted by adding "/S" to the original Q-number (i.e., Q18003/S001). An amendment, or any additional information relevant to the original Q-Sub like meeting minutes, is denoted by adding a "/A" to the original Q-number (i.e., Q18003/A001).

3.1.5 Today's Medical Device Meetings with FDA

Although meetings with the FDA are not required, Sponsors developing devices should embrace the opportunity to obtain FDA feedback to answer any specific questions that they may have on *their* device. This includes feedback on product development and classification, the requirements and/or design of nonclinical evaluations, the requirements and/or design of clinical studies, the requirements and/or design of HF/usability evaluations, or other requirements prior to making a device submission. The FDA strongly encourages device Sponsors to come to them with questions, particularly in situations when specific questions arise that do not appear to be covered by current regulation or guidance. All meetings are requested via the Q-Sub process, must be in English, and must comply with the eCopy program [6]. A CDRH Premarket Review Submission Cover Sheet should accompany the meeting request, unless it is submitted using the eSTAR (Electronic Submission Template; note that the eSTAR is only an option for certain Q-Subs as of this writing, so be sure to check FDA's website for the most current eSTAR options, as FDA is continuously updating this process).

One important note on talking with the FDA: You should only request to meet with the FDA as much as is necessary to continue to advance development or keep a product on the market. Requests just to "meet" with the FDA with no clear purpose should be avoided.

Every submission to the FDA should be clear and concise. It is often "easy" to be either overly verbose or very scant with information; both of these approaches can be very frustrating to FDA reviewers. The FDA wishes to help you get your device to the market, so help them help you. A balance must be struck in the amount of information you provide—you need to provide enough information so the agency can provide useful and meaningful answers; however, you do *not* need to provide excruciating detail for every little nuance. As we will discuss in Chap. 4, the same theory applies to your submission package.

Anytime you submit information to the FDA, it must be accompanied by a cover letter. The cover letter you include with your meeting request should include contact information for the submitter, the Q-Sub type (only 1 type per submission), the requested method of feedback (e.g., written feedback only or a meeting). If the Q-Sub request includes an option for a meeting, the cover letter should also include the requested meeting duration (in general, CDRH will limit meetings to one hour because if longer is needed, the scope of the Q-Sub likely needs to be limited), a proposed agenda, meeting format being requested (in person, videoconference, or teleconference), proposed dates and times you will be available for a meeting, and list of Sponsor attendees. If you would like specific FDA personnel to attend the meeting (i.e., staff from CBER or CDER), you will also need to indicate this in the cover letter. Do note that the FDA requests a minimum of three date/time options—it is wise to make these as open as possible. A good example would be "anytime during November 3–7" to allow the FDA to coordinate their schedules too.

Along with the cover letter, there are certain aspects of your request that should be clearly identified within your Q-Sub. The purpose of the meeting, along with your goal(s) related to the outcome, should be clearly defined. A fairly detailed description of the device or product (combination products), including how it functions and the basic science behind the device concept. If the manufacturing process could impact safety or efficacy, then it could also impact any recommendations from the agency and should therefore be included in your Q-Sub request. You should also include the proposed indications for use (or intended use), as well as a listing of your regulatory history (previous communications from the FDA).

Additional requirements based on meeting type are provided in the respective section below. Of note, the guidance provides a detailed summary of the required content and review process by submission type. The guidance also provides the current timelines for when Sponsors can anticipate receiving feedback or determinations. As always, review the most recent guidance update for specifics related to agency expectations and timelines.

Importantly, for all Q-Sub meetings, the submitter of the Q-Sub is responsible for drafting and submitting meeting minutes to FDA as an amendment to the Q-Sub within 15 days of the meeting. If FDA has no edits to the submitter's meeting minutes, they will be considered final; FDA will communicate their acceptance via email. If FDA does edit the submitter's meeting minutes, the revised version will be emailed to the submitter within 30 days; FDA's edited meeting minutes are considered final 15 days after they are sent to the submitter, unless the submitter lets FDA know there is a disagreement.

3.1.5.1 Pre-submissions

A Pre-Submission, or "Pre-Sub" meeting enables medical device Sponsors to obtain formal FDA feedback on specific questions to guide product development and/or application preparation prior to an intended submission of a marketing application. Sponsors can use this type of meeting to hear the FDA's viewpoint on specific questions. Questions can include those related to proposed nonclinical studies and clinical study design/acceptance criteria. The FDA strongly recommends the Sponsor engage in Pre-Sub meetings *prior* to executing planned studies. For a Pre-Sub, a Sponsor can request written feedback only or a meeting (written feedback will be sent by FDA prior to the meeting).

Per FDA guidance, the contents of the Pre-Sub should include the planned future submission type (i.e., IDE, IND, marketing submission, etc.), an overview of the development history of the device, and the list of Sponsor questions. The Pre-Sub document must include all relevant background information for FDA to be able to answer the Sponsor's questions. In general, only a study synopsis is needed for FDA to provide meaningful feedback on your proposed studies. However, FDA does permit Sponsors to submit a full protocol for their review and comment prior to the Sponsor starting the study if the Pre-Sub pertains to a nonsignificant risk (NSR) device, an IDE exempt device, or a study that is planned to be conducted OUS. (outside of the United States).

The guidance provides information for the review process. Within fifteen (15) days of receipt of a valid eCopy or eSTAR, the FDA will conduct an acceptance review of the meeting request. The checklist the FDA uses is provided in Appendix 1 of the guidance—it is a very helpful resource to ensure you are requesting the correct meeting type.

The FDA's written comments can be expected seventy (70) days after the request is received or five (5) days prior to a scheduled meeting, whichever comes first. If the FDA grants medical device Sponsors a meeting (in person, videoconference, or teleconference), the meeting is typically scheduled to occur within seventy (70) to seventy-five (75) days of the FDA's receipt of the meeting request.

3.1.5.2 Submission Issue Requests

The purpose of a Submission Issue Request (SIR) is exactly as one might expect: to seek feedback from the FDA on deficiencies identified by the FDA in the Sponsor's 510(k), De Novo, IDE, HDE, PMA, Investigational New Drug (IND) application, New Drug Application (NDA), or Biologics License Application (BLA) (including associated amendments or supplements, i.e., "FDA's submission letter," as shortened in subsequent paragraphs).

A SIR is intended to facilitate interactions between the Sponsor and the FDA to quickly resolve or clarify the identified issues so a program can move forward. As such, a SIR should contain any specific questions the Sponsor has relevant to their planned response. If additional data is required, the Sponsor should provide a protocol along with the rationale for the chosen approach. For SIRs, the Sponsor must choose whether they want a meeting with FDA *or* written feedback.

Note that Sponsors are expected to formally respond to any FDA submission letters within the requested timeline, regardless of whether they submit a SIR.

SIRs should include a cover letter that identifies the submission number that is prompting the SIR, the CDRH Premarket Review Submission Cover Sheet (FDA Form 3514), the purpose of the meeting, applicable background information, preferred feedback format (i.e., written feedback or a meeting; if a meeting, type requested [in person, videoconference, or teleconference]), and the questions for which the Sponsor is seeking FDA's feedback.

Because a SIR is directly related to a submission review, a response to a SIR is prioritized. If a Sponsor provides their SIR to FDA within sixty (60) days of receiving FDA's submission letter, FDA will strive to provide feedback within twenty-one (21) days. If the SIR is provided to FDA more than 60 days after receiving FDA's submission letter, FDA will strive to provide feedback within seventy (70) days.

Note No Sponsor wants to have to request a Submission Issue meeting, but do not shy away from this type of meeting if the FDA has indicated that there are deficiencies in what they have received. Earnest attempts to provide the FDA with what they require to continue to do their work will be met with sincere appreciation. **Happy reviewers often make the most efficient reviewers!**

3.1.5.3 Study Risk Determinations

Medical device studies are classified into one of three risk categories: significant risk (SR), nonsignificant risk (NSR), and exempt from IDE regulations [7]. For studies that are not exempt, Sponsors are responsible for making the initial risk determination (SR or NSR) and presenting it to the Institutional Review Board (IRB). When an IDE is submitted, the FDA determines whether a device study is SR or NSR. However, Sponsors (but also clinical investigators and IRBs, if they do not agree with the Sponsor's determination) can request the FDA's ruling on study risk [8].

Example Our coronary stent and CardioColander examples would very obviously be classified as significant risk. However, it may not be obvious whether our blood pressure monitor would be classified as nonsignificant risk. Therefore, we may submit a study risk determination to the FDA prior to beginning a clinical study.

A study risk determination (SRD) should include a description of the device and how it will be used, a full protocol for the study in question, and a description of the intended patient population.

There is no meeting with FDA to discuss study risk determinations. The FDA will issue a formal, binding letter to the Sponsor indicating the study risk classification within ninety (90) days of receiving the request. Because the FDA is the final decision-maker on study risk, IRBs do not need to conduct separate risk assessments if FDA has issued a decision.

3.1.5.4 Informational Meetings

Device Sponsors can request a FDA meeting as a way of introducing their device and sharing information about development plans *without* the expectation of feedback. Informational meetings are appropriate to set the collaborative stage to help the FDA learn about the device being developed and be notified of the submissions that the company plans to make to FDA in the near term (6–12 months). These meetings can be particularly helpful if the device is complicated and/or significantly differs from current device technologies. They also can help the FDA plan their resources. As such, although these meetings occur infrequently, there can be significant value to meeting with the FDA under these circumstances, even though the FDA participates in a "listen only" mode (i.e., they may ask questions for clarification, but they will not be providing feedback to the Sponsor). The FDA encourages these meeting requests but has stated that they will honor them only when FDA's resources allow. Informational meetings are typically scheduled within ninety (90) days of the FDA's receipt of the meeting request, again, as resources permit.

Example With respect to our product, CardioColander, because it is such a novel device, we would be wise to request an informational meeting with the FDA. During this meeting, we would show the FDA our schematics and ideally a prototype of the device to familiarize the agency with our device concept and our intended use. The FDA will not make recommendations about our concept or development program, but they will be familiar with what we are developing and why, which will smooth the process when we want to interact with the FDA down the road.

3.1.5.5 Other Q-Subs

There are other interactions a Sponsor can have with the FDA that will be given a Q-number even though they are not technically Q-Sub meetings. As always, check with current guidance for information related to any interaction you would like to have with the FDA.

PMA Day 100 Meetings

Sponsors of PMA applications are entitled to a Day 100 meeting during which the FDA will discuss any deficiencies in the application that have been identified to date during their review. This meeting type is elaborated upon in the FDA guidance, *Requests for Feedback and Meetings for Medical Device Submissions: The Q-Submission Program* [5] (and also see *Cybersecurity in Medical Devices: Quality System Considerations and Content of Premarket Submissions* [9]). Per regulation, the PMA Day 100 meeting must occur no later than one hundred (100) days after the FDA has officially filed the PMA. Prior to the meeting, FDA will provide the Sponsor with a description of the deficiencies; note that the list of deficiencies will be sent by FDA regardless of whether a Day 100 Meeting was requested by the Sponsor. Best practice dictates requesting a Day 100 meeting in the cover letter of the initial

PMA submission. It is also possible to submit a separate Q-Submission, no later than 70 days after the filing of the PMA.

During this meeting, the FDA will also discuss an estimated time to complete their review, they will indicate whether an advisory panel will be invoked, they will mention the need for any additional premarketing data, and (if applicable) will discuss any postmarketing obligations that will be imposed on the Sponsor.

Let's assume the following scenario for our CardioColander medical device. During development, we chose, for business reasons, not to have a Determination Meeting with the FDA to discuss our development plan and clinical protocol. We did not want to spend the time waiting for the FDA's feedback, nor did we want to spend the money to talk with the FDA in person. In fact, we were so excited about our product and so confident we were studying it correctly, we did not meet with the FDA at all during development.

Just before we submitted our PMA, we hired a new Head of Regulatory Affairs who reviewed our application. She was particularly concerned about some gaps she felt she saw in the PMA (e.g., we did not conduct human factors testing on our device). She reluctantly agreed to submit the PMA, but insisted that we request a Day 100 meeting because she wants the opportunity to fully understand any issues the FDA identifies during their review as early as possible. Although we would strongly recommend meeting with the FDA throughout the development process, this scenario provides an excellent example of a situation where a Day 100 meeting may be particularly useful.

Formal Early Collaboration Meetings

As described above, there are two types of early collaboration meetings: Determination Meetings and Agreement Meetings. Both of these early collaboration meetings are described in detail in the FDA guidance *Early Collaboration Meetings Under the FDA Modernization Act (FDAMA); Final Guidance for Industry and CDRH Staff* [10]. For medical device Sponsors who plan to submit a PMA or PDP (product development protocol), a Determination Meeting should be requested to talk with the FDA about the type and amount of scientific evidence needed to ensure the device is effective for its intended use [11]. The two major objectives for Sponsors engaging with the FDA in a Determination Meeting should be to obtain feedback from FDA on whether clinical studies are required to establish device efficacy, and to agree upon the most efficient development program to establish device efficacy that has a reasonable chance of succeeding. A successful Determination Meeting is one where the Sponsor exits with a clear development pathway that is (at least at a high level) acceptable to FDA.

Determination Meetings are scheduled on a date that is mutually agreed upon between FDA and the company, within thirty (30) days of the FDA's receipt of the meeting request. FDA will meet with the Sponsor either in person, via videoconference, or via teleconference. The applicant and FDA will both draft meeting minutes; the applicant should provide their meeting minutes to FDA within seven (7) days of the meeting, for FDA's review and comment. Within thirty (30) days of the meeting, FDA will provide the applicant with the memorandum of agency determination that reflects the outcome of the meeting and the agency's binding determination.

An Agreement Meeting is an option for anyone planning to investigate the safety and efficacy of any implant or Class III medical device. Sponsors of Class III medical devices or implants that are intended to be submitted via the 510(k) pathway are not eligible for a Determination Meeting and thus must request an Agreement Meeting to solicit FDA feedback. The purpose of the Agreement Meeting is to achieve consensus with FDA on the critical components of an investigational plan including a clinical protocol to investigate the device's efficacy and/or safety [12, 13].

Agreement meetings are scheduled within thirty (30) days of FDA's receipt of the meeting request or on a date that is mutually agreed upon between FDA and the company. Like the Determination Meeting, FDA will meet with the Sponsor either in person, or via teleconference. The applicant and FDA will both draft meeting minutes; the applicant should provide their meeting minutes to FDA within seven (7) days of the meeting, for FDA's review and comment. Within thirty (30) days of the meeting, FDA will provide the applicant with the memorandum of agency agreement that reflects the outcome of the meeting and the agency's binding determination.

For both Determination and Agreement Meetings, any determinations and agreements made during the meeting are documented in writing by FDA, shared with the meeting applicant within thirty (30) days of the meeting, and are binding.

That is, the FDA's determination regarding necessary scientific evidence cannot legally be changed unless FDA establishes that adhering to the determination could be contrary to public health. Similarly, FDA's agreement on major aspects of the Sponsor's investigational plan in the establishment of device efficacy and safety is binding. It can only be changed after the Sponsor has had the opportunity to meet with FDA to discuss the issues.

The Sponsor is also bound by these determinations/agreements and cannot change the basic tenets of the determination/agreement (e.g., intended use and indications, product design, investigational plan, clinical study protocol, etc.). Significant changes by the Sponsor may nullify formal determinations/agreements made with the agency. Any changes that the Sponsor wishes to make that impact the determinations/agreements should be discussed with FDA in advance of those changes being implemented.

Using our medical device examples, we would be very smart to engage FDA in a Determination Meeting to discuss the development plan and clinical testing protocol for CardioColander. We would prepare our meeting request, which would include the details of our development plan (testing already done and proposed nonclinical and clinical testing), and a synopsis or full clinical protocol for FDA to consider. We could expect to meet with FDA in approximately thirty (30) days of the FDA receiving our meeting request. Ideally, we would want to meet face-to-face (either in person or via videoconference) with FDA to discuss our ideas and gain consensus. After the meeting, we would submit Sponsor meeting minutes within the seven (7) days requested per the guidance document and would wait approximately thirty (30) days for FDA to issue their formal and binding determination.

Breakthrough Devices Submissions

Submissions associated with the Breakthrough Devices Program, such as requests for inclusion in the program and requests for FDA feedback, are tracked via the Q-Sub program. Additional information regarding the Breakthrough Devices Program can be found on FDA's website and in their 2024 guidance document, *Breakthrough Devices Program* [14].

Safer Technologies Program Submissions

Submissions associated with the Safer Technologies Program (STeP), such as requests for inclusion in the program and requests for FDA feedback, are tracked via the Q-Sub program. Additional information regarding STeP can be found on FDA's website and in their 2021 guidance document, *Safer Technologies Program for Medical Devices* [15].

Accessory Classification Request Submissions

Accessory Classification Requests, for a new or existing accessory type, are tracked via the Q-Sub program. Additional information regarding Accessory Classification Requests can be found on FDA's website and in their 2017 guidance document, *Medical Device Accessories—Describing Accessories and Classification Pathways* [16].

3.1.6 Who Should Attend an FDA Meeting?

Sponsors have different philosophies on who should attend FDA meetings. Our general guidance is that *only* those who will be speaking or might need to answer FDA questions during the meeting should attend. FDA meetings are not a time to train staff, educate investors, or leverage politicians with an agenda. FDA wants to hear from company experts from each discipline primarily, followed by experts in the field (including regulatory affairs experts), but only to the extent that each attendee contributes to a productive meeting. It may be appropriate to have the CEO of a company attend if they are also a scientific expert or is well versed in the details of the device application. However, we would discourage a CEO from attending an FDA meeting if the only reason for attendance was simply "to see what an FDA meeting is like," "to show company solidarity," or "to intimidate using a C-class executive."

> If you are granted an FDA meeting, you should make every effort to attend.

These kinds of interactions are invaluable in helping to solidify a good working relationship with FDA. You will certainly receive valuable information regarding the answers to your questions with context that would otherwise not be available in written comments. More importantly, it is often the case that you will also receive unsolicited guidance (or even "off the record guidance") that can be helpful in decision-making.

Everyone who attends the FDA meeting should take notes during the meeting so that Sponsor meeting minutes (documentation of what the Sponsor believes was discussed and agreed to during the meeting) can be submitted to FDA.

> **Expect** that if you bring your corporate lawyer (for the sole purpose of corporate legal support, i.e., the lawyer does not have any regulatory or product/question-specific contributions), the FDA may also bring their lawyers and the meeting may be less collegial and productive.

In general, FDA CDRH meeting attendees include the lead reviewer, branch chief and at least one representative from each functional discipline to ensure that they have the adequate staff to address any of the Sponsor's questions during the meeting. However, some junior reviewers, trainees, and other FDA staff may attend, depending on their availability and interest.

If there are specific FDA personnel you wish to have attend your meeting, you may identify those individuals in your meeting request. For example, if you have a drug-device combination product that is being regulated as a device, you might request a consult from the Office of New Drugs if you have questions about the drug component of your product. Similarly, if your product is designed for use in both adults and children, you may request a pediatric device expert to attend the meeting to address specific questions related to use in children.

> **Important Note**: If you are attending an FDA meeting in person at an FDA facility, security screening for all participants is required. Be sure to arrive at least thirty (30) minutes in advance to ensure you have plenty of time. Foreign visitors must receive special security clearance <u>in advance</u> of the meeting, or they will not be permitted entry to the facility.

Consult FDA's website or with your lead reviewer for detailed instructions, as the procedures for entry can change.

3.1.7 Guidelines for Interacting with the FDA

FDA meetings can be in person, via videoconference, or teleconference, or the FDA may provide written comments only to meeting requests. Additionally, phone calls and emails may occur throughout the lifecycle of a product and provide important touchpoints of interaction. In all cases, polite interaction is necessary. Prior to all meetings, it is common courtesy to provide the lead reviewer with a final meeting agenda, via email, which includes the meeting details (date, time, location), a final list of Sponsor attendees, and a list of discussion points with rough time estimates for each topic. This is especially helpful (and often requested by FDA) when only certain topics will be discussed because other topics were resolved as a result of receiving FDA's written feedback prior to the meeting.

During the meeting, respect all speakers and maintain a collaborative stance throughout the discussions. Listen carefully to what the FDA has to say—after all, that is why you are there in the first place. Do not interrupt the FDA or each other, do not raise your voice, do not threaten, "school" or "teach" the FDA, and do not talk in a condescending manner.

While not as common currently (as of this writing), due to FDA attendees being in separate locations from one another, sometimes on videoconferences or teleconferences the FDA may place you on a brief hold to enable them to discuss a topic in private prior to providing their final guidance during the meeting. During this time, be sure to remain silent so that the FDA can discuss. As Sponsor, you may also put FDA on a brief hold if you need to have a quick internal discussion prior to providing a response to FDA. The time on hold should be kept to the shortest period possible. Be certain that you are indeed muted/on hold (there have been FDA meetings where the Sponsor later learned that the FDA could hear everything that was said!). Additionally, if you can hear FDA talking when they indicated they would go on hold, respectfully interrupt and let them know you can hear them talking. This builds trust and demonstrates solid ethics.

At the end of the meeting with FDA, summarize any action items and be certain to thank FDA for their time. They spent as much time, if not more, preparing for the meeting than the Sponsor has.

3.1.8 What Types of Questions Should Be Asked (and Not Asked)?

Composing the right questions in a meeting request is one of the biggest challenges of the entire process. It is important that the questions not be "half-baked." FDA will not answer open-ended questions, nor will they do what they perceive the Sponsor's job to be. They are not going to plan your device development pathway for you. They also typically do not offer answers to questions that have not been asked.

For example, let us consider our CardioColander medical device. We would like to engage FDA in a discussion about the clinical requirements to establish definitive efficacy and safety of our device.

> A poor question would be: "What kind of clinical study does the FDA require to establish efficacy of CardioColander?"

> A better question would be: "We have submitted a protocol for the proposed definitive efficacy and safety study of CardioColander for the purpose of lowering LDL cholesterol levels in the blood. Does the FDA agree with the specified primary efficacy endpoint of LDL (measured as FF [total cholesterol {TC} minus high-density lipoprotein {HDL}-cholesterol minus triglycerides {TGs}/5 in mg/dL]) at Week 12 after CardioColander has been implanted?"

Additionally, we have chosen to use a prototype device made of stainless steel rather than titanium because it is more cost-effective for us at this time.

> A poor question would be: "Titanium is too expensive to use in our pig proof-of-concept trial. We have a prototype made out of stainless steel. Does the FDA agree that we can use the stainless-steel device in the planned pig study?"

> A better question would be: "Our finish device will be comprised of a titanium casing. However, our current prototype is comprised of stainless steel. We have tested the properties of both materials and have concluded based on the data contained herein that stainless steel would be an adequate material for testing proof-of-concept efficacy and safety of our device in our planned pig study. Does the FDA agree with our conclusion?"

It is also a good rule of thumb to consider each aspect of product development (device controls, and nonclinical, HF, and clinical testing) when coming up with questions for the FDA. Each aspect of development must progress toward a final, commercializable device with evidence to support its use. You do not want any elements to lag behind. If you are not sure whether a particular aspect of your program is sound, it is better to ask FDA than to be surprised during application review. Rework is costly, time-consuming, and entirely avoidable with the right interactions with FDA.

Do not assume that your plans for development are, or will be, acceptable to FDA. It is also important to realize that FDA does not care if your company can afford to do what the FDA requires to progress development. The FDA's purview is to ensure that the device being developed or maintained on the market is safe and effective for its intended use. In general, FDA is also not concerned with reimbursement or commercial considerations once the product is being marketed. Whether a company succeeds or fails (or makes money or goes bankrupt) is outside their jurisdiction.

Finally, in every interaction with FDA, we would recommend you ask FDA specifically one "catch-all" question: "Does FDA have any additional guidance they are willing to offer regarding our program?" This question signals to FDA a Sponsor's openness to the process and to FDA's guidance. We have seen many times where FDA will offer up valuable insights, or points to consider, that dramatically change development. If FDA does provide guidance, be sure to follow it or have a very good justification for why you are not.

3.2 The 510(k) Pathway: Predicates and Substantial Equivalence

3.2.1 Establishing the Predicate

As mentioned earlier, on FDA's website, there are several databases that are helpful to the medical device Sponsor. One such database is the FDA 510(k) database that contains detailed information about all of the medical devices that have been cleared to date by FDA under the 510(k) regulatory route. This database is searchable using a variety of device product characteristics, and the FDA provides detailed instructions on how to best search the database to find the information that you seek. If you are trying to determine how your medical device would be classified more broadly or are looking for a device that is similar to yours to use as a predicate device, this database is a good place to start.

The goal when selecting a predicate device is to identify a legally marketed device (as defined in Sect. 2.4.3.1) that is similar in the intended use and in the technological characteristics to the device in which you want to market in the U.S. ("new" device). For purposes of this discussion, the intended use of a device encompasses the general purpose of the device or its function. It also encompasses the indications for use (the "indication," or the disease or condition the device will diagnose, treat, prevent, cure or mitigate, including a description of the patient population for which the device is intended), as we have discussed previously in Chap. 2.

Although applications with a single predicate device are most simple and are preferred by the FDA, do not feel that you must find an exact match to a single cleared device. In fact, it is rather common for medical device Sponsors to use more than one predicate device to help establish substantial equivalence. However, if a Sponsor does choose to use multiple predicate devices in an application, the FDA requires that a primary predicate—a legally marketed device that has an indication and technological characteristics that are most similar to the new device—be identified. The guidance document referenced below is a good resource for understanding how to establish a predicate device.

It should be noted that the use of a "split predicate" is not permitted. That is a situation where a medical device Sponsor attempts to use one predicate to try to establish substantial equivalence for the indication and an altogether different predicate to try to establish equivalence for the technological characteristics in a cleared device that has a different intended use.

Whether one or multiple predicate devices are used, a Sponsor establishes the basic similarities between a new device and a predicate device (or devices) by comparing the specifications, materials, and technology between the two devices. The FDA has gone on record indicating that very few 510(k) submissions rely solely on descriptive information but include a package of data derived from nonclinical performance testing. Some even include clinical data. The nature and amount of data needed to support substantial equivalence of both the intended use and technological characteristics depends on the type of medical device.

We recommend talking with the FDA prior to submitting your 510(k) application to gain agreement with the FDA on the extent of evidence that will be required (see Sect. 3.1.5 of this chapter for more information about formal meetings with the FDA). Additionally, you may find helpful information in the FDA's guidance on how to select a predicate (see *The 510(k) Program: Evaluating Substantial Equivalence in Premarket Notifications [510(k)]*) [17], which also includes details around the use of multiple predicates, how the FDA evaluates the intended use of devices, and how the FDA compares the technological characteristics between two or more devices. Lastly, there is an FDA guidance in draft form (issued September 2023), *Best Practices for Selecting a Predicate Device to support a Premarket Notification [510(k)] Submission* [18], that is helpful to use in conjunction with the aforementioned guidance.

3.2.2 *Establishing Equivalence*

As indicated in Sect. 2.4.3.1, according to FDA regulations, a new device is considered substantially equivalent to a predicate device if it has the same intended use *and* the same technological characteristics as a legally marketed device OR has the same intended use as the predicate, *and* has different technological characteristics but does not raise different questions of safety and effectiveness, *and* the information submitted to FDA demonstrates that the device is as safe and effective as the predicate [19].

What if the new device has different technological characteristics (e.g., it is made of different materials or has different design features)?

Sponsors can establish equivalence for new devices that have either a slightly different indications for use or different technological characteristics to the predicate by including two additional pieces of evidence:

(1) Arguments and data establishing that the new device shows the same safety and efficacy profile as the marketed predicate device.
(2) Arguments and data demonstrating that the new device does not result in different questions related to human safety and efficacy when the device is used as intended as compared to the predicate device.

If the FDA determines through the review of a 510(k) application (discussed earlier in this chapter) that the device is substantially equivalent to a legally marketed medical device, it is placed in the same class as the predicate device. However, if the FDA determines that the device is not substantially equivalent, it is termed "NSE" and is automatically placed into Class III.

FDA's July 2014 guidance, *The 510(k) Program: Evaluating Substantial Equivalence in Premarket Notifications [510(k)]* [17], contains a decision flowchart for helping Sponsors determine whether their new device is SE to a chosen predicate. Additionally, the guidance document illustrates these steps through various examples, such as when your new device has a slightly different indications for use or has some technological differences compared to the predicate. Another FDA guidance would also be helpful for instances of technological characteristics differing from the predicate, *Benefit-Risk Factors to Consider When Determining Substantial Equivalence in Premarket Notifications (510(k)) with Different Technological Characteristics* [20].

Note that FDA issued a guidance document in January 2022, *Principles of Premarket Pathways for Combination Products* [21], indicating that new device-led combination products (i.e., device-drug or device-biologic) will need to use a device-led combination product as the predicate in order to demonstrate substantial equivalence. FDA explains that this is because the addition of a biologic or drug to a device would likely result in a new intended use and/or different technological characteristics that raise different questions of safety and effectiveness compared to a device-only predicate. Similarly, a new combination product with a different active ingredient from a predicate would differ significantly in materials and/or design, which would also likely raise different questions of safety and effectiveness.

Similarly, devices containing AI may need a predicate that also contains AI (otherwise, depending on level of risk, they may need to pursue the De Novo or PMA pathways), in order to demonstrate substantial equivalence. This is because the use of AI in the new device, when compared to a similar predicate device that does not use AI, may result in a new intended use and/or different technological characteristics that raise different questions of safety and effectiveness compared to the predicate device that does not contain AI [22].

3.3 The HDE, De Novo, and PMA Pathways: Benefit-Risk Determinations

The FD&C Act (sections 513 and 520(m)(2)(C)) allows for a degree of uncertainty in terms of the benefits and risks of devices to the patient (and user, when different from the patient), in the context of evaluating devices for premarket authorization. Devices are required to have a reasonable assurance of safety and effectiveness, not an absolute assurance. The appropriate amount of uncertainty in the device's benefit and risk depends on the circumstances as well as the breadth of information about the device. In terms of benefit, FDA considers the magnitude and type, as well as the probability of the patient experiencing benefit and the duration of such benefits. In terms of risk, FDA considers the extent of probable harm/risk by assessing individual and collective: device and procedure-related adverse events (serious and non-serious) and complications; duration and probability of harmful events; and risks from false-positive or negative results (for diagnostics).

To provide clarity on the key factors FDA considers during the review process of certain device submissions, in terms of the benefit-risk determination of the device or device-led combination product, FDA issued two guidance documents in August 2019, *Consideration of Uncertainty in Making Benefit-Risk Determinations in Medical Device Premarket Approvals, De Novo Classifications, and Humanitarian Device Exemptions* [23] and *Factors to Consider When Making Benefit-Risk Determinations in Medical Device Premarket Approval and De Novo Classifications* [24].

3.4 Nonclinical Research

It would be rare indeed for a Class I or even Class II medical device application to include data from animal studies, although sometimes the FDA requests it for special Class II devices. In contrast, most Class III devices regulated via the PMA pathway, as well as all combination products, must include evidence of safety and efficacy of the novel device (or combination product) in their application. For FDA to permit testing in humans, comprehensive nonclinical testing must be performed. Animal testing may be done without submitting an application to the FDA; however, it is always prudent to discuss with the FDA any nonclinical testing prior to conducting animal studies to be sure that the proposed study/studies will meet regulatory requirements. We provided information on how to engage the FDA in those discussions earlier in this chapter.

Nonclinical testing can involve everything from bench testing of mechanical engineering performance (e.g., device materials fatigue, wear, tensile strength, mechanical, and electrical) and biological performance (e.g., in vitro [typically cell lines], ex vivo [animal and human tissue samples], in situ animal carcass or human cadaver testing) and animal testing. These kinds of testing are used for things like software and hardware/device verification and validation, reprocessing and sterilization validation, human factors, and biocompatibility.

Generally speaking, nonclinical testing should be conducted in a controlled and systematic manner and defined *a priori* in a nonclinical study protocol. The protocol should include a description of the device, the proposed indications for use, and details regarding the methods, procedures, and analysis of the study. The design of the nonclinical study should simulate the clinical use of the device and meaningful time points should be sampled. FDA's December 2019 guidance, *Recommended Content and Format of Non-Clinical Bench Performance Testing Information in Premarket Submissions* [25], is a helpful tool for determining relevant information to include in protocols, summaries, reports, etc., specific to nonclinical bench performance testing conducted by a third-party facility or the device manufacturer.

A description and justification of the animal model being used in the study is important to help the FDA interpret the results from the study. The selection of the animal model must align with the device type and purpose. When it comes to the selection of the best animal model, keep in mind there is no perfect animal model for a human disease, so even the best model will have its limitations. However, the best model will be one that allows for the best assessment of function (efficacy) and safety. This information will also support the laboratory bench engineering and biological studies that are used to set and measure that the device meets the design specifications and control. That is to say, the device will perform adequately for the indication in the intended environment of operation. Continuing with the example of the coronary stent, swine have typically been the animal of choice for coronary stent evaluation, but the rabbit iliac artery model could be an appropriate primary or complementary model to assess stent efficacy and safety.

The biological evaluation of sterile and non-sterile medical devices that come into direct, indirect, and sometime transient contact with human body is often conducted in accordance with International organization of Standardization ISO 10993 as described in the guidance document—*Use of International Standard ISO 10993-1, "Biological evaluation of medical devices—Part 1: Evaluation and testing within a risk management process"* [26]. This guidance document provides a useful framework for what biocompatibility endpoints to evaluate based on the nature of how the device is in contact with the body. In addition to ISO 10993, there are other standards that apply to biocompatibility testing, such as ICH, OECD, USP, and ASTM. For example, ASTM F2901-19 is a standard for the selection of testing that should be conducted to evaluate the potential for neurotoxicity caused by medical devices that come in contact with nervous tissue. A list of FDA-recognized standards and test methods as well as recommended premarket submission contents regarding these standards and tests are provided in the draft guidance—*Biocompatibility Testing of Medical Devices—Standards Specific Information for the Accreditation Scheme for Conformity Assessment (ASCA) Program* [27]. Screening for and characterizing extractables and leachables (e.g., plasticizers and metal ions) from device components that come in contact with the human body is used to identify what biocompatibility endpoints need to be characterized or to demonstrate chemical equivalency to a device or device component that has previously been demonstrated to be biocompatible. Recommendations for collecting and reporting of chemical characterization data are provided in the draft guidance—*Chemical Analysis for Biocompatibility Assessment of Medical Devices* [28].

In the nonclinical study protocol, safety and efficacy (if appropriate) outcomes should be defined and measured objectively. In particular, all possible risks associated with the device should be measured, including performance and handling of the device. In the case of animal studies clinical and histological pathology data should be collected, including high quality images. The things that need to be considered in the design and conduct of animal studies to assess how a medical device interacts with biological systems, including physiological, pathological, and toxicological effects of the device, and how the biological system may affect the device are outlined in the guidance—*General Considerations for Animal Studies Intended to Evaluate Medical Devices* [29].

The nonclinical study protocol should be finalized prior to study initiation. The FDA strongly prefers that nonclinical studies be conducted under Good Laboratory Procedures (GLPs, see 21 C.F.R. § 58). A final study report, which summarizes the actual study procedures, data analysis, results, and interpretation, should be prepared according to 21 C.F.R. § 58.185. The FDA will expect to receive the final report body, but also will want to receive copies of all raw data associated with the report for their own review and analyses. Because the regulatory submission requirements are evolving rapidly (particularly with respect to electronic or electronic common technical document (eCTD) submissions), be sure to review the electronic submissions regulations and guidances, and possibly consult with the FDA on the technical requirements for submitting these data to ensure that your company complies prior to submission.

Of particular relevance to our cardiovascular device theme, the FDA issued a specific guidance in 2010 for cardiovascular stent manufacturers, which provides detailed recommendations of what should be investigated in nonclinical testing (see *Non-Clinical Engineering Tests and Recommended Labeling for Intravascular Stents and Associated Delivery Systems* [30]). FDA has published many topical guidances that outline the agency's recommendations for development of devices for specific uses. In addition, there are other standards that are available for specific devices or uses of a device, for example ISO 25539-2:2020—Cardiovascular implants—Endovascular devices Part 2: Vascular stents. A wise Sponsor will explore FDA's guidance documents for all guidances that may be applicable to their specific product prior to initiation of nonclinical research.

As a best practice, conduct only as many nonclinical studies as is necessary to understand the device and how it will perform in vivo so that you can take your device into the clinic as quickly as possible. Although scientific rigor and a basic understanding of device function are needed, companies can get bogged down in so much nonclinical testing that they lose sight of the end goal. Consider a risk-based approach where you conduct basic animal research, then talk with the FDA about whether you have enough to support clinical investigation. Be prepared to continue to test aspects of your device in animals at the same time that you are testing the device in humans. This will help ensure that you continue to make efficient progress in the regulatory space without breaking the budget. The FDA wants this too.

When considering how many nonclinical studies to conduct, do not forget to see what information is available in the published peer-reviewed scientific literature. Building on the coronary stent example, consider developing a new drug-eluting stent (DES). Up to 40% of patients receiving coronary stents suffer what is called restenosis of the coronary arteries where the stents are placed. This is due to vascular tissue lining the artery growing in and around the stent resulting in the artery being blocked again. One approach to dealing with this is to add a drug to the stent that when released slowly and in a controlled manner, inhibits the tissue growth thus preventing or slowing the rate of restenosis.

If one is developing a new DES, the chances are the metal stent itself may be the same as a stent that has already been approved. It is also possible that the polymer (the substances that hold the drug to the stent and controls the release of the drug) could be the same as used for another DES. In some cases, there are studies published in the scientific literature regarding the efficacy and safety of the stent and/or the polymer. If one is lucky or smart, the drug in the new stent is one that has been approved for another indication but was found to potentially inhibit restenosis. In this case, much of the safety information about the stent, the polymer, and the drug can be obtained from the literature and/or an FDA-approved label for the drug being used for another indication. This could greatly reduce the number of nonclinical bench and animal studies (and time) needed to get the new DES to the clinic.

3.5 Investigational Device Exemption Application (IDE)

Once nonclinical research has been completed and you have solicited guidance from the FDA on the acceptability of your nonclinical coverage to support clinical use (soliciting guidance is optional but highly recommended), can a Sponsor simply design and execute a clinical study? The answer to that question all depends on the *risk* of the device. Devices that do not pose a significant risk to humans are considered nonsignificant risk (NSR) and devices that are classified as "exempt" do not require any special application to the FDA to start clinical studies. How do I know if my device is exempt, you ask? Well, your device is exempt if it complies with 21 C.F.R. § 812.2(c). Specifically, your device is exempt if it is:

- A legally marketed device used in accordance with its labeling.
- A diagnostic device that complies with the labeling requirements in 21 C.F.R. § 809.10(c) and if the clinical testing:
  - Is noninvasive.
  - Does not require an invasive sampling procedure that presents significant risk;
  - Does not by design or intention introduce energy into a subject,
  - Is not used as a diagnostic procedure without confirmation by another medically established diagnostic product or procedure,
- Consumer preference testing, testing of a modification, or testing of a combination of devices if the device(s) are legally marketed device(s) AND if the testing is not for the purpose of determining safety or effectiveness and does not put subjects at risk.
- A device intended solely for veterinary use.
- A device shipped solely for research with laboratory animals and contains the labeling "CAUTION—Device for investigational use in laboratory animals or other tests that do not involve human subjects."
- A custom device within the meaning of section 520(b) of the Federal Food, Drug, and Cosmetic Act, unless the device is being used to determine the safety or effectiveness for commercial distribution.

In most cases, the Sponsor makes the initial risk determination and the IRB where the proposed clinical study is being conducted evaluates that determination. If the IRB agrees with the Sponsor that the device is NSR/exempt, you may still need to obtain IRB review and approval of the protocol and consent participants under 21 C.F.R. § 56 and 50, respectively (see 21 C.F.R. § 812.2(c) for more information about these requirements). In these cases, the IRB often serves as the proxy for the FDA in their review, approval, and continuing review of device studies for medical devices that are classified as NSR/exempt. The Sponsor also must comply with the abbreviated IDE requirements under 21 C.F.R. § 812.2 (b), including adequate labeling, study monitoring, records retention, and safety reporting.

In contrast, devices are classified as significant risk (SR) when the medical device presents a potential for serious risk to the health, safety, and welfare of humans [31]. By regulation, these devices are:

- Implants.
- Used in supporting or sustaining human life.
- Of substantial importance in diagnosing, curing, mitigating, or treating disease, or preventing impairment of human health.

For device Sponsors whose medical device is classified as having significant risk (SR; or not exempt), an Investigational Device Exemption (IDE) application must be prepared, submitted, and FDA approval must be obtained before starting a clinical study.

Why is an IDE required for SR devices, you ask? Well, in the U.S., it is against the law to ship both drugs and non-exempt medical devices across state lines or to use drugs or medical devices that are unapproved by the FDA for investigational use or not authorized by FDA for commercial sale. For drugs, an open IND permits a company to legally import, ship, and administer an investigational drug for use in clinical trials in the U.S. The medical device equivalent of the IND is the IDE, and the requirements for the IDE are outlined in 21 C.F.R. § 812.1. An approved IDE allows an SR/non-exempt investigational device Sponsor, that otherwise would be required to comply with performance standards or have premarket authorization (clearance/grant/approval), to lawfully import and ship the device across state lines for the purpose of conducting clinical investigations. Approved IDEs are also needed for evaluation of certain modifications or new uses of a legally marketed device. For a NSR device, an IDE is considered approved for conducting studies once Sponsors meet the abbreviated requirements found in 21 C.F.R. § 812.1(b) and obtain approval from the Institutional Review Boards (IRBs) where the studies are conducted. Exempt studies are those that are exempt from most of 21 C.F.R. Part 812, with the exception of Part 812.119; however, an IRB must review an exempt study before it begins, in accordance with IRB regulations. Advice on SR and NSR studies can be found in the guidance document, *Information Sheet Guidance For IRBs, Clinical Investigators, and Sponsors Significant Risk and Nonsignificant Risk Medical Device Studies*. Additional information on exempt studies can be found in the Information Sheet Guidance entitled, *Frequently Asked Questions About Medical Devices* [32].

As we discussed previously in this chapter, if you are not sure about the risk classification of your medical device, you can hold a meeting with the FDA to have them assign a classification for you.

Regulation 21 C.F.R. § 812.20 establishes the content included in an IDE application:

- The name and address of the Sponsor.
- A summary of all prior investigations of the device and a summary of the investigational plan as described in 812.25(a) through (e).
- A description of the methods, facilities, and controls used for the manufacturing, processing, packing, storage, and, where appropriate, installation of the device.
- A list of the names and addresses of all the investigators planned to participate in the clinical trial and an example of the investigator commitment letter obligating them to comply with U.S. laws.

- A certification that all participating investigators have signed the commitment letter, a list of all the investigators participating in the investigation, and a statement that no investigators will be allowed to participate in the study until they have signed the commitment letter.
- A list of the name, address, and chairperson of the IRB (or each IRB, if multiple are used) that is overseeing the clinical study and a certification of the action(s) taken by each IRB.
- The name and address of any institution where the study is being conducted.
- If the device is to be sold, the amount to be charged and an explanation of why sale does not constitute commercialization of the device.
- A claim for categorical exclusion under 21 C.F.R. § 25.30 or 25.34 or an environmental assessment under 21 C.F.R. § 25.40 (showing how the device affects different aspects (e.g., air, water, soil) of the environment).
- A copy of all the labeling for the device (see 21 C.F.R. § 812.5).
- A copy of all forms and informational materials to be provided to the research participants, including the informed consent form(s).

Once the FDA receives an IDE application, FDA is to review and act on an IDE application within thirty (30) days of receipt. If the Sponsor does not receive a decision from FDA thirty (30) days following FDA's receipt of their IDE, they can begin their study, once they've obtained IRB approval. While this is legally an option, we recommend awaiting FDA's feedback before starting your study or contacting FDA to obtain the status of their feedback if nothing was received after thirty (30) days.

On or before Day 30 (calendar days), the FDA will contact the Sponsor with one of three possible outcomes:

- Approval—the Sponsor can proceed with the proposed clinical study.
- Approval with Conditions—the Sponsor can proceed, although conditions have been placed on the Sponsor.
- Disapproval—When the FDA does not permit a Sponsor to proceed, usually because the risk outweighs the benefits, and/or the study is not designed adequately to inform the Sponsor about a reasonable scientific question.

If the Sponsor receives a disapproval notification, the FDA will indicate the reason(s) for the determination. If approval is garnered, the Sponsor can begin the clinical trial as long as IRB approval has also been granted. If approval with conditions is issued, the Sponsor may proceed, but has up to forty-five (45) calendar days to address the issues identified by the FDA, unless stated otherwise in the decision letter. See FDA's Guidance, *FDA Decisions for Investigational Device Exemption Clinical Investigations* [33] for more information, including some excellent examples.

According to the 2014 guidance, the FDA also has the flexibility to grant "staged approval" or "staged approval with conditions." Similar to partial clinical holds in the world of drug IND regulation, these subset classifications allow the FDA to grant IDE approval or approval with conditions for a portion of the proposed study, while not permitting other aspects of the study to proceed until additional information is submitted to and evaluated by the FDA. Such gating allows the Sponsor to proceed with some data collection on the medical device in a way that protects human subjects from harm. This can often occur in situations where information about the safety of the product is obtained in a small subset (or cohort) of patients before exposing a larger group (or more vulnerable set of patients) to the device.

Consider an example with CardioColander: we propose in our IDE to expose 10,000 patients to our novel medical device in our first in human study. Given the size of that proposed patient population, it is likely that the FDA would want us to test a smaller group of subjects (e.g., 50 patients) and learn about the safety of the device in humans on a much smaller scale before opening up the study to so many subjects. Likewise, assume we wanted to test subjects who were 18-years-old up to 90-years-old. The FDA might consider restricting our initial IDE exposure to subjects who are 18-years-old to 65-years-old to evaluate initial safety before exposing geriatric subjects.

3.6 Clinical Research

Similar to nonclinical research requirements, it would be rare for a Class I or most Class II medical device applications to include clinical study data. Clinical studies are almost always conducted to support a device or combination product regulated via a PMA, but a few 510(k) devices do require clinical data to support the application to demonstrate device efficacy and safety [7]. Similar to nonclinical research, clinical research requirements may be met or at least partially met through published literature or information in the public domain.

Like the framework for clinical trials for drugs, clinical development for medical devices typically starts with pilot studies and culminates in pivotal (confirmatory) clinical trials. Pilot studies are conducted to establish initial safety in humans, device feasibility, and inform later-stage trials. Pilot studies should be small, ranging in size from exposure to a few subjects to less than 100. The goals of these pilot studies range from establishing initial safety to proof-of-concept efficacy.

Confirmatory, later-stage, pivotal studies are based on results obtained in pilot studies and are designed to specifically demonstrate efficacy and safety of the device as it is intended to be used in the marketplace. Pivotal studies can range in size from 100 subjects to thousands, depending on the type of device and purpose of the study, and are typically powered using standard statistical techniques. See the FDA's guidance entitled, *Design Considerations for Pivotal Clinical Investigations for Medical Devices Guidance for Industry, Clinical Investigators, Institutional Review Boards and Food and Drug Administration Staff* [34] as well as the guidance, *Acceptance of Clinical Data to Support Medical Device Applications and Submissions: Frequently Asked Questions* [35].

In this guidance, the FDA discusses standards for designing, conducting, recording, and reporting studies that involve the participation of human subjects, as well as requirements related to conducting studies under Good Clinical Practice

(GCP). It provides a wealth of knowledge for the medical device Sponsor to follow when embarking on clinical development work. In evaluating a PMA, the FDA relies on valid scientific evidence to determine whether there is reasonable assurance that a device is safe and effective [36]. This typically involves data from at least one well-controlled clinical trial that has clinically significant results. The guidance also discusses the necessity for device Sponsors to understand the purpose of their device and to design trials that most appropriately test the function and utility of the device for its intended use. The FDA also urges Sponsors, to address device functionality (regardless of the device's purpose), user skill and training, learning how to use the device, and human factors (or how the user interacts with the device). The next section provides more details on human factor testing.

The ethical and technical considerations of clinical studies of medical device products involving children are discussed in the 2022 draft FDA guidance (*Ethical Considerations for Clinical Investigations of Medical Products Involving Children*) [37]. In this guidance, FDA recommends that any available clinical data from published studies and reports, and/or actual use information be considered when designing a clinical trial to maximize the information obtained from and minimize the number of subjects involved in the trial. If the medical device indication(s) involve both children and adults, a properly designed clinical trial that includes both children and adult subjects could provide adequate sample sizes for both pediatric and adult populations and possibly reduce the number of trials needed.

Similar to nonclinical study planning, it is important to have a detailed protocol completed prior to initiating any clinical study. The protocol should clearly define the patient population, intended use of the device, detailed methods and procedures, and specifics regarding the analysis of the efficacy and safety data collected. The design of the clinical study should reflect the clinical use of the device in the marketplace. Similar to nonclinical testing protocols, all possible risks associated with the device should be measured, including performance and handling of the device.

Federal regulations require Sponsors to post clinical trial information for some studies online at Clinicaltrials.gov [38]. Specifically, all clinical trials of devices with health outcomes, all clinical trials other than small feasibility studies, and any pediatric postmarketing surveillance studies must be listed on this public site. As a result, Clinicaltrials.gov can be an outstanding resource for the medical device Sponsor for competitive intelligence, regulatory strategy, and concept development and modeling for future clinical study designs.

Note that just as every device is unique, so is the clinical development plan. Not all PMAs will include data from all phases of clinical development. It is best to consult with the FDA about the necessary clinical trials that must be conducted to support your device PMA.

In today's healthcare climate, clinical trials are conducted all over the world. The FDA will accept clinical data from medical device trials conducted OUS when specific conditions are met (see FDA's guidance entitled, *Acceptance of Medical Device Clinical Data from Studies Conducted Outside the United States* [39] for details). OUS trials should be prospectively defined and should follow the same requirements as US-based trials [40].

An important benchmark for FDA for their willingness to accept OUS clinical trial data is whether the agency can reasonably conclude the safety and effectiveness of a device under its conditions of use. FDA evaluates whether differences in the regulatory landscape, medical treatment, and patient population that underlie data obtained in an OUS trial would be predictive of device efficacy and/or safety in U.S. patients. The greater the differences between the OUS and U.S. climate, the less FDA feels comfortable extrapolating study results. In such cases, the FDA is likely to ask for confirmatory evidence from a U.S. clinical trial. Finally, and most importantly, FDA's expectations are that clinical trials conducted OUS will follow the same standards as trials conducted in the US. All trials, regardless of where they are conducted, must follow GCP, ensure protection of human subjects, and have investigational review board or ethics committee oversight.

The most common scenarios for when clinical data may be required in a 510(k) submission to demonstrate substantial equivalence (SE) to the predicate device are described in the July 2014 guidance, *The 510(k) Program: Evaluating Substantial Equivalence in Premarket Notifications [510(k)]* [17] and the 2023 draft guidance *Recommendations for the Use of Clinical Data in Premarket Notification [510(k)] Submissions* [41]. The most common scenarios used to determine the necessity of clinical data to demonstrate SE are:

(1) There are differences between the indications for use of the new device and the predicate device.
(2) There are differences between technological characteristics of the new device and the predicate device.
(3) SE between the new device and the predicate device cannot be determined by nonclinical testing (analytical, bench, and/or animal).
(4) A newly identified or increased risk for the predicate device suggests clinical data may be needed for the new device.

AI is starting to be used in clinical trials to collect data, conduct quality control checks, develop clinical protocols, review tables, listings, and figures (TLFs) and generate patient narratives. AI-enabled technology and devices that use patient data to diagnose disease have also come to the forefront. Examples include devices/systems to improve image quality, detect electrocardiogram (ECG) abnormalities, and assist in the diagnosis of a given disease or condition. FDA is starting to regulate this rapidly changing technology. There are two recently (January 2025) released guidance documents on the subject, *Considerations for the Use of Artificial Intelligence to Support Regulatory Decision-making for Drug and Biological Products* [42] and *Artificial Intelligence-Enabled Device Software Functions: Lifecycle Management and Marketing Submission Recommendations* [43]. For more examples of AI in assisting clinical diagnosis please refer to Sects. 2.4.3.2.1 and 2.4.3.3.1.

3.7 Human Factor and Usability Engineering

Human factor engineering (HFE) and usability engineering (UE) data are rarely reviewed by the agency for Class I and Class II devices. However, HFE and UE testing are almost always conducted to support a device or combination product; it is considered part of a robust design control development program [2, 7, 44]. The FDA has four useful guidance documents related to HFE/UE data collection and submission:

- The first, *Applying Human Factors and Usability Engineering to Medical Devices* [45], provides recommendations on appropriate HFE/UE testing to increase the likelihood that your device will be safe and effective when used commercially.
- The second, *List of Highest Priority Devices for Human Factor Review: Draft Guidance for Industry and FDA Staff* [46], provides guidance for the device types required to provide HFE/UE data in their submission package. If your device type is not in this list, it does not mean you do not need to perform HFE/UE testing. It means the agency may not require the data to review as part of your submission package.
- The third, *Purpose and Content of Use-Related Risk Analyses for Drugs, Biological Products, and Combination Products* [47], provides recommendations on the purpose of a use-related risk analysis (URRA) and how it can be used to help determine what HF data is needed.
- The fourth, *Application of Human Factors Engineering Principles for Combination Products: Questions and Answers* [48], covers the application of human factors engineering (HFE) principles to the development of combination products as defined under 21 C.F.R. part 3. This guidance also serves as the final version of the draft guidance document entitled *Human Factors Studies and Related Clinical Study Considerations in Combination Product Design and Development* [49]. This guidance document conveys information on the unique aspects of combination products that affect HFE process. For example, a key consideration in the HFE process is the inclusion in the design control of a URRA of the device, the drug, and the combination product. As there may be use-related risks associated with the device, the drug/biologic, and combination product that may not exist with the device or the drug/biologic alone.

FDA recognizes that different industry standards can provide useful information to help guide device design and testing. These include HE75, which is a hefty design tool and standard that provides processes to implement, and elements of HFE/UE testing. Another, ANSI/AAMI/IEC 62366, focuses on the safety aspects of HFE/UE testing by helping to develop and evaluate the usability of a device. Lastly, ANSI/AAMI/ISO 14971, a risk management standard, helps identify hazards associated with device use.

It is important to note that although these industry standards are recognized as useful documents, they only provide information on the process to implement HFE/

UE. These standards do not replace the FDA requirements for acceptability. In addition, indicating "our amazing device conforms to ANSI/AAMI 62366" in your application really only means that your evaluation process was possibly valid, but it does not ensure the device will be safe to use upon release to the market. FDA will ask for your data, the safety-related considerations and issues discovered during development, and the solutions implemented to ensure safety.

3.7.1 What You Need to Consider

So, what are the HFE/UE requirements? Glad you asked! In a nutshell, you need to account for the "human factor" during your design development. You need to consider the impact of:

- Each user type (i.e., healthcare provider, patient, caregiver and associated characteristics).
- Possible use environments (i.e., hospital, home, office and associated noise, lighting, temperature).
- Every step (i.e., not done, performed incorrectly).

You need to ensure that each human (user), no matter the education and experience level, or the comfort level with using a device or technology, is capable of using your device as intended in a "typical" environment. We no longer live in an environment where we categorize everything as user error and assume no responsibility. User error is defined as when an action (or lack of action), different from what the manufacturer expected, elicited a response that was different than the user expected; was not caused solely by device failure, but did or could result in harm. Your goal as the device manufacturer is to limit the possibility of user error as much as possible. As you will note in Chap. 5, "blaming" user error does not eliminate or even reduce the Sponsor's responsibility when a device is on the market.

Example Suppose you designed a blood pressure monitor for home use. The user was to secure the blood pressure cuff around his/her upper arm, but instead put the device on his/her leg. This "user error" was not a device failure per se, but was a failure of the user to follow the device instructions and likely resulted in an inaccurate blood pressure reading.

Right now, you are probably thinking—you cannot engineer around *every possible* scenario. You are correct, and the agency knows this to be true. This is where the user interface (UI) specific to your device will be of assistance. The UI includes all points of user interaction. Engineers typically think of physical controls, display elements, and the practical interface as the only options to correct a problem. However, UI also includes packaging, labeling, and device training; all of which are possible sources for solutions.

3.7.1.1 Who Will Use This Device?

While developing your device, you need to remain cognizant of who the users will be. What is special about this group of people that requires consideration during development? What level of training will be necessary to use this device? Careful consideration must go into characterizing the use population. There is always the possibility of vast differences within your user group. The FDA expects you to know the characteristics of your user group that could or will influence design.

Consider our blood pressure monitor. Who will use this device? Physicians and nurses… caregivers… patients (probably adults, but not necessarily only adults) … the health-conscious and athletes, just to name a few.

What is special about this group of people and how does knowing who makes up this group of users influence device design? There are many differences between the various users':

- Education levels—college graduates to the possibility of less than a high school education.
- Ages—which implies differing physical abilities, different comfort levels with devices and technology, and some users may have vision impairments (i.e., they wear glasses, are blind, or are color blind).
- Comorbidities—a patient battling cancer may be weaker than a healthy athlete.
- Technological abilities—the older generation, in general, may not be as tech-savvy or as comfortable with certain types of technology.
- Body types—smaller or larger diameter arms.
- There are others…

From the short list above, we see a few common threads:

- Education level and technology abilities will affect several aspects of the UI. The practical interface needs to be simple. The user manual should be written such that someone with less than a high school education understands what you want them to do. For example, placement of the cuff to get a correct reading should probably include a diagram/picture along with words.
- Different ages and comorbidities will also influence several aspects of the UI. The hook and loop fastener used cannot grip too tightly to itself (may be difficult for people to release); however, it cannot grip too loosely (in which case the device does not provide an accurate reading). The hook and loop fastener strip needs to be long enough to accommodate smaller arms as well as big, muscular arms. Also, the display should use text as well as indicator lights, and the text needs to be large enough to be easily read without blocking any other information on the display.

In contrast, consider a coronary stent. Who will use this device? The patient is just the host; they do not interact with the device, so they would not be considered users. The only user will be the surgeon implanting the stent. What is different about this group of users and how does it influence device design?

- Education levels—all college graduates with high levels of education.
- Ages—which implies a difference in experience.

From this list, you can see that your user manual and device training should be directed at a certain education level and technical expertise. For example, you can indicate in medical terms the procedure required and location for stent placement.

As you can tell from this short exercise, as your list of possible users increases, so does the list of items for which you need to be aware.

3.7.1.2 Where Will Your Device Be Used?

The Use Environment, or where the device is used, is also a major consideration that can impact device design. As we saw with the user types discussed in the previous section, the Use Environment can vary greatly. There are many differences between the various locations where a device can be used:

- Surgical suites—have good lighting; sterile environment; physicians are gowned (including gloves and protective eyewear); intense but concentrated effort; exposure to bodily fluids as a result of surgery.
- Hospital rooms—have adequate lighting; can be very noisy with many distractions; nurses and physicians wear gloves for procedures and may use protective eyewear; not a sterile environment. Design features may need to be added to prevent tampering/misuse in this less controlled environment.
- Doctor's office— similar to hospital rooms.
- Patient's home—lighting could be poor or bright; can be very noisy with many distractions; may or may not wear gloves (depending on what they are doing).
- Patient's work environment—similar to home environment, but could be "dirty" (e.g., factory), possible desire for "discrete use."
- Restaurant—lighting could be poor; very noisy with many distractions; possibly cramped for space; possible desire for "discrete use."

From this short list above, you can see that the Use Environment can vary greatly, and therefore the needs for your device will as well. If a user will need to wear gloves, a touch screen may or may not be the best idea. Display brightness necessity can be affected by the Use Environment's lighting. If your device will be used in a sterile environment, the outside will need to be designed such that it can be sterilized.

Our blood pressure monitor example could be used in the majority of the Use Environments listed above, and possibly others (e.g., pharmacies, the mall, etc.). However, our stent example is only used in the surgical suite. Very different Use Environments—very different design needs.

3.7.1.3 What Are the Critical Tasks and Use-Related Hazards?

Throughout development, you should keep an eye toward "what could go wrong with this step, and could it have a bad outcome." If a step is performed incorrectly or not at all, and the outcome would or *could* result in compromised medical care or serious harm, then that step is considered a Critical Task. You already know that if something is designated as a "hazard," it means that there is the potential for harm, even possibly serious harm. A Use-Related Hazard is the possibility for harm brought about through the use (either correct or incorrect) of a device.

> A simple non-cardiac example: insufficient insulin delivery (Use-Related Hazard: compromised medical care which could result in serious harm) because the thumbscrew was hard to push (Critical Task: pushing the button all the way down to the resting position).

Tools such as failure mode effects analysis (FMEA) and fault tree analysis will help define which tasks or steps should be considered critical. Whether or not you have a predicate device (Sect. 3.2), information about similar devices, or devices with similar components, can be very useful for determining Critical Tasks. The Manufacturer and User Facility Device Experience (MAUDE) database, the Medical Product Safety Network (MedSun) adverse event reporting program, medical device recall information, and the FDA Review Summary for a predicate/similar device component, all found on the FDA website, are good potential sources of information for determining Critical Tasks. Another good source, albeit unusual, is online video sources—people do the unexpected and they like to video themselves doing it!

In addition, analytical or empirical approaches to assess user interaction with your device are also very helpful for determining Critical Tasks. Analytical approaches can include task or heuristic analyses. Empirical data can be obtained through inquiry, interviews, or formative evaluations (i.e., simulated use).

Clearly there are a variety of ways to determine how people will use your device—both correctly and incorrectly. While knowing that someone is using your device correctly is great—knowing what people are doing incorrectly and *why* is extremely useful. If the directions were not clear; the buttons are hard to push; the display is difficult to read in low lighting; "my thumb isn't that long." These are all examples of useful feedback to help you make adjustments so your user can correctly use your device in a typical Use Environment and reduce the possibility of Use-Related Hazards.

One thing to keep in mind—in our fast-paced, ever-changing world, you need to ensure your information is current. Device development can take years, but even after your device is on the market, you will probably make changes. As an example, consider our blood pressure monitoring device again. When first developed, the only User Environments were medical. Life evolved, we became more cognizant of the possibility of heart trouble, and people started to monitor their blood pressure at home. Now, blood pressure monitors are available at the mall. Different users and Use Environments, as well as possibly updated Critical Tasks and Use-Related Hazards, must be considered when making device improvements.

3.7.2 *Reducing Risks—Mitigation Strategies for Identified Risks*

You now know what is special about the people who will use your device, possible issues in the environment where it will be used, and what tasks could result in harm. The next step is to ensure you have taken all measures possible to eliminate or reduce the risk of harm and ensure correct use.

> Design modifications are the best way to eliminate Use-Related Hazards. Inherent Safety by Design involves the use of design features to ensure a hazard never happens or is greatly minimized. As implied by the name, these features are a permanent, inherent element of the device that cannot be "turned off," deactivated for a period of time, or circumvented. An example would be a port design such that only the correct component can be plugged into that port.

If you cannot use design features to avoid a hazardous situation, the next step is to develop protective measures. These will allow a fault to happen, but not become hazardous. An example of this would be the "are you sure" boxes… are you sure you want to do that step, because it could be harmful? Alternatively, establishing a maximum input or a confirmation screen to ensure data was entered correctly would be other examples.

Unfortunately, as we acknowledged earlier, one simply cannot design for *every possible* scenario especially in the medical arena with so many different users. Again, this is why the FDA allows the use of packaging, labeling, and device training as part of the UI package. If all else fails, you can provide information to ensure the safe and effective use of your device. The obvious example would be training materials, but this also includes warning stickers or signs.

There is, of course, a caveat to the execution of reducing risks. It is critical to ensure that the modifications you made or information you provided does not introduce new or worse Use-Related Hazards. The exercises outlined in Sect. 3.7.1.3 will need repeating throughout development.

3.7.3 HF Validation Testing

Once you have identified your Critical Tasks, established mitigation strategies, and have a final device design, you are ready to begin Human Factor (HF) Validation Testing. These tests are designed to demonstrate that your entire UI package appropriately addresses Use-Related Hazards while considering all intended users in expected Use Environments. In addition, prior to initiating HF testing, you need a plan. That plan is called a protocol, and it should clearly define several elements of your test (similar to what is required for Clinical Studies, as discussed in Sect. 3.5). Developing a protocol to ensure collection of the data you need in the best way possible is very challenging. The agency is willing to review a draft protocol to help guide you. We strongly advise you to take advantage of this.

As you will recall, we discussed different types of users (Sect. 3.7.1.1) and their Use Environments (Sect. 3.7.1.2). From each of these sections, we learned:

- As your list of possible users increases, so does the list of items for which you need to be aware.
- The User Environment can vary greatly, and therefore the needs for your device will as well.

This also means that you will need to test each user in simulated conditions realistic to each User Environment (e.g., lighting, distractions). This may require that several protocols are developed and executed in order to evaluate just one Critical Task. Of note, if actual use testing is required (e.g., insulin delivery devices evaluated during at-home use), then an IDE will be required prior to performing the tests (Sect. 3.4). If actual use testing will be performed, you may need to obtain Institutional Review Board (IRB) approval before initiating the study.

By guidance, you are required to evaluate a minimum of 15 participants from each distinctive user group. Each Critical Task and use scenario must be presented in a logical sequence. Although you are allowed to group tasks during testing, you must ensure they follow a "natural workflow" for your device. It is important to note that even though you assigned a very low probability of a Critical Task occurring, you are still required to evaluate this task during HF testing. Also of note, you cannot include employees or partners as part of your testing population, as this could bias the results.

You will also need to provide a clear definition of success for this user group, in this Use Environment, performing this Critical Task. If you want to evaluate the Instructions for Use (IFU), you can only do so in an indirect manner. During the HF testing, you cannot require a test participant to consult the IFU—it is entirely their choice whether or not to use the IFU. If you are evaluating required training programs, guidance requires a minimum of one (1) hour between training and use in order to account for training decay. If training decay is a possible source for an identified Use-Related Hazard, you will need to increase the amount of time between training and use (e.g., one (1) or two (2) days). The protocol must contain the details of testing, training content and delivery, and time requirements to evaluate training decay.

The types of data collected during HF testing includes both objective and subjective. Examples of objective data collected includes observing participants perform under the critical use scenario, and any responses to knowledge-based questions. Other feedback given on the device is often considered subjective data, for example specific concerns voiced by participants related to difficulties with tasks, "I'm confused," or if you ask a participant their opinion on why a step was executed incorrectly.

Both the objective and subjective data you collect are extremely valuable. It allows you to confirm appropriate mitigation strategies are in place. If a mitigation strategy did not work, or if other, unexpected Use-Related Hazards were introduced, it is "back to the drawing board" for risk mitigation (Sect. 3.7.2) and then (re-)validating via additional HF testing. The study protocol should include information on how to detect previously unanticipated errors.

It is important to note, you cannot simply "cover everything in the IFU or required training" and be okay. As we all know, there are people who never read instructions. If an update to the IFU or training is warranted, and it mitigates the risk, this will need to be proven through additional HF testing.

A source of frustration is that HF testing can be very cyclical; evaluate data, develop new or updated mitigate strategies, perform new HF testing to confirm strategies work, evaluate new data. Once the cycle stops, and you have mitigated everything to the best of your abilities, you will then need to develop a report. In the HFE/UE report, you should provide not only the details of your HF testing (including data and conclusions) with identified Critical Tasks, but also details about your identified users, Use Environments, and mitigation strategies. The guidance *Applying Human Factors and Usability Engineering to Medical Devices: Guidance for Industry and FDA Staff* [44] provides a nice outline with bullet point indications of what to provide in each section.

As mentioned previously, the guidance document, *List of Highest Priority Devices for Human Factor Review* [45], provides guidance for the device types required to provide HFE/UE report and data in their submission package. As of this writing, this was still a draft guidance, so changes to the FDA's recommendations may be incorporated when they issue the final guidance. As mentioned previously, this guidance does not indicate HF testing is not required; it means the agency may not require the data to review as part of your submission package. In addition to the device types listed in the draft guidance, there are situations where the agency will want to see HF testing data.

Examples include updating the user interface due to safety concerns, including different users than indicated with the predicate device, modifications due to information from recall reports. Basically, if there is a clear safety issue, or the possibility of a safety issue due to possible Use-Related Hazards, HF testing data may be required in your submission. The need to include an HFE/UE report (HF testing data) in your submission is a discussion you should have with the agency (Sect. 3.1.3).

References

1. FDA's Webpage, Medical Device Databases. Available online at https://www.fda.gov/medical-devices/device-advice-comprehensive-regulatory-assistance/medical-device-databases. Accessed20 July 2025.
2. C.F.R. § 820.
3. C.F.R. § 812.18(a)
4. FDA's Webpage, U.S. Agents. Available online at https://www.fda.gov/medical-devices/device-registration-and-listing/us-agents. Updated 23 December 2017. Accessed 20 July 2025.
5. FDA's guidance for Industry and Food and Drug Administration Staff: *Requests for Feedback and Meetings for Medical Device Submissions: The Q-Submission Program* – FINAL GUIDANCE. Available online at https://www.fda.gov/regulatory-information/search-fda-guidance-documents/requests-feedback-and-meetings-medical-device-submissions-q-submission-program. Accessed 12 August 2025.
6. FD&C Act, section 745(A)(b).
7. C.F.R. § 812.
8. C.F.R. § 812.2(b)(1).
9. FDA's guidance for Industry and Food and Drug Administration Staff: *Cybersecurity in Medical Devices: Quality System Considerations and Content of Premarket Submissions.* Available online at https://www.fda.gov/regulatory-information/search-fda-guidance-documents/cybersecurity-medical-devices-quality-system-considerations-and-content-premarket-submissions. Accessed 20 August 2025.
10. FDA's guidance for Industry and CDRH Staff: *Early Collaboration Meetings Under the FDA Modernization Act (FDAMA); Final Guidance for Industry and CDRH Staff* – FINAL GUIDANCE (May 2001). Available online at https://www.fda.gov/regulatory-information/search-fda-guidance-documents/early-collaboration-meetings-under-fda-modernization-act-fdama-final-guidance-industry-and-cdrh. Accessed 20 July 2025.
11. FD&C Act, section 513(a)(3)(D).
12. FD&C Act, section 520(g)(7).
13. C.F.R. § 812.25.
14. FDA's Webpage, Breakthrough Devices Program. Available online at https://www.fda.gov/medical-devices/how-study-and-market-your-device/breakthrough-devices-program. Updated 7 November 2024. Accessed 20 July 2025.
15. FDA's Webpage, Safer Technologies Program for Medical Devices. Available online at https://www.fda.gov/medical-devices/how-study-and-market-your-device/safer-technologies-program-step-medical-devices. Updated 15 March 2021. Accessed 20 July 2025.
16. FDA's guidance for Industry and FDA Staff: *Medical Device Accessories – Describing Accessories and Classification Pathways* - FINAL GUIDANCE (December 2017). Available online at https://www.fda.gov/regulatory-information/search-fda-guidance-documents/medical-device-accessories-describing-accessories-and-classification-pathways. Accessed 20 July 2025.
17. FDA's guidance for Industry and Food and Drug Administration Staff: *The 510(k) Program: Evaluating Substantial Equivalence in Premarket Notifications [510(k)]* – FINAL GUIDANCE (July 2014). Available online at https://www.fda.gov/regulatory-information/search-fda-guidance-documents/510k-program-evaluating-substantial-equivalence-premarket-notifications-510k. Accessed 20 July 2025.

18. FDA's draft guidance for Industry and Food and Drug Administration Staff: *Best Practices for Selecting a Predicate Device to support a Premarket Notification [510(k)] Submission* – DRAFT GUIDANCE (September 2023). Available online at https://www.fda.gov/regulatory-information/search-fda-guidance-documents/best-practices-selecting-predicate-device-support-premarket-notification-510k-submission. Accessed 20 July 2025.
19. C.F.R. § 807.100(b).
20. FDA's guidance for Industry and Food and Drug Administration Staff: *Benefit-Risk Factors to Consider When Determining Substantial Equivalence in Premarket Notifications (510(k)) with Different Technological Characteristics*—FINAL GUIDANCE (September 2018). Available online at https://www.fda.gov/regulatory-information/search-fda-guidance-documents/benefit-risk-factors-consider-when-determining-substantial-equivalence-premarket-notifications-510k. Accessed 20 July 2025.
21. FDA's guidance for Industry and FDA Staff: *Principles of Premarket Pathways for Combination Products*—FINAL GUIDANCE (January 2022). Available online at https://www.fda.gov/regulatory-information/search-fda-guidance-documents/principles-premarket-pathways-combination-products. Accessed 20 July 2025.
22. Executive Summary for the Digital Health Advisory Committee Meeting: Total Product Lifecycle Considerations for Generative AI-Enabled Devices, November 20–21, 2024. Available online at https://www.fda.gov/media/182871/download. Accessed 20 July 2025.
23. FDA's guidance for Industry and Food and Drug Administration Staff: *Consideration of Uncertainty in Making Benefit-Risk Determinations in Medical Device Premarket Approvals, De Novo Classifications, and Humanitarian Device Exemptions*—FINAL GUIDANCE (August 2019). Available online at https://www.fda.gov/regulatory-information/search-fda-guidance-documents/consideration-uncertainty-making-benefit-risk-determinations-medical-device-premarket-approvals-de. Accessed 20 July 2025.
24. FDA's guidance for Industry and Food and Drug Administration Staff: *Factors to Consider When Making Benefit-Risk Determinations in Medical Device Premarket Approval and De Novo Classifications*—FINAL GUIDANCE (August 2019). Available online at https://www.fda.gov/regulatory-information/search-fda-guidance-documents/factors-consider-when-making-benefit-risk-determinations-medical-device-premarket-approval-and-de. Accessed 20 July 2025.
25. FDA's guidance for Industry and Food and Drug Administration Staff: *Recommended Content and Format of Non-Clinical Bench Performance Testing Information in Premarket Submissions*—FINAL GUIDANCE (December 2019). Available online at https://www.fda.gov/regulatory-information/search-fda-guidance-documents/recommended-content-and-format-non-clinical-bench-performance-testing-information-premarket. Accessed 20 July 2025.
26. FDA's guidance for Industry and Food and Drug Administration Staff: *Use of International Standard ISO 10993-1, "Biological evaluation of medical devices—Part 1: Evaluation and testing within a risk management process"*—FINAL GUIDANCE (September 2023). Available online at https://www.fda.gov/regulatory-information/search-fda-guidance-documents/use-international-standard-iso-10993-1-biological-evaluation-medical-devices-part-1-evaluation-and. Accessed 20 July 2025.
27. FDA's draft guidance for Industry, Accreditation Bodies, Testing Laboratories, and Food and Drug Administration Staff: *Biocompatibility Testing of Medical Devices—Standards Specific Information for the Accreditation Scheme for Conformity Assessment (ASCA) Program*—DRAFT GUIDANCE (September 2024). Available online at https://www.fda.gov/regulatory-information/search-fda-guidance-documents/biocompatibility-testing-medical-devices-standards-specific-information-accreditation-scheme-0. Accessed 20 July 2025.

28. FDA's draft guidance for Industry and Food and Drug Administration Staff: *Chemical Analysis for Biocompatibility Assessment of Medical Devices*—DRAFT GUIDANCE (September 2024). Available online at https://www.fda.gov/regulatory-information/search-fda-guidance-documents/chemical-analysis-biocompatibility-assessment-medical-devices. Accessed 20 July 2025.
29. FDA's guidance for Industry and Food and Drug Administration Staff: *General Considerations for Animal Studies Intended to Evaluate Medical Devices*—FINAL GUIDANCE (March 2023). Available online at https://www.fda.gov/regulatory-information/search-fda-guidance-documents/general-considerations-animal-studies-intended-evaluate-medical-devices. Accessed 20 July 2025.
30. FDA's guidance for Industry and FDA Staff: *Non-Clinical Engineering Tests and Recommended Labeling for Intravascular Stents and Associated Delivery Systems*—FINAL GUIDACNE (April 2010). Available online at https://www.fda.gov/regulatory-information/search-fda-guidance-documents/non-clinical-engineering-tests-and-recommended-labeling-intravascular-stents-and-associated-delivery. Accessed 20 July 2025.
31. C.F.R. § 812.3(m).
32. *Information Sheet Guidance For IRBs, Clinical Investigators, and Sponsors Significant Risk and Nonsignificant Risk Medical Device Studies.* Additional information on exempt studies can be found in the Information Sheet Guidance entitled, *Frequently Asked Questions About Medical Devices* (January 2006). Available online at https://www.fda.gov/media/75459/download. Accessed 20 July 2025.
33. FDA's guidance for Sponsors, Clinical Investigators, Institutional Review Boards, and Food and Drug Administration Staff: *FDA Decisions for Investigational Device Exemption Clinical Investigations*—FINAL GUIDANCE (April 2014). Available online at https://www.fda.gov/regulatory-information/search-fda-guidance-documents/fda-decisions-investigational-device-exemption-clinical-investigations. Accessed 20 July 2025.
34. FDA's guidance for Industry, Clinical Investigators, Institutional Review Boards, and FDA Staff: *Design Considerations for Pivotal Clinical Investigations for Medical Devices Guidance for Industry, Clinical Investigators, Institutional Review Boards and Food and Drug Administration Staff*—FIAL GUIDANCE (November 2013). Available online at https://www.fda.gov/regulatory-information/search-fda-guidance-documents/design-considerations-pivotal-clinical-investigations-medical-devices. Accessed 20 July 2025.
35. FDA's guidance for Industry and Food and Drug Administration Staff: *Acceptance of Clinical Data to Support Medical Device Applications and Submissions: Frequently Asked Questions*—FINAL GUIDANCE (February 2018). Available online at https://www.fda.gov/regulatory-information/search-fda-guidance-documents/acceptance-clinical-data-support-medical-device-applications-and-submissions-frequently-asked. Accessed 20 July 2025.
36. C.F.R. § 860.7(c)(1).
37. FDA's draft guidance for Industry, Sponsors, and IBRs: *Ethical Considerations for Clinical Investigations of Medical Products Involving Children*—DRAFT GUIDANCE (September 2022). Available online at https://www.fda.gov/regulatory-information/search-fda-guidance-documents/ethical-considerations-clinical-investigations-medical-products-involving-children. Accessed 20 July 2025.
38. C.F.R. § 11.

39. FDA's guidance for Industry and Food and Drug Administration Staff: *Acceptance of Medical Device Clinical Data from Studies Conducted Outside the United States*—FINAL GUIDANCE (February 2018). Available online at https://www.fda.gov/regulatory-information/search-fda-guidance-documents/acceptance-clinical-data-support-medical-device-applications-and-submissions-frequently-asked. Accessed 20 July 2025.
40. C.F.R. § 860.7(c)(2).
41. FDA's draft guidance for Industry and Food and Drug Administration Staff: *Recommendations for the Use of Clinical Data in Premarket Notification [510(k)] Submissions*—DRAFT GUIDANCE (September 2023). Available online at https://www.fda.gov/regulatory-information/search-fda-guidance-documents/recommendations-use-clinical-data-premarket-notification-510k-submissions. Accessed 20 July 2025.
42. FDA's draft guidance for Industry and Other Interested Parties: *Considerations for the Use of Artificial Intelligence to Support Regulatory Decision-making for Drug and Biological Products*—DRAFT GUIDANCE (January 2025). Available online at https://www.fda.gov/regulatory-information/search-fda-guidance-documents/considerations-use-artificial-intelligence-support-regulatory-decision-making-drug-and-biological. Accessed 20 July 2025.
43. FDA's draft guidance for Industry and Food and Drug Administration Staff: *Artificial Intelligence-Enabled Device Software Functions: Lifecycle Management and Marketing Submission Recommendations*—DRAFT GUIDANCE (January 2025). Available online at https://www.fda.gov/regulatory-information/search-fda-guidance-documents/artificial-intelligence-enabled-device-software-functions-lifecycle-management-and-marketing. Accessed 20 July 2025.
44. C.F.R. § 808.
45. FDA's guidance for Industry and Food and Drug Administration Staff: *Applying Human Factors and Usability Engineering to Medical Devices*—FINAL GUIDANCE (February 2016). Available online at https://www.fda.gov/regulatory-information/search-fda-guidance-documents/applying-human-factors-and-usability-engineering-medical-devices. Accessed 20 July 2025.
46. FDA's draft guidance for Industry and Food and Drug Administration Staff: *List of Highest Priority Devices for Human Factor Review: Draft Guidance for Industry and FDA Staff*—DRAFT GUIDANCE (February 2016). Available online at https://www.fda.gov/regulatory-information/search-fda-guidance-documents/list-highest-priority-devices-human-factors-review. Accessed 20 July 2025.
47. FDA's draft guidance for Industry and Food and Drug Administration Staff: *Purpose and Content of Use-Related Risk Analyses for Drugs, Biological Products, and Combination Products*—DRAFT GUIDANCE (July 2024). Available online at https://www.fda.gov/regulatory-information/search-fda-guidance-documents/purpose-and-content-use-related-risk-analyses-drugs-biological-products-and-combination-products. Accessed 20 July 2025.
48. FDA's guidance for Industry and FDA Staff: *Application of Human Factors Engineering Principles for Combination Products: Questions and Answers*—FINAL GUIDANCE (September 2023). Available online at https://www.fda.gov/regulatory-information/search-fda-guidance-documents/application-human-factors-engineering-principles-combination-products-questions-and-answers. Accessed on 20 July 2025.
49. FDA's draft guidance for Industry and FDA Staff: *Human Factors Studies and Related Clinical Study Considerations in Combination Product Design and Development*—DRAFT GUIDANCE (February 2016). Available online at https://www.fda.gov/files/about%20fda/published/Human-Factors-Studies-and-Related-Clinical-Study-Considerations-in-Combination-Product-Design-and-Development.pdf. Accessed 20 July 2025.

Chapter 4
Compiling and Submitting Your Device Application

4.1 Compiling Your Submission Package

You have traveled a long journey—you know the type and class of your device (Chap. 2), and you have evaluated it extensively (Chap. 3). Importantly, you have discussed your device, and the development plans, tests/studies, and results along the way with the FDA (Chap. 3). The next big step is preparing and submitting your package of information (your regulatory application) to the FDA. Here, we will provide additional details to help you on your way to market.

The basics of what to include are directly related to regulation and the medical device itself. More information must be provided as the risk of the medical device increases (e.g., higher class of device). The level of detail, complexity, and thoroughness of the information provided in a medical device application can also vary as a function of other factors unrelated to the regulatory requirements (e.g., company requirements, level of regulatory knowledge and expertise, and author variations). Some Sponsors include everything they have done related to device development, no matter how small, because they feel it's required; they want to give the FDA everything "just in case," or they simply don't know what they must include. In contrast, some Sponsors include so little information that the FDA is left to fill in the blanks or go back to the Sponsor repeatedly to get missing information. Both approaches result in frustrated reviewers and slower reviews. A poor-quality application will result in delays in getting your device to market.

The overarching goal of every Sponsor should be to strive to achieve a balance between a thorough application and speed to submission.

Happy FDA reviewers = quickest path to market

G. B. Halt Jr. et al., *FDA and Intellectual Property Strategies for Medical Device Technologies*, https://doi.org/10.1007/978-3-032-12649-8_4

You do not need to tell the FDA *every* detail of *every* aspect of your program. That could take years to document and would result in many thousands of pages and very large electronic files (and very overwhelmed reviewers!). However, if not enough information or the wrong information is provided, the FDA reviewer is left wanting.

To better understand FDA's expectation of the level of detail to be included in premarket applications, it would be ideal to read the guidance, *The Least Burdensome Provisions: Concept and Principles* [1]. This FDA guidance elaborates upon "least burdensome," which is defined as "the minimum amount of information necessary to adequately address a relevant regulatory question or issue through the most efficient manner at the right time." FDA applies this concept to all devices throughout the total product lifecycle (i.e., premarket and postmarket). The least burdensome approach encompasses the varying approaches to generating data and providing information, the type of data and information, and the timepoint(s) in the device's lifecycle when data should be generated and/or submitted to FDA.

The trick is to provide the FDA with only the <u>necessary and required information</u> couched in a <u>clear and concise story</u> so that they can easily review the data, (safety, efficacy, human factors, etc.) and readily make a sound regulatory decision.

The length of a regulatory application can vary, sometimes dramatically, based on the type of submission, the type of device, the complexity of the device, and the data needed to support a regulatory decision. Submitting a well-written, succinct package eases FDA's review and can result in getting your device to market faster. Remember, the FDA wants to get beneficial new and innovative devices into the hands of healthcare providers and patients as quickly as possible.

In the following sections, the necessary 510(k), PMA, De Novo, and combination product application contents (including labeling information) will be discussed. The 2024–2025 Medical Device User Fee Amendments (MDUFA) fees associated with submitting medical device applications for the FDA's review are also outlined in the content that follows. Unfortunately, it's not free to submit these applications like it is to submit Q-Submissions!

4.1.1 General Guidelines

The following information applies to all medical device applications that are submitted to FDA. It is important to adhere to these guidelines to ensure the FDA accepts your application and can perform an efficient review:

- All submission content must be provided in English. If the source information is not in English, the document must be translated and both the original and translated document must be submitted in your application.
- All pages of the submission must be numbered sequentially. There can be no skips or duplicates.

Please note that at any time FDA may request that the Sponsor send them a sample of the device. Be sure you have one ready to send to the FDA. Alternatively, you can proactively include a sample device with your application. If you cannot send your device to the FDA (e.g., it cannot be reasonably shipped due to size limitations), you will need to make a device available for the FDA to travel to see it.

4.1.2 Online Submission of CDRH Premarket Applications: CDRH Portal

On October 3, 2022, FDA launched the CDRH Customer Collaboration Portal (CDRH Portal) for the online submission of eCopy and eSTAR premarket applications. The CDRH Portal, as the name suggests, is only for CDRH (and CDRH-led combination products) submissions; biologics and devices used in blood establishments (such as blood banks, blood/plasma donor centers, blood product manufacturers, etc.) must be submitted to CBER using the FDA Electronic Submissions Gateway (ESG). The CDRH Portal is the *mechanism* for submitting the regulatory application, whereas the eSTAR (described in Sect. 4.1.2.1) and eCopy (described in Sect. 4.1.2.2) are the forms the actual application must be in, to submit the application through the CDRH Portal.

Anyone can register for a CDRH Portal account (it's free!), which will then be used to submit or track CDRH premarket applications. Each user of the CDRH Portal should have their own account. Accounts are not intended to be shared by multiple people or departments or transferred from one person to another. After 11 months of inactivity, the registered CDRH Portal user will receive an email. If the user does not log into the account, the account will be disabled after one (1) year of no use.

It is important that you **keep a copy of your application** you submitted, for record-keeping purposes. The CDRH Portal is NOT a repository for submissions. You will also want original copies so you can answer any questions that the FDA may have during review (Chap. 5 provides more information about the review process).

All submissions should be uploaded into the CDRH Portal in a single zip (for eCopy) or pdf file (for eSTAR or eCopy); a submission should not be uploaded in multiple parts. The CDRH Portal *does* have file size limitations. Individual submission files cannot exceed 4 GB and attachments (within pdf files) cannot exceed 1 GB. Electronic files exceeding these limits will need to be mailed to CDRH's Document Control Center. Thought should be put into naming the application that gets submitted through the CDRH Portal, as this can't be edited in the future. For example, to distinguish between submissions listed in the CDRH Portal once you submit more in the future, you may want to consider including the product name (for a 510(k), De Novo, or PMA, for example) or the main discussion topic (for a Q-Sub, for example).

Anything sent via the CDRH Portal that is received before 4 PM Eastern Time on a business day (Monday through Friday, except for U.S. Federal holidays) will be processed by FDA that same day. Anything received after 4 PM Eastern Time will be processed the next business day.

Near real-time submission status is displayed in the CDRH Portal progress tracker for the applications that are identified in the blue box in Fig. 4.1 (and will be displayed in the future for the application listed in the gray box in Fig. 4.1). Although progress can be tracked in the CDRH Portal for some applications, for *all* applications FDA will formally notify the person identified (in the premarket submission) as the official correspondent of official actions and requests.

FDA's Webpage, Send and Track Medical Device Premarket Submissions: CDRH Portal [2] is a helpful resource and one to reference regularly as you prepare your premarket applications. This webpage contains the link for accessing the CDRH Portal as well as update announcements and links to pertinent information. FDA is committed to improving upon and maintaining the CDRH Portal, as indicated in the FDA draft guidance, *MDUFA Performance Goals and Procedures, Fiscal Years 2023 through 2027* [3]. Lastly, the CDRH Portal contains Portal Help articles. If your

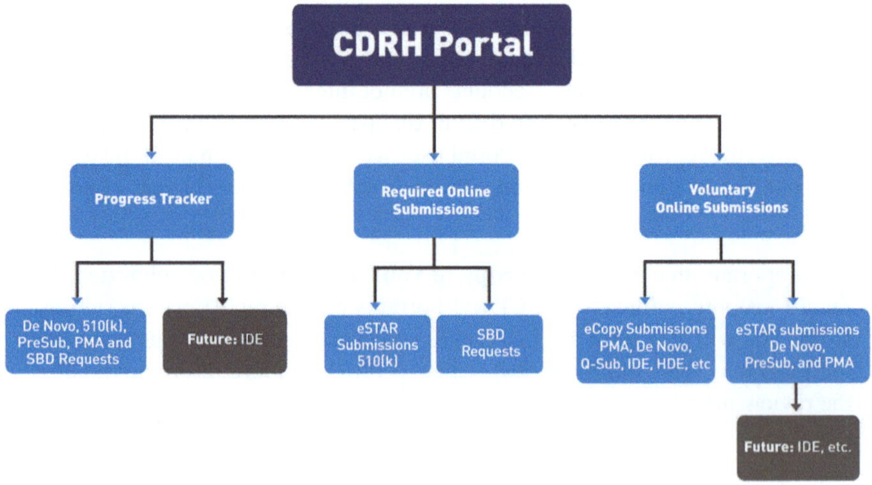

Fig. 4.1 Flow chart outlining CDRH portal capabilities [2]

CDRH Portal questions are not addressed in those articles, you can send an email to ccp@fda.hhs.gov to get more assistance.

4.1.2.1 ESTAR

CDRH created an interactive pdf form, the electronic Submission Template and Resource (eSTAR), for the submission of regulatory applications. The uses of this template continue to evolve and the template itself is periodically updated. Thus, once you're ready to begin drafting your regulatory application, it would be ideal to consult FDA's webpage, eSTAR Program [4], to download the current version of the eSTAR. Additionally, the webpage contains current information regarding which submissions *must* be composed/submitted using the eSTAR, which can voluntarily be composed/submitted using the eSTAR, and dates for those submissions that will be required to use the eSTAR in the future. Most of this information is captured as it exists as of this writing in Fig. 4.1. Lastly, the webpage contains helpful tips for managing eSTAR attachments, information regarding acceptable file attachments, and other composition- and submission-related details for the eSTAR. Once you've reviewed FDA's webpage, should you still have eSTAR-specific questions for FDA, you can send an email to eSubPilot@fda.hhs.gov.

FDA's webpages, eSTAR Program [4] and Send and Track Medical Device Premarket Submissions: CDRH Portal [2], list the submission types that require an eSTAR, but should be noted that the eSTAR could make a reviewer's life significantly easier (again, Happy FDA Reviewer = Quickest Path to Market). Thus, if it is possible to use the eSTAR for your regulatory application, we strongly recommend it, as it will significantly benefit your team *and* FDA.

The eSTAR is a huge advancement for CDRH and is very beneficial to submitters of regulatory applications *and* to FDA reviewers. The eSTAR standardizes the format of the information submitted, which allows for consistency between applications and easier accessibility for FDA reviewers and the submitter. The format and components of the eSTAR are aligned with FDA's internal review template, to ensure FDA reviewers are receiving the necessary and expected information from submitters. The eSTAR is an interactive pdf, guiding submitters through the information needed for the specific application they are preparing, reducing the risk of omitting important details needed for the review of their application. There are also fields in the eSTAR that auto-populate information that was entered in earlier fields in the form, to avoid errors in duplicating information and eliminating the need to enter the same information multiple times. Note that the eSTAR has a combination of fields requiring the submitter to enter information and sections requiring attachment of documents prepared by the submitter (documents that would have historically been submitted via paper or eCopy). A helpful feature of the eSTAR is that it has built-in forms that would traditionally (when sent via paper or eCopy) have been standalone documents (i.e., Declaration of Conformity, Form 3514, Truthful and Accuracy Statement, Indications for Use Form, and 510(k) Summary).

As of this writing, the eSTAR is required for all medical device 510(k) submissions (unless exempted), including device-led combination product 510(k) submissions and Dual 510(k)/CLIA Waiver IVD submissions. The eSTAR is voluntary for De Novo submissions and 513(g) requests as well as for PMA submissions and Pre-Submissions (a type of Q-Submission) sent to CDRH.

The eSTAR eliminates the need to Refuse to Accept (RTA) review by FDA.

No more RTA holds!

Importantly, use of the eSTAR eliminates the need for a Refuse to Accept (RTA) review by FDA and thus RTA holds, because the eSTAR's design ensures all necessary components are present and complete when finalizing the eSTAR for submission. In other words, if you have completed all fields and sections required for the selected type of application, which are marked red in the eSTAR until they are complete (once complete, they turn green), "STATUS: eSTAR COMPLETE" will display in green at the top of the eSTAR. Note, however, that there is still the possibility of an eCopy Hold (the same language is used for the eSTAR and eCopy), for instances where submitters send an eSTAR containing the message "eSTAR INCOMPLETE" through the CDRH Portal. There is also the possibility of an early hold in the review of your application if your responses to any of the prompts in the eSTAR are inaccurate (i.e., inaccurate responses to drop down menus may result in omitting subsequent sections or information).

While the *submission* of many regulatory applications (elaborated upon in other chapters) requires the submitter to pay a user fee, the *use* of the eSTAR is free. In other words, you can download and prepare the eSTAR for free and only need to pay the applicable user fee in order to submit the regulatory application to CDRH (or CBER) for FDA review through the CDRH Portal (or ESG for CBER).

4.1.2.2 eCopy

Nearly all medical device submissions now need to be sent to FDA in electronic format (called eCopy). There is an FDA guidance, *eCopy Program for Medical Device Submissions* [5], that provides requirements and recommendations for an electronic application. FDA's webpage, eCopy Medical Device Submissions [6], is also a helpful resource.

There are a few things to note when preparing your electronic application:

- The only font types allowed are Times New Roman, Verdana, Arial, Tahoma, or Helvetica.

  - The recommended font size is 12 (smaller sizes may be acceptable, especially in tables and figures).
  - The font color should be black (if hyperlinks and bookmarks are used, they can be blue).
  - You should use one font type throughout your application.

- It is best to create your PDF versions from your original document—scanned copies do not have searchable text.

  - If you must scan a document (it happens), use optical character recognition (OCR) so the text is searchable. OCR documents will be rejected.
  - Be certain to check the entire document after you render it to PDF. Sometimes content does not render properly and you get smiley faces, triangles, and other surprising characters instead of the text and symbols you were expecting.

- Using Adobe plug-ins is risky—reviewers may not be able to review your content if they do not have access to the specific plug-in.
- An eCopy can be as simple as a single PDF file. Check the guidance for the rules for required naming conventions.
- Each new submission is a new eCopy. For example, if you are responding to a deficiency letter, that will be its own eCopy submission (not cumulative, with what was previously submitted).
- There is no maximum total submission size restriction. However, to avoid possible delays in the submission process, it is recommended that the total eCopy be $\leq$ 1 GB (refer to Sect. 4.1.2 for size restrictions on submitting through the CDRH Portal).

  - While there is no limit on total eCopy size, each pdf file included in your eCopy must be $\leq$ 50 MB.

- Although hyperlinks and bookmarks are not required, their use is recommended because they facilitate FDA review. For example, you indicate "refer to Sect. 1. 2.3.4 for additional information" … the "Sect. 1.2.3.4" would be hyperlinked so the reviewer simply clicks and quickly hops to that section to continue their review. Bookmarks within the application allow the reviewer to review the TOC from anywhere in the document. This is very convenient if you include attachments or appendices—again, one click and the reviewer can easily continue their review. There are, of course, rules about how to create and denote links in your document.

It is strongly recommended that you check the hyperlinks and bookmarks before submitting your application, to ensure they go to the correct locations—bad directions never get you to your destination and only increase frustration for FDA and for your team!

Before preparing your eCopy, be sure to read, understand, and comply with the FDA's guidance entitled, *eCopy Program for Medical Device Submissions* [5]. Additionally, FDA's webpage, eCopy Medical Device Submissions [6], contains common eCopy errors and information on how to avoid eCopy hold. If you cannot prepare your eCopy in-house, there are expert consulting organizations who do this day in and day out and will ensure compliance with the FDA's eCopy requirements.

If you receive an eCopy hold letter, the letter will provide specific reasons why your eCopy failed to load and directions for submitting a replacement eCopy. The guidance from FDA should be straightforward. Additionally, FDA's webpage, eCopy Medical Device Submissions [6], contains information on responding to an eCopy hold. However, should the reason(s) why your eCopy failed to load be outside the bandwidth of your team, consider consulting with electronic submissions experts who can quickly and easily tackle such technical issues. An eCopy Hold means FDA's review of your submission is delayed until the issue(s) are resolved so you'll want to work quickly to get FDA's review back on track. If after reviewing FDA's guidance document and webpage you still have eCopy-specific questions, you can email FDA at CDRH-eCopyinfo@fda.hhs.gov.

4.1.3 Premarket Notification 510(k)

Sponsors who want to market a medical device in the United States must submit a 510(k) to the FDA, unless: the device is exempt, the device is subject to PMA regulations (refer to Sect. 4.1.4 for the components of a PMA), or there is no predicate device and thus the subject device will be submitted through the De Novo process (refer to Sect. 4.1.5 for the components of a De Novo). For those devices requiring a 510(k), the device cannot be marketed until the Sponsor has received a letter from FDA indicating that the device is substantially equivalent to the selected predicate (i.e., an SE letter); the SE decision, as discussed throughout this book, is based on the information submitted by the Sponsor.

As of October 1, 2023, all 510(k) submissions must be submitted through the CDRH Portal using the eSTAR, unless exempted, as described in FDA's guidance, *Electronic Submission Template for Medical Device 510(k) Submissions* [7].

There are three (3) types of 510(k) applications available to medical device Sponsors to demonstrate substantial equivalence: traditional, abbreviated, and special. The type of 510(k) is driven by the information used to establish substantial equivalence between the new device and the predicate(s).

(1) A Traditional 510(k) is the most common of all device applications and can be used for any original 510(k) or modification to a device previously cleared under a 510(k).
(2) An Abbreviated 510(k) is an option when the submission relies on conforming to FDA-recognized and/or voluntary consensus standards, compliance with special controls established by regulation for the device type, and/or FDA guidance

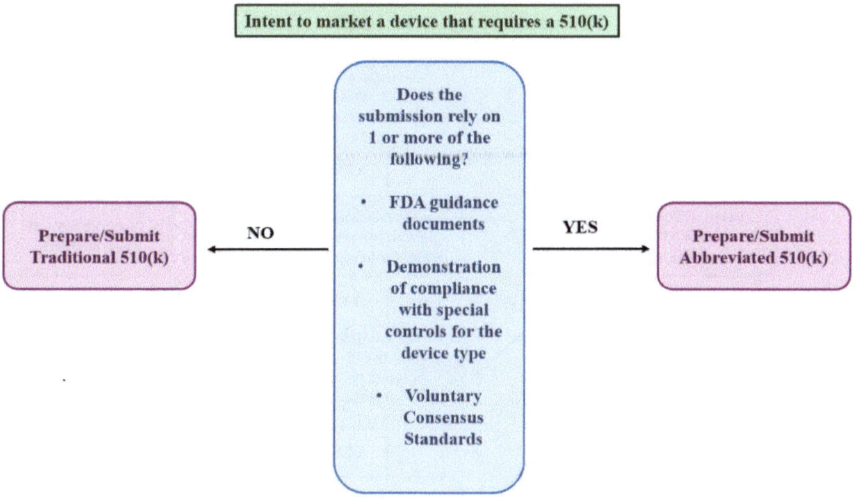

Fig. 4.2 Flow chater to determine whether an Abbreviated 510(k) is appropriate

documents. Refer to FDA's guidance, *The Abbreviated 510(k) Program* [8], as well as Fig. 4.2 (created from information in the aforementioned guidance) for help in determining whether an Abbreviated 510(k) is appropriate for your device.

(3) A Special 510(k) is applicable for Sponsors making a change(s) to their own device previously cleared under a 510(k), provided that the methods to evaluate the change(s) are well established and the results of these methods can be sufficiently reviewed in a risk analysis or summary format. Refer to FDA's guidance, *The Special 510(k) Program* [9], as well as Fig. 4.3 (adapted from the aforementioned guidance) for help in determining whether a Special 510(k) is appropriate for your device.

As for making changes to a device that has already received FDA clearance, note that some changes may be implemented without submitting a new 510(k) submission for FDA's review and clearance. FDA's webpage, "Is a new 510(k) required for a modification to the device?" [10], should be used to evaluate changes to a cleared device. FDA has also issued 2 helpful guidance documents on this topic, *Deciding When to Submit a 510(k) for a Change to an Existing Device* [10] and *Deciding When to Submit a 510(k) for a Software Change to an Existing Device* [11].

As outlined in 21 C.F.R. 807.87 and FDA's guidance, *Electronic Submission Template for Medical Device 510(k) Submissions* [7], the basic elements of all 510(k) submissions are:

- Identification of the submission type (i.e., Traditional, Special, Abbreviated, Dual 510(k)/CLIA Waiver, etc.).
- Cover letter.

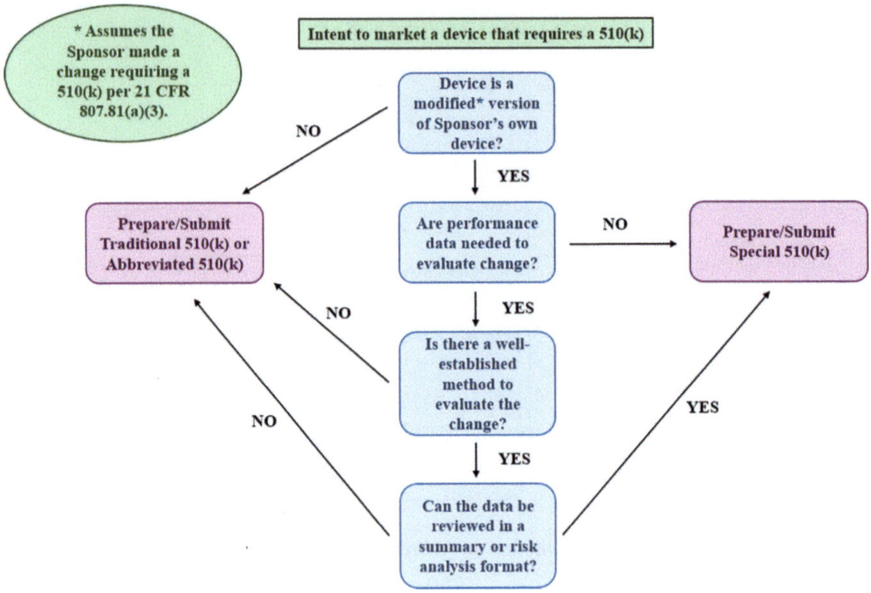

Fig. 4.3 Flow chart to determine whether a Special 510(k) is appropriate

- Letters of reference, as applicable.
- Applicant and correspondent information.
- Pre-Submission Correspondence & Previous Regulatory Interaction(s).
- Consensus standards, as applicable.
- Device name (including trade or proprietary name and common or usual name or classification name of the device).
- Device description.
- Proposed indications for use.
- Classification.
- Predicate and substantial equivalence.
- Design/special controls, risks to health, & mitigation measures (applicable to Special 510(k)s only).
- Proposed labeling.
- Reprocessing, as applicable.
- Sterility, as applicable.
- Shelf life.
- Biocompatibility.
- Software/firmware, as applicable.
- Cybersecurity/interoperability, as applicable.
- Electromagnetic compatibility (EMC), & electrical, mechanical, wireless, & thermal safety, as applicable.
- Performance testing (i.e., bench, animal, clinical, etc.).
- References, as applicable.

- Administrative documentation (i.e., Executive summary, truthful and accuracy statement, 510(k) summary or 510(k) statement, financial certification/disclosure, etc.).

Appendix A of *The Special 510(k) Program* [9] contains FDA guidance on the content and format of a Special 510(k) while *The Abbreviated 510(k) Program* [8] contains FDA guidance on the content and format of an Abbreviated 510(k).

4.1.4 Premarket Approval (PMA)

For companies who wish to market a Class III device, they must submit a PMA application to the FDA for review and approval. As mentioned earlier in the chapter, the eSTAR may be used to compose and submit a PMA. As of this writing, FDA has not posted a date for mandatory use of the eSTAR for PMAs. Regulations dictating the elements of a PMA are provided in 21 C.F.R. § 814.20. In addition, the FDA provides a detailed summary of requirements along with links to current guidance documents on their premarket approval (PMA) webpage [12]. FDA's 2019 guidance document, *Acceptance and Filing Reviews for Premarket Approval Applications (PMAs)* [13], is a helpful resource to consider prior to compiling your PMA.

In summary, the basic elements of a PMA include, but are not limited to:

- Premarket Review Submission Cover Sheet (FDA Form 3514 for eCopy only; if using the eSTAR, form is built into PDF).
- Clinicaltrials.gov Certification (form FDA-3674).
- Cover letter (signed and dated).
- Sponsor identification information.
- TOC (applicable for eCopy submissions only).
- Identification of the trade secret or confidential commercial or financial information related to the medical device (e.g., patents, copyrights, trademarks).
- Executive summary, approximately ten (10) to fifteen (15) pages in length.
    - The summary should provide the reviewer with an overall understanding of the main features of the device and the supporting data and include the indications for use, device description, marketing history, labeling claims linked to corresponding studies' conclusions, and summaries of conducted studies.
    - A helpful reminder provided by FDA on their PMA webpage [12] is that the summary section, when well written and appropriately developed, can serve as the starting point for FDA's Summary of Safety and Effectiveness Data (SSED; note that this is publicly posted upon PMA approval) and will also support FDA's (and as applicable a review panel [a.k.a. advisory committee]) review process.
- Complete description of the device (refer to the FDA's guidance, *Quality System Information for Certain Premarket Application Reviews*).

- List of all standards used. Adequate information must be provided to demonstrate how the subject device meets the referenced mandatory and voluntary standards. A justification must be provided for a deviation(s) from mandatory standards; an explanation must be provided for a deviation(s) from voluntary standards.
- Technical sections. The technical sections provide the "meat" of the application and are divided into two sections: Nonclinical studies and clinical investigations. You must provide sufficient detail about each study and investigation so the reviewer understands "who, what, where, when, why, and how" and can interpret the results appropriately. Each type of study and investigation has certain details that must be included (e.g., good laboratory practices information; informed consent; investigational review board information, etc.).
- Samples of the device and device components, if requested by FDA, or the location where FDA may examine and test one or more devices.
- Copies of all proposed labeling for the device.
- Environmental assessment. 21 C.F.R. § 814.20(b)(11) states that an environmental assessment in accordance with 21 C.F.R. § 25 must be included in the PMA submission (see 21 C.F.R. § 25.34(d) for more information). If you are not sure what you need to include in your PMA, consult the FDA.
- Omissions. If a Sponsor thinks that any of the information required under 21 C.F.R. 814.20(b) does not apply to the subject device and thus does not include this information in their PMA, they must include a statement in the PMA (listed in the TOC) that identifies and justifies the omission. FDA will let the Sponsor know if they do not accept the justification.

Three additional things to note regarding your PMA application. First, you must periodically update your pending application with new safety and/or efficacy information as it becomes available from ongoing or completed studies. This is particularly important if emerging data suggest that there needs to be a change to the labeling for the device (such as warnings, precautions, contraindications, and adverse reactions). Such updates are submitted as amendments to a pending PMA. Some amendments can affect review timelines (see 21 C.F.R. § 814.37(c)(1)). More information about the potential impact of amendments to the PMA on FDA review timelines is provided in Chap. 5.

Secondly, you can greatly facilitate the FDA's review processes when preparing the executive summary (the 10–15 page document previously mentioned) required in 21 C.F.R. § 814.20(b)(3) by presenting the information in an order that the FDA will use to create their SSED, using the following Table of Contents (note that you can refer to approved PMAs to see the format of the SSEDs):

(1) General information including but not limited to, generic and trade names, product codes, and applicant name and address. Note, some items will be left blank, like the date of panel recommendation, PMA number, etc.
(2) Indications for use.
(3) Contraindications.
(4) Warnings and precautions.
(5) Device description.

(6) Alternative practices and procedures.
(7) Marketing history.
(8) Potential adverse effects of the device on health.
(9) Summary of nonclinical studies, including laboratory, animal, and other studies.
(10) Summary of clinical studies.
(11) Financial disclosure.
(12) Summary of supplemental clinical information.
(13) Conclusions drawn from nonclinical and clinical studies (both effectiveness and safety conclusions) including a benefit-risk determination and overall conclusion.

Thirdly, while the list we have provided in this chapter is quite extensive and comes from FDA's PMA webpage [12] as well as the regulations, note that it is not a comprehensive list. If you are developing a PMA device, whether in the early stages of development or approaching the time of compiling your submission, we recommend reviewing the eSTAR, as it *is* comprehensive. Further, it indicates which fields are required (based on a green, red, gray color coding) in the context of your product based on questions answered as you complete the interactive form. It also contains descriptions of fields that need to be completed as well as help guides throughout, where additional background is provided for the given section.

A final point to mention regarding the submission of a PMA is the option for a modular review. This approach allows the Sponsor to submit nonclinical data and manufacturing information while still collecting, analyzing, and compiling the clinical data on the subject device. Using this feature, FDA can review your PMA in sections or "modules," which are submitted at different times to become a complete application. The goal of the PMA modular review is to potentially resolve deficiencies earlier in the process and to increase review efficiency. FDA released a guidance in January 2025, *Premarket Approval Application and Humanitarian Device Exemption Modular Review* [14], that is a good resource for additional details on this option. Note that, as indicated in the title of the guidance, this is an option for HDEs as well as PMAs.

4.1.5 De Novo Classification Request (De Novo)

As previously indicated, FDA conducts a risk-based evaluation of the device that is subject of a De Novo Classification Request to determine if the subject device can be classified as a Class I or II device. A De Novo Classification Request can be submitted to FDA for two different reasons:

(1) A submitter received an NSE determination from FDA on their 510(k) because either their device has different technological characteristics (compared to their selected predicate) that raised different questions of safety and effectiveness,

their device has a new intended use (compared to their selected predicate), or their selected predicate was not actually a viable predicate; or

(2) The submitter did not previously submit a 510(k) and receive an NSE determination from FDA because they had already determined that there was no legally marketed device to use as a predicate for making their justification of substantial equivalence. FDA recommends that Sponsors intending to submit a De Novo consider first submitting a Pre-Sub to obtain feedback from the applicable CDRH premarket division.

FDA's October 2021 guidance, *De Novo Classification Process (Evaluation of Automatic Class III Designation)* [15], can be helpful in better understanding the process for submitting a De Novo and FDA's review of your submission. FDA's other October 2021 guidance, *Acceptance Review for De Novo Classification Requests* [16], can provide additional insight into FDA's criteria for evaluating your submission. Appendix A of the guidance lists all of the content that should be included in a De Novo submission.

At a high level, the required elements are as follows (note that some of these components are only applicable for an eCopy submission as administrative-type elements are built into the eSTAR):

- Coversheet identifying the request a "Request for Evaluation of Automatic Class III Designation."
- Cover letter.
- Proposed indications for use.
- Over-the-counter or prescription or both.
- Device description (including components, principles of operation, pictures, diagrams, etc.).
- Classification information and supporting data
    - Submitter's proposed classification and justification.
    - Discussion of why general controls only or general and special controls provide reasonable assurance of the device's safety and effectiveness.
    - A list of the special controls (if proposing a Class II designation) that would allow FDA to conclude that there is reasonable assurance your device is safe and effective for its intended use.
    - Clinical data (as applicable) to support the reasonable assurance of your device's safety and effectiveness.
    - Nonclinical (bench, animal) performance data.
    - Information on all of the applicable topics: biocompatibility; animal study(ies); shelf life; reprocessing and sterilization; electrical, wireless, mechanical, and thermal safety and electromagnetic compatibility; software/firmware and cybersecurity/interoperability; and literature.
    - Description of the probable benefits of your device compared to the anticipated or probable risks when your device is used as intended.
    - Proposed labeling (i.e., labels, labeling, advertisements, instructions for use, etc.).

- General summary of submission/executive summary (not required but very helpful for FDA's review of the submission).

As of October 1, 2025, De Novo applications *must* be composed and submitted using the eSTAR (prior to that date, use of eSTAR is voluntary). FDA's guidance, *Electronic Submission Template for Medical Device De Novo Requests* [17], is a helpful resource for particulars on the use of the eSTAR for De Novo applications.

Aside from the guidance documents cited in this chapter, it is always ideal to search FDA's Guidance Database to check for any new ones released or updates to existing ones, as well as guidances on your type of device. Additionally, FDA's webpage, De Novo Classification Request [18], contains current information on this submission pathway.

4.1.6 Combination Product Applications

If you are preparing a combination product application, you will need to consult with the reviewing centers at the FDA to ensure you are submitting the right information in the correct format. Typically, the format of the application will be dictated by the lead reviewing Center. If the product is a drug-device combination product wherein the primary mode of action (or PMOA) of the combination product is chemical in nature, the format of the application will typically follow that of drugs (e.g., 505(b)(1) or 505(b)(2)) and contain a device component in the submission. CDRH can be asked to provide consultation on the device components of such an application. If the product is a device-drug combination product wherein the PMOA of the combination product is *not* chemical in nature, the format of the application will typically follow that of device (e.g., 510(k), De Novo, or PMA) and contain a drug component in the submission. CDER or CBER can be asked to provide consultation on the drug-biologic components of such an application. Alternatively, some companies choose to submit a separate device application (when possible) and cross-refer to that application in their drug application. This is done via a letter of reference.

> Whether to submit the device information in one combination product application or to submit the device information in a separate application is a strategic regulatory decision that has pros and cons which must be weighed carefully. If you are not sure which approach to take or want more information on the benefits and risks of each approach, consult with a regulatory medical device expert.

The 21st Century Cures Act requires Sponsors to clearly identify combination product applications for FDA review as a "combination product application" (see 21 C.F.R. § 503(g)(8)(C)(v)). Submissions for device-drug combination products that are being regulated as devices (based on the PMOA) must also include patent certification(s) or a statement as described in section 505(b)(2) and provide notice as described in section 505(b)(3) if the combination product contains as a constituent part an approved drug.

4.2 Device Labeling Requirements

As a device Sponsor, you would never simply distribute your device without accompanying information that explains what the device does, how to use it properly, and what the risks of your device may be. However, every company may have different information that they would like to present about their device. Some companies may want to include a great deal of information whereas others may want to keep the labeling "short and sweet." Not surprisingly, the FDA regulates what information must be included with a device. Many of the earliest guidances on this topic still remain in effect: *Labeling Regulatory Requirements for Medical Devices* [19], but also see *General Program Memorandum #G91-1* [20], *Guidance on Medical Device Patient Labeling* [21], and *Guidance for Industry: Alternative to Certain Prescription Device Labeling Requirements* [22]. Be sure to check the FDA's website for current requirements because the regulations pertaining to medical device labeling are evolving (FDA's webpage, Device Labeling [23], is a good place to start). Additionally, sometimes specific types of devices will have their own labeling guidelines and requirements, which are usually conveyed in the product-specific FDA guidance.

The Federal Food, Drug, and Cosmetics Act defines a "label" as:

> …a display of written, printed, or graphic matter
>
> (1) upon any article or any of its containers or wrappers, or
> (2) accompanying such article at any time while a device is held for sale after shipment or delivery for shipment in interstate commerce.

Importantly, the term "accompanying" in the regulation is interpreted by the FDA very broadly. Any printed material that is provided in association with or pertains to the device, including materials shipped with the device, is considered by the FDA to be labeling. This includes items that one would expect (e.g., healthcare provider/patient instruction on the proper use of the device and the carton and container in which the device is shipped), but it also includes items that may be of surprise (e.g., website materials, advertisements, promotional notices, emails, posters, sales force training materials, press releases, conference handouts, etc.). All labeling materials are subject to specific FDA requirements and failure to meet these requirements can result in the FDA misbranding the medical device (as defined in Section 502 of the Federal Food, Drug and Cosmetic Act; more information is provided in Sect. 4.2.5 of this chapter).

In the sections that follow, we will focus on the printed instructions that accompany a medical device that is meant for use by healthcare providers (also called the physician's package insert or physician's label) and the printed instructions that accompany a medical device that is meant for use by patients (also called the patient package insert or patient label). In Chap. 5, we will briefly touch on advertising and promotional labeling.

4.2.1 *Standard Requirements for Medical Device Labeling (21 C.F.R. § 801)*

General regulations that apply to medical device labeling are provided in 21 C.F.R. § 801. In summary, labeling for your medical device must contain language that:

- Clearly and conspicuously identifies the name and place of business (city, state, and zip code) of the manufacturer, packager, or distributor of the device. For situations where the medical device is manufactured and distributed by the Sponsor, labeling should include the statement, "Manufactured and distributed by [Sponsor Name]." For situations where the medical device is being manufactured or distributed by someone other than the Sponsor, labeling should include "Manufactured for [Sponsor Name]," and/or "Distributed by [Sponsor Name]."
- Describes the adequate directions for use, including (but not limited to):
  - Instructions provided in layman's terms on how to use the device so that it can be used safely and as intended.
  - Information about the dose, frequency of use, and duration of use should be provided. Any variations in dose or frequency or duration of use based on patient characteristics (e.g., age, gender, disease conditions, etc.) should be noted.
  - Special instructions related to preparation for use (e.g., use on an empty stomach, clean treatment "area" with alcohol before using this device, etc.) should be included.
  - Route of administration (e.g., adhere to skin, inject, etc.).
  - Warnings against use in certain patients (e.g., pediatric or geriatric, etc.) where its use may be dangerous to their health.
  - Warnings against unsafe dosage or methods or duration of administration or application.

Note: There are exemptions from the adequate directions of use requirements (see *Subpart D—Exemptions From Adequate Directions for Use*).

Per 21 C.F.R. § 801, medical device labeling must also:

- Display the required information in a legible font, entirely contained on the article of labeling (e.g., it cannot run off the side of a carton or be obscured by folds in packaging).
- Be provided in English or in the predominant local language if the local language is other than English. For example, if the medical device is prescription only

and is only marketed in Puerto Rico, it is permissible to use the local language exclusively.
- Present any expiration date, date of manufacture, or any other date notifying the user of important information in YYYY-MM-DD format. For example, an expiration date of October 15, 2020, must be presented as 2020-10-15. Not all labels will include date information.

The use of symbols in labeling is not required. However, there are special requirements associated with the use of symbols in medical device labeling. Generally speaking, symbols must be accompanied by explanatory text, or be standard FDA-recognized symbols, or adhere to a standard developed by a standards development organization, or SDO, but there are exceptions (see 21 C.F.R. § 801.15 for details and the FDA's guidance, *Use of Symbols in Labeling* [24]). If you are using symbols, be sure to check the FDA's website for current requirements because the regulations pertaining to the use of symbols in medical device labeling are evolving.

As indicated previously, the FDA can misbrand a medical device if it does not adhere to mandatory labeling requirements. The FDA can also misbrand a medical device if:

- The brand name or proprietary name of the device appears twice the size or larger than the established name, name in an official compendium, or common name on any labeling.
- The device is subject to a performance standard and the labeling does not clearly specify that standard.
- There is any language that gives the impression of FDA clearance, grant, or approval if the device has not been cleared, granted, or approved by the FDA.
- The medical device does not comply with the color additives provisions listed under Section 706 of the FFDCA.

4.2.2 Combination Product Labeling

There are no combination-product-specific regulations or guidances that address product labeling. If you are developing a combination product, you will likely be required to submit labeling that conforms, in spirit, to all medical device labeling requirements as well as the Physician's Labeling Rule (content and formatting requirements for Physician's Labeling Rule are outlined in 21 C.F.R. § 201.56 and 201.57). It is important to discuss the content and format of combination product labeling with the FDA as early as possible and throughout development. It is also important to search FDA's guidance document database, as there are often product-specific guidance documents (on certain device types as well as on certain types of combination products), and such guidances may include information pertaining to labeling.

One great example of labeling for a device-drug combination product is provided in the PMA submitted by Medtronic, Inc., for IN.PACT Admiral Paclitaxel-Coated

Percutaneous-Transluminal Angioplasty Balloon Catheter, which was approved by the FDA on April, 19, 2018. The product labeling can be found on FDA's website at https://www.accessdata.fda.gov/cdrh_docs/pdf14/P140010S037C.pdf (accessed 5 August 2025).

4.2.3 Unique Device Identification (21 C.F.R. § 830)

Since early 2000, the FDA has issued very few additional recommendations to Sponsors on the content and format of medical device labeling even though medical device innovation has grown at an incredible rate. Because of the paucity of guidance in this area, there is enormous variability in medical device labeling from product to product. Unlike for FDA-approved drug products, there is no FDA-endorsed template for labeling and no single repository for cleared, granted, and approved medical device labels. That is bad news for healthcare providers, patients, and medical device Sponsors and is a place where enhanced transparency would be welcomed.

Unlike pharmaceutical drug and biologic products, medical devices often remain in the marketplace for years or even decades. Thus, a challenge unique to devices is that the instructions for use and warning labels can get lost, may be damaged, or rendered illegible after repeated use. If these labels are unavailable or illegible, medical devices may not be used as intended and patients can be harmed as a result.

There is recognition at the FDA that a lack of device labeling standardization and other online tools may be harming patient and that *change is needed.*

In a Federal Register announcement dated January 7, 2013, the FDA stated that there is a "growing need for medical device labeling to be delivered in a clear, concise and readily accessible format so that patients, caregivers, and healthcare providers may access and utilize device labeling as efficiently and effectively as possible [25]." In 2018, then FDA commissioner Scott Gottlieb furthered this stating that the FDA is now focused on ways to enhance and modernize medical device safety and innovation [26].

A final rule was issued by FDA at the end of 2013 to establish a system to adequately identify devices through distribution and use as well as to require medical device labels to include a unique device identifier (UDI; except where the final rule provides for an alternative placement or exception). Per the final rule, unless subject to an alternative or exception, labelers of medical devices must submit information about the devices to FDA's Global Unique Device Identification Database (GUDID). Additionally, the final rule requires the device package and label of every medical device to include a UDI and requires that each UDI be provided in a plain-text version, in a form that uses automatic identification and data capture (AIDC) technology. In cases where the device is intended to be used more than once and intended to be reprocessed before each use, the final rule requires that the UDI be directly marked on the device itself.

To further assist labelers of medical devices in understanding and complying with UDI requirements, FDA issued two final guidances, *Unique Device Identification: Direct Marking of Devices* [27] and *Unique Device Identification System: Form and Content of the Unique Device Identifier (UDI)* [28].

The requirements outlined in the final rule and in the UDI guidances go a long way toward standardizing the tracking of specific device usage and safety events associated with every unique device.

In June 2025, FDA released a draft guidance, *Unique Device Identifier (UDI) Requirements for Combination Products* [29], to provide clarification on how the UDI requirements apply to a combination product with a device component. The guidance provides examples of when the device component needs to contain UDI versus when it does not. In general, the guidance stipulates that most combination products with one or more device components are subject to UDI requirements. However, the UDI requirement does not apply to investigational devices and Class I devices exempted from good manufacturing practice (GMP) requirements (i.e., 21 C.F.R. § 820). As with all draft guidance documents, the contents may vary once the draft is finalized, so be sure to check for the most current version when referring to draft guidance, in the event it was finalized since the first time you referenced it.

Sponsors' mandatory use of the FDA's electronic Global Unique Device Identification Database (GUDID) will facilitate the documentation of medical device use, accurate data collection on device adverse events, track recalls, and avoid device misuse and counterfeiting. A guidance document was released in December 2024, *Global Unique Device Identification Database (GUDID)* [30], that includes FDA's recommendations on the information that labelers should submit to the GUDID.

4.2.4 Device Labeling Examples

The fact that there is little guidance from the FDA on what information should and should not be included in a healthcare provider's and patient's package insert poses interesting challenges for the medical device Sponsor. It is worth pointing out that the content of medical device applications is proprietary and not published by the FDA (aside from summary reviews). Therefore, the healthcare provider and patient labeling are the only public-facing documentation available once the device is marketed. It is important to note that labeling is not just a regulatory document, but a legal one as well.

As a guiding principle, a Sponsor should strive to prepare labeling that is truthful, instructive on the proper use of the device, balanced (weighing equally the benefits and risks of the device), and meets all the aforementioned FDA labeling requirements. Keep in mind, labeling must be tailored with the end user in mind (e.g., comprehension level and appropriately wording it for the end user [medical professional versus patient]). Labeling comprehension studies may be requested by the FDA if your patient labeling appears to be too technical in nature for the audience.

Physician's labeling can be written in technical and scientific language whereas the FDA requires that patient labeling must be written in a way that someone with a 6th grade education can read and understand.

Remember your label can and will be used in the competitive marketplace, so each labeling statement (or claim) must be considered carefully. Reimbursement is often tied to specific labeling claims. In sum, when you propose labeling to the FDA, you will want to propose language that is your ideal from a regulatory, legal, and competitive marketing perspective. Once your product is cleared, granted, or approved, making labeling changes can be expensive and difficult (and may require additional regulatory submissions), so it is important to go after the language that you want at the outset.

4.2.4.1 Label Content Comparison of Five Marketed Products

As case in point, let's look at a few product labels for marketed devices and evaluate the similarities and differences. Consider the labeling for three self-expanding stents:

- **Boston Scientific—Innova™ Vascular OVER-THE-WIRE Self-Expanding Stent System** [31].
- **Covidien—EverFlex™ Self-Expanding Peripheral Stent System** [32].
  - Note: this label also includes language for the Protégé™ [33] EverFlex™ Self-expanding Biliary Stent System, a similar device.
- **Terumo Medical Corporation—Misago®, Peripheral Self-Expanding Stent System** [34].

Each label conforms to the FDA's medical device labeling requirements and contains all mandatory elements, but they also contain additional information that is intended to aid in the proper use of the medical device. The Boston Scientific and Terumo labels have a TOC whereas the Covidien label does not. The Boston Scientific label begins with a high-level summary of warnings about the device followed by a device description in text and then the intended use, contraindications, warnings, and precautions. The Covidien and Terumo labels present a device description first, including both text and images, followed by the indication(s) for use, followed by contraindications, warnings, and precautions.

All these devices have similar intended uses; however, the description in each package insert is different:

- Boston Scientific—Innova™ Vascular OVER-THE-WIRE Self-Expanding Stent System is indicated to improve luminal diameter in the treatment of symptomatic de novo or restenotic lesions in the native superficial femoral artery (SFA) and/or proximal popliteal artery (PPA), with reference vessel diameters of 4.0–7.0 mm and lesion lengths up to 190 mm.
- Covidien—EverFlex™ Self-Expanding Peripheral Stent System is intended to improve luminal diameter in the treatment of symptomatic de novo or restenotic lesions up to 180 mm in length in the native Superficial Femoral Artery (SFA) and/or proximal popliteal arteries, with reference vessel diameters ranging from 4.5 to 7.5 mm.
- Terumo Medical Corporation—Misago®, Peripheral Self-Expanding Stent System is indicated to improve luminal diameter in symptomatic patients with de novo or restenotic native lesions or occlusions of the superficial femoral artery (SFA) and/or proximal popliteal artery, with reference vessel diameters of 4.0–7.0 mm and lesion length up to 150 mm.

Also note that the description of the clinical studies varies dramatically between the inserts. This is partially due to variations in clinical programs run by each company, but it is also a function of how each company chose to describe the data.

Finally, note that on the last page of the Covidien and Terumo labeling, a glossary of symbols and their meanings is provided, according to regulation. Symbols are used in the Boston Scientific label, but unlike the other labels, there is no official symbol legend. The symbols are simply provided on the last page with accompanying descriptive text. Although seemingly different, the symbol notations in the Boston Scientific labeling are entirely compliant with FDA regulation 21 C.F.R. § 801.15.

Now let's consider the labeling for a different medical device. The AngelMed Guardian® [35] System is an implantable cardiac monitor intended to detect and alert patients of a potential heart attack. The FDA approved Angel Medical Systems, Inc.'s PMA for this device on April 9, 2018, and the labeling can be found on their website at the following link: https://www.accessdata.fda.gov/cdrh_docs/pdf15/P150009C.pdf (accessed 5 August 2025).

Although this device is quite different from the self-expanding stents mentioned above, simply by virtue of the different types of devices and device functions, the overall content and organization of the labeling is strikingly different. In the Angel Medical Systems labeling for this device, there is an introduction to the device that provides an overview of the device purpose and component parts. The indications for use appear on page 3 of the label and are followed immediately by contraindications, followed by an extensive discussion of the details of the IMD device description. Precautions and adverse effects begin on page 14 of the label. Details regarding the clinical data do not appear until page 53 of the label, but more detail is provided in this label about the design aspects of the clinical program than in the other labels we have discussed. Interestingly, an index is provided at the end of both the Users Guides (i.e., labeling for healthcare providers) and Patient Manuals to aid in navigation and is intended to supplement the TOC. Patient labeling begins on page 102 of the label. Note that it is written in easy-to-understand language, with the patient in mind.

Let's look at the labeling for one more product—Defibtech, LLC's automated external defibrillator, Lifeline/ReviveR DDU-100 and associated models. It is a portable, battery-operated defibrillator that will deliver lifesaving electrical shocks to people experiencing sudden cardiac arrest. This product was approved by the FDA on February 1, 2018, and the labeling is provided on the FDA's website at: https://www.accessdata.fda.gov/cdrh_docs/pdf16/P160032C.pdf (accessed 5 August 2025).

The labeling for this product shows pictures of the device on the first page followed by graphics of the key use features of the device on page iii of the label. A TOC is provided for ease of navigation to key information in the labeling. In contrast to the other labels summarized here, a considerable amount of technical information is provided about the device and device operation up front. The clinical study information is much less prominent and de-emphasized as compared with other labels we have summarized (see page 53 of the label). Again, all the basic information required to comply with FDA regulations appears in this labeling, but this is yet another example showing the vast differences in medical device product labeling.

4.2.5 Misbranding Ramifications

The FDA's medical device labeling requirements are intended to ensure proper medical device use and to keep patients safe, but what happens if you do not adhere to FDA labeling requirements?

Typically, the first action that the FDA will take will be to send you a letter (either "untitled" or a "warning letter") that points out the specific areas of noncompliance in your labeling. Both types of letters are also posted on the FDA's website for transparency. In the letter, they will require you to update your labeling to adhere to U.S. law and guidance within a specified timeframe (typically fifteen (15) days). The FDA will also indicate in the letter that you must respond formally to their letter describing in detail how you are modifying your current labeling to correct the violation(s).

If you fail to update the labeling, the FDA has the authority to remove the device from interstate commerce (seizure). They may also place an injunction against the Sponsor which is a "cease and desist" court order preventing further violation of federal law. Fines and imprisonment can also result if the Sponsor continues to market the misbranded medical device. The fines for each offense for violating the Food, Drug, and Cosmetics Act, as of the February 09, 2018, report on Congress.gov [36], are:

- Up to $100,000 for a misdemeanor by an individual that does not result in death.
- Up to $200,000 for a misdemeanor by a corporation that does not result in death.
- Up to $250,000 for a misdemeanor by an individual that results in death, or a felony.

- Up to $500,000 for a misdemeanor by a corporation that results in death, or a felony.
- The maximum imprisonment for a misdemeanor under the Act remains a year for each offense.

However, since 2018, there has been no update to the website where these fines and penalties were posted and more current information does not appear to be publicly available.

Be sure to **adhere to U.S. labeling regulations!**

Consult with the FDA and/or an experienced regulatory consultant if you have questions about whether you adhere to all labeling requirements for your medical device.

4.3 User Fees for Device Applications

U.S. law permits the FDA to charge a fee to review medical device applications. Proof of the payment of fees is required at the time of the submission of the application for review, unless a waiver of the fee has been secured. If the fee has not been waived, and the FDA does not receive the fee for an application, the FDA will Refuse to Accept the application. For all applications, a Medical Device User Fee Cover Sheet (FDA Form 3601, generated in the FDA's Customer Collaboration Portal) is required to be submitted. The FDA has published several application-specific user fee guidances:

- *User Fees and Refunds for Premarket Notification Submissions (510(k)s)* [37].
- *User Fees and Refunds for Premarket Approval Applications and Device Biologics License Applications* [38].
- *User Fees and Refunds for De Novo Classification Requests* [39].

Every year, the FDA publishes the fiscal year (October 1 through September 30) fees in the Federal Register and on their website. Traditional, Special, and Abbreviated 510(k)s are charged the same user fee. The Fiscal year (FY) 2026 medical device user fees are provided in Table 4.1.

Do NOT send the user fee with the application.

Table 4.1 FY 2026 medical device user fees (In USD) [40]

Type of submission	Fee for small business	Standard fee
510(k)	$6517	$26,067
PMA	$144,818	$579,272
De Novo	$43,446	$173,782

Information on how to submit the user fee can be found on the FDA's website.

If you are a small business, user fees may be waived or reduced. See the FDA's guidance, *Medical Device User Fee Small Business Qualification and Determination: Guidance for Industry, Food and Drug Administration Staff and Foreign Governments* [41].

There are certain user fee requirements based on the particular application type and category that is submitted to the FDA. More information about medical device user fee requirements for other categories and application types can be found online on the FDA's website at https://www.fda.gov/medical-devices/premarket-submissions-selecting-and-preparing-correct-submission/medical-device-user-fees#fees (accessed 5 August 2025).

If your device is a combination product, what can you expect in terms of user fees? To aid Sponsors, the FDA issued a guidance entitled, *Application User Fees for Combination Products*, which outlines how user fees are assessed for different types of combination products. In addition, the FDA guidance entitled, *User Fees and Refunds for Premarket Approval Applications and Device Biologics License Applications* [38], summarizes user fees by PMA type and provides information and links to additional information as well as information related to possible user fee refunds. In sum, if you submit a single application for a combination product, you will be subject to the fee associated with the type of application the FDA is reviewing.

Recall our discussion in Sect. 4.1.6 about how the primary mode of action (PMOA) dictates the content and format of combination product applications. If the product is a device drug combination wherein the PMOA is mechanical in nature, the FDA will regulate the application as a medical device (thus a 510(k), PMA, or De Novo will be submitted), and medical device user fees will apply (according to the submission type, i.e., 510(k), De Novo, or PMA). Alternatively, if the product is a drug-device or biologic-device combination wherein the PMOA is chemical in nature, the FDA will regulate the application as a drug or biologic (thus an NDA or BLA will be submitted), respectively, and prescription drug user fees apply (according to the submission type, i.e., an NDA or BLA with clinical data or/and NDA or BLA without clinical data). Given the amount of money involved, should there be any uncertainty, it may be ideal to call the FDA's User Fee Helpdesk at (301) 796–7200 or email userfees@fda.gov to confirm fees for your particular application.

References

1. FDA's guidance for Industry and FDA Staff: *The Least Burdensome Provisions: Concept and Principles* – FINAL GUIDANCE (February 2019). Available online at https://www.fda.gov/regulatory-information/search-fda-guidance-documents/least-burdensome-provisions-concept-and-principles. Accessed 26 July 2025.
2. FDA's Webpage, Send and Track Medical Device Premarket Submissions Online: CDRH Portal. Available online at https://www.fda.gov/medical-devices/industry-medical-devices/send-and-track-medical-device-premarket-submissions-online-cdrh-portal. Updated 15 November 2024. Accessed 26 July 2025.
3. FDA's *MDUFA Performance Goals and Procedures, Fiscal Years 2023 through 2027*- DRAFT GUIDANCE. Available online at https://www.fda.gov/media/157074/download. Accessed 26 July 2025.
4. FDA's Webpage, eSTAR Program. Available online at https://www.fda.gov/medical-devices/how-study-and-market-your-device/estar-program. Updated 21 November 2024. Accessed 26 July 2025.
5. FDA's guidance for Industry and Food and Drug Administration Staff: *eCopy Program for Medical Device Submissions* – FINAL GUIDANCE (April 2020). Available online at https://www.fda.gov/regulatory-information/search-fda-guidance-documents/ecopy-program-medical-device-submissions. Accessed 26 July 2025.
6. FDA's Webpage, eCopy Medical Device Submissions. https://www.fda.gov/medical-devices/how-study-and-market-your-device/ecopy-medical-device-submissions. Updated 3 October 2022. Accessed 26 July 2025.
7. FDA's guidance for Industry and Food and Drug Administration Staff: *Electronic Submission Template for Medical Device 510(k) Submissions* – FINAL GUIDANCE (October 2023). Available online at https://www.fda.gov/regulatory-information/search-fda-guidance-documents/electronic-submission-template-medical-device-510k-submissions. Accessed 26 July 2025.
8. FDA's guidance for Industry and Food and Drug Administration Staff: *The Abbreviated 510(k) Program* – FINAL GUIDANCE (September 2019). Available online at https://www.fda.gov/regulatory-information/search-fda-guidance-documents/abbreviated-510k-program. Accessed 26 July 2025.
9. FDA's guidance for Industry and Food and Drug Administration Staff: *The Special 510(k) Program* – FINAL GUIDANCE (September 2019). Available online at https://www.fda.gov/regulatory-information/search-fda-guidance-documents/special-510k-program. Accessed 26 July 2025.
10. FDA's guidance for Industry and Food and Drug Administration Staff: *Deciding When to Submit a 510(k) for a Change to an Existing Device* – FINAL GUIDANCE (October 2017). Available online at https://www.fda.gov/regulatory-information/search-fda-guidance-documents/deciding-when-submit-510k-change-existing-device. Accessed 26 July 2025.
11. FDA's guidance for Industry and Food and Drug Administration Staff: *Deciding When to Submit a 510(k) for a Software Change to an Existing Device* – FINAL GUIDANCE (October 2017). Available online at https://www.fda.gov/regulatory-information/search-fda-guidance-documents/deciding-when-submit-510k-change-existing-device. Accessed 26 July 2025.
12. FDA's Webpage, Premarket Approval (PMA). Available online at https://www.fda.gov/medical-devices/premarket-submissions-selecting-and-preparing-correct-submission/premarket-approval-pma. Updated 16 May 2019. Accessed 26 July 2025.
13. FDA's guidance for Industry and Food and Drug Administration Staff: *Acceptance and Filing Reviews for Premarket Approval Applications (PMAs)* – FINAL GUIDNACE (December 2019). Available online at https://www.fda.gov/regulatory-information/search-fda-guidance-documents/acceptance-and-filing-reviews-premarket-approval-applications-pmas. Accessed 26 July 2025.
14. FDA's guidance for Industry and FDA Staff: *Premarket Approval Application and Humanitarian Device Exemption Modular Review* – FINAL GUIDANCE (January 2025). Available

online at https://www.fda.gov/regulatory-information/search-fda-guidance-documents/premarket-approval-application-and-humanitarian-device-exemption-modular-review. Accessed 5 August 2025.
15. FDA's guidance for Industry and Food and Drug Administration Staff: *De Novo Classification Process (Evaluation of Automatic Class III Designation)* – FINAL GUIDANCE (October 2021). Available online at https://www.fda.gov/regulatory-information/search-fda-guidance-documents/de-novo-classification-process-evaluation-automatic-class-iii-designation. Accessed 5 August 2025.
16. FDA's guidance for Industry and Food and Drug Administration Staff: *Acceptance Review for De Novo Classification Requests* – FINAL GUIDANCE (October 2021). Available online at https://www.fda.gov/regulatory-information/search-fda-guidance-documents/acceptance-review-de-novo-classification-requests. Accessed 5 August 2025.
17. FDA's guidance for Industry and Food and Drug Administration Staff: *Electronic Submission Template for Medical Device De Novo Requests* – FINAL GUIDANCE (August 2024). Available online at https://www.fda.gov/regulatory-information/search-fda-guidance-documents/electronic-submission-template-medical-device-de-novo-requests. Accessed 5 August 2025.
18. FDA's Webpage, De Novo Classification Request. Available online at https://www.fda.gov/medical-devices/premarket-submissions-selecting-and-preparing-correct-submission/de-novo-classification-request. Updated October 4, 2022. Accessed 5 August 2025.
19. *Labeling Regulatory Requirements for Medical Devices.* Available online at https://www.fda.gov/files/medical%20devices/published/Labeling---Regulatory-Requirements-for-Medical-Devices-(FDA-89-4203).pdf. Accessed 5 August 2025.
20. *Device Labeling Guidance #GP1–1 (Blue Book Memo) - General Program Memorandum #G91–1* – FINAL GUIDANCE (March 1991). Available online at https://www.fda.gov/regulatory-information/search-fda-guidance-documents/device-labeling-guidance-g91-1-blue-book-memo. Accessed 5 August 2025.
21. FDA's guidance for Industry and Food and Drug Administration Staff: *Guidance on Medical Device Patient Labeling* – FINAL GUIDANCE (April 2001). Available online at https://www.fda.gov/regulatory-information/search-fda-guidance-documents/guidance-medical-device-patient-labeling. Accessed 5 August 2025.
22. Federal Register Notice: *Guidance for Industry: Alternative to Certain Prescription Device Labeling Requirements.* Vol 65, No. 14, page 3463, 21 January 2000. Available online at https://www.federalregister.gov/documents/2000/01/21/00-1407/guidance-for-industry-alternative-to-certain-prescription-device-labeling-requirements-availability. Accessed 5 August 2025.
23. FDA's Webpage, Device Labeling. https://www.fda.gov/medical-devices/overview-device-regulation/device-labeling. Updated 23 October 2020. Accessed 5 August 2025.
24. *Use of Symbols in Labeling.* Available online at https://www.fda.gov/medical-devices/device-labeling/use-symbols-labeling. Accessed 5 August 2025.
25. Federal Register Notice: *Accessible Medical Device Labeling in a Standard Content and Format Public Workshop; Request for Comments.* Vol. 78, No. 4, page 951, January 7, 2013. Available online at https://www.govinfo.gov/content/pkg/FR-2013-01-07/pdf/2013-00003.pdf. Accessed 5 August 2025.
26. Statement from FDA Commissioner Scott Gottlieb, M.D., on new efforts to enhance and modernize the FDA's approach to medical device safety and innovation, April 17, 2018. Available online at https://www.fda.gov/news-events/fda-brief/fda-brief-fda-brings-additional-efficiency-and-modernization-regulation-digital-health-part-digital#:~:text=In%20this%20case%2C%20the%20FDA,and%20for%20regulating%20tobacco%20products. Accessed 5 August 2025.
27. FDA's guidance for Industry and Food and Drug Administration Staff: *Unique Device Identification: Direct Marking of Devices* – FINAL GUIDANCE (November 2017). Available online at https://www.fda.gov/regulatory-information/search-fda-guidance-documents/unique-device-identification-direct-marking-devices. Accessed 5 August 2025.

28. FDA's guidance for Industry and Food and Drug Administration Staff: *Unique Device Identification System: Form and Content of the Unique Device Identifier (UDI)* – FINAL GUIDANCE (July 2021). Available online at https://www.fda.gov/regulatory-information/search-fda-guidance-documents/unique-device-identification-system-form-and-content-unique-device-identifier-udi. Accessed 5 August 2025.
29. FDA's draft guidance for Industry and Food and Drug Administration Staff: *Unique Device Identifier (UDI) Requirements for Combination Products* – FINAL GUIDANCE (June 2025). Available online at https://www.fda.gov/regulatory-information/search-fda-guidance-documents/unique-device-identifier-requirements-combination-products. Accessed 5 August 2025.
30. FDA's Webpage, *Global Unique Device Identification Database (GUDID).* Available online at https://www.fda.gov/medical-devices/unique-device-identification-system-udi-system/global-unique-device-identification-database-gudid. Updated 17 December 2024. Accessed 5 August 2025.
31. U.S. Registration No. 4,914,126. Owner: Boston Scientific Scimed, Inc.
32. U.S. Registration No. 3,270,214. Owner: Covidien LP.
33. U.S. Registration No. 2,660,323. Owner: Covidien LP.
34. U.S. Registration No. 4,021,057. Owner: Terumo Corporation.
35. U.S. Registration No. 2,738,542. Owner: Angel Medical Systems, Inc.
36. Enforcement of the Food, Drug, and Cosmetic Act: Select Legal Issues, February 09, 2018. Available online at https://www.congress.gov/crs-product/R43609#:~:text=Consequently%2C%20for%20FD%26C%20Act%20misdemeanors,for%20an%20individual%20is%20%24250%2C000. Accessed 5 August 2025.
37. FDA's guidance for Industry and Food and Drug Administration Staff: *User Fees and Refunds for Premarket Notification Submissions (510(k)s)* – FINAL GUIDANCE (October 2022). Available online at https://www.fda.gov/regulatory-information/search-fda-guidance-documents/user-fees-and-refunds-premarket-notification-submissions-510ks. Accessed 5 August 2025.
38. FDA's guidance for Industry and Food and Drug Administration Staff: *User Fees and Refunds for Premarket Approval Applications and Device Biologics License Applications* – FINAL GUIDANCE (October 2022). Available online at https://www.fda.gov/regulatory-information/search-fda-guidance-documents/user-fees-and-refunds-premarket-approval-applications-and-device-biologics-license-applications. Accessed 5 August 2025.
39. FDA's guidance for Industry and Food and Drug Administration Staff: *User Fees and Refunds for De Novo Classification Requests* – FINAL GUIDANCE (October 2022). Available online at https://www.fda.gov/regulatory-information/search-fda-guidance-documents/user-fees-and-refunds-de-novo-classification-requests. Accessed 5 August 2025.
40. U.S. FDA, FY 2025 MDUFA User Fees. Available online at https://www.fda.gov/industry/fda-user-fee-programs/medical-device-user-fee-amendments-mdufa-fees. Accessed 20 August 2025.
41. FDA's guidance for Industry, Food and Drug Administration Staff and Foreign Governments: *Medical Device User Fee Small Business Qualification and Determination Requests* – FINAL GUIDANCE (July 2025). Available online at https://www.fda.gov/regulatory-information/search-fda-guidance-documents/medical-device-user-fee-small-business-qualification-and-determination. Accessed 5 August 2025.

Chapter 5
During and After Review

5.1 FDA Review of Your Device Application

It is important to understand the steps taken by the FDA after you submit your application. Being prepared to interact with the FDA if they have questions on the information submitted will help to ensure that the review of your application continues smoothly. As indicated in Chap. 3, be sure to keep a copy of the application you submitted. It will facilitate your communications with the FDA and allow you to look at exactly what they are seeing if, for example, they call you to discuss the schematic presented in a specific figure of your application.

As a general recommendation, you should make a list of all the gaps and limitations in your application prior to submission (no submission is perfect and there are probably things that are keeping you up at night!). Take the time to prepare preliminary responses in case the FDA asks about these gaps or limitations. This advance effort will allow you to respond to the FDA questions with rapid and well-formulated, responses. A timely, well thought-out response will make all the difference in allowing the FDA to finish their review and get your product to market quicker.

There are times that important information becomes available after you submit your application. In some circumstances, you may be able to update your application during review without penalty. However, the FDA will only review content submitted in the original application or information the Sponsor submitted in response to an FDA request during review. Any other (non-FDA requested) updates to your application may not be part of the FDA's substantive (scientific) review. As always, if you are not sure if an update will impact the timing of the FDA's review, consult with the Agency.

As for the timing related to reviews, the FDA routinely updates their performance goals. The timelines provided herein are from *MDUFA [Medical Device User Fee Amendments (MDUFA)] Performance Goals and Procedures, Fiscal Years 2023 Through 2027* (sometimes referred to as MDUFA V) [1] and provides the Agencies'

G. B. Halt Jr. et al., *FDA and Intellectual Property Strategies for Medical Device Technologies*, https://doi.org/10.1007/978-3-032-12649-8_5

goal (in percentages) for turnaround of the submissions received. The specific timeline for your application could have shorter or longer durations than those indicated below. As always, please check the FDA website for current expectations as these are ever changing.

5.1.1 FDA Review Process

Whether your submission is a 510(k), PMA, De Novo, or Combination Product, the FDA generally follows the same basic steps during review (Fig. 5.1). Your likely first questions about the whole process are: What happens if the application has deficiencies, gaps, or holes of which you may or may not be aware? Will the FDA refuse to accept my submission? Maybe … Generally, the first communication the Sponsor will receive from the FDA will indicate whether or not the application was accepted for review.

PMA and De Novo medical device submissions are submitted electronically using the electronic copy (eCOPY) program. The process is explained in the guidance document *eCopy Program for Medical Device Submissions, Guidance for Industry and*

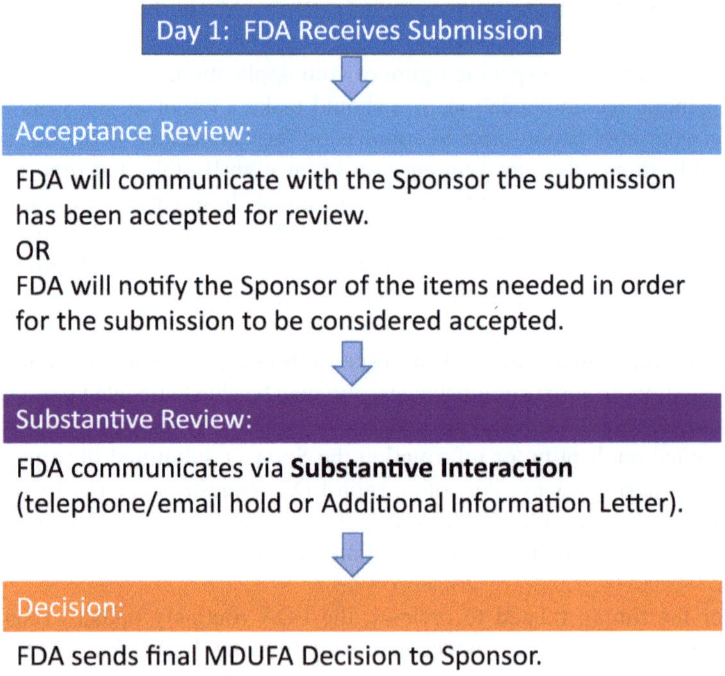

Fig. 5.1 Overview of the FDA review process

Food and Drug Administration Staff, issued in April 2020 [2]. Additional submission types for which an eCOPY is required include Q-subs, Investigational Device Exemptions (IDEs), Humanitarian Device Exemptions (HDEs), and others.

All 510(k) submissions are required to be electronically submitted through eSTAR, unless exempted. Correct completion of the eSTAR submission will give a status message "eSTAR Complete" which can them be submitted for review. If the status message indicates "eSTAR Incomplete" and is submitted for review, the FDA will place your eSTAR on "eCOPY Hold." The "eCOPY HOLD" notification email will list the incomplete sections in the submitted eSTAR. The use of the eSTAR eliminates the need for a Refuse to Accept (RTA) review by FDA and thus RTA holds (refer to Sect. 4.1.2.1). The substantive review will be conducted according to the procedures found in FDA guidance *The 510(k) Program: Evaluating Substantial Equivalence in Premarket Notifications [510(k)]* [3].

Currently, the use of eSTAR is voluntary for PMA and De Novo submissions, although the latter must be submitted via eSTAR starting October 1, 2025. The FDA will communicate (via fax, email, or other written communication) with the Sponsor to identify the assigned FDA reviewer or lead and to either accept or reject a submission after a preliminary review of the information provided. Keep in mind, the review performed at this stage is *very* preliminary. If you receive notice that your submission was accepted, it *does not* mean that your submission was "perfect;" it simply means you complied with the minimum requirements for submission.

During this time, the FDA evaluates the application per a checklist(s) specific to the application type. If a submission is refused for filing, the FDA will provide the Sponsor with the checklist indicating the items that are missing. If something is missing, the FDA may request the Sponsor submit the missing information, depending on what it is and the timeframe the FDA thinks is required to assemble and submit it.

Let's consider our coronary stent example. Fictional company Medicus Societas has a coronary stent that is equivalent to a cleared stent with only slight modifications. Somehow, while compiling the 510(k) package for submission, the Truthful and Accuracy Statement (21 C.F.R. § 807.87) was inadvertently excluded. This is a fairly "small" item and the FDA will probably request it be submitted (quickly—possibly within 24 h) so the substantial review can start. However, as mentioned in Chap. 2, there are guidances specifically applicable to coronary stents. For various reasons, this company chose to not follow guidance; however, they didn't explicitly explain what they did and why. This has the potential to be a "big" item—one that could take a long time to address. It is possible that the FDA will not ask for this information in order to start the substantial review; they will simply indicate in the checklist that the information was not provided in the initial package and refuse to accept (RTA) the submission.

Once your application is accepted for review, the FDA will begin a substantial review of the entire submission package. In Chap. 2 we covered *all* the information that goes into your submission package—it can be a lot of data. The FDA will review everything and compile their own list of gaps, holes, and deficiencies found during review. The FDA will let the Sponsor know what deficiencies (or questions) they find during review and will issue an Additional Information (AI) letter (Sect. 5.1.1.1).

It is also possible for the FDA to initiate an Interactive Review during their substantial review of the submission package. A helpful guidance from the FDA entitled *Types of Communication During the Review of Medical Device Submissions* [4] is a good resource for not only the types of communication a Sponsor may receive, but also the purpose and intent behind these communications.

The specific review processes and timelines for 510(k), PMA, and De Novo submissions are further discussed in Sects. 5.1.2, 5.1.3, and 5.1.4, respectively.

5.1.1.1 Additional Information Letter (Deficiencies Noted During Review)

If you receive an AI letter indicating deficiencies in your application, it is not an attack on you or your company. An AI letter does not indicate your product will never make it to market. It could simply mean that sufficient information was not provided about "a concern" in your original submission. You may already have the information the FDA needs and just didn't include enough detail in your submission. If so, great! You will have the opportunity to provide that additional detail, via a formal response to the FDA.

> It is important to note that all Additional Information letters are reviewed by supervision at the FDA. You cannot "fight" a decision by going to the next level of management at the FDA.

An AI letter may contain a few or many questions, and the number of questions is typically related to the quality of the application. The better the quality, the fewer questions the FDA will have for the Sponsor. In the AI letter, the FDA will provide the basis for the deficiency (e.g., reference to a specific guidance, recognized standard, etc.) or other support (i.e., a scientific issue with information to support the FDA's position) to assist the Sponsor with understanding.

If you get an AI letter, read it carefully. There will be a date by which the agency expects to receive your answers, and you must comply with that timeline. Try to understand what the FDA is asking of you and provide the appropriate information in your response. As with your original submission, provide a clear and concise response that truly addresses the concern, or deficiency, the FDA identified. Any subsequent deficiencies identified by the FDA during review will be mainly limited to any issues raised by the additional information you provide.

As the Sponsor, you have the right to not provide any additional information as requested by the FDA. If such a decision is made, you should provide a written response to the FDA declining to submit the requested information.

Additionally, the Sponsor can withdraw an application at any time during the review. Again, you should provide written correspondence indicating withdrawal of the application.

5.1.2 510(k) Review Process

Figure 5.2 provides an overview of the FDA 510(k) review process as of the guidance update in October 2023. Upon receipt of the submission the FDA gives the submission a unique control number, often referred to as the "510(k)" or "K" number. The first step will occur within fifteen (15) calendar days of the FDA's receipt of the submission. This first pass check of the application is done to ensure that User Fees have been paid (or were waived) and to ensure that the application is reviewable (an eCopy is provided, per regulation). Additionally, the FDA will determine if, at a high level, the submission contains all the necessary information for a substantive review.

During this time, the FDA evaluates the 510(k) application per a checklist (the specific checklist used depends on the type of 510(k) application). The FDA guidance *Refuse to Accept Policy for 510(k)s* [5] provides a checklist of the preliminary questions the FDA will seek to answer before determining if an application will be accepted or rejected.

Assuming that there are no RTA issues, within sixty (60) calendar days of receipt of the submission, the FDA will provide a Substantive Interaction communication (telephone call/email to place the review on hold, or an AI letter).

By Day 90, the FDA will issue a decision. If a decision is not made by Day 100, you will receive written feedback with all outstanding issues preventing the FDA from making a decision concerning your device. This letter will include action items for both the FDA and the Sponsor, as appropriate, with estimated dates for completion by each party. The issues identified may require discussion during a meeting (either in person, via videoconference, or via teleconference) or may be resolved through written correspondence.

For all 510(k) types, the FDA's commitment to industry is to review and provide final action on 95% of all submitted 510(k) within ninety (90) FDA days (refer to the FDA's MDUFA Commitment Letter). Possible FDA decisions include an order declaring a device substantially equivalent (SE) to a legally marketed predicate device (i.e., FDA clearance); an order declaring a device is not SE to a legally marketed predicate device (i.e., an FDA NSE letter); an AI letter; or advise the Sponsor that a 510(k) submission is not required (i.e., product is not regulated as a device). Additional information on these actions is provided in the FDA guidance *FDA and Industry Actions on Premarket Notification (510(k)) Submissions: Effect on FDA Review Clock and Goals* [6].

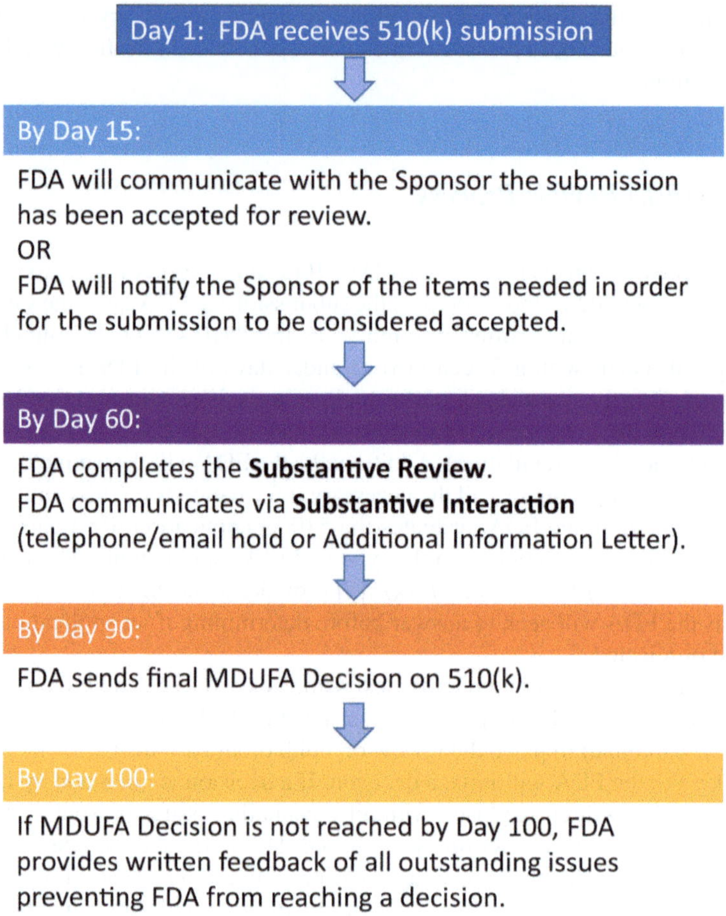

Fig. 5.2 Review process for 510(k) applications

5.1.3 PMA Review Process

Figure 5.3 provides an overview of the FDA PMA review process as of the guidance update in December 2019. The acceptance and filing reviews begin at the time the PMA is submitted and are completed within the first fifteen (15) and forty-five (45) calendar days after the FDA receives the application, respectively. The purpose of these reviews is to allow the FDA to establish whether the application contains all the required elements in the correct format to permit the review team to proceed with an unhindered scientific review. The FDA's guidance *Acceptance and Filing Reviews for Premarket Approval Applications (PMAs)* [7] provides details on acceptance and filing, including an easy-to-follow checklist.

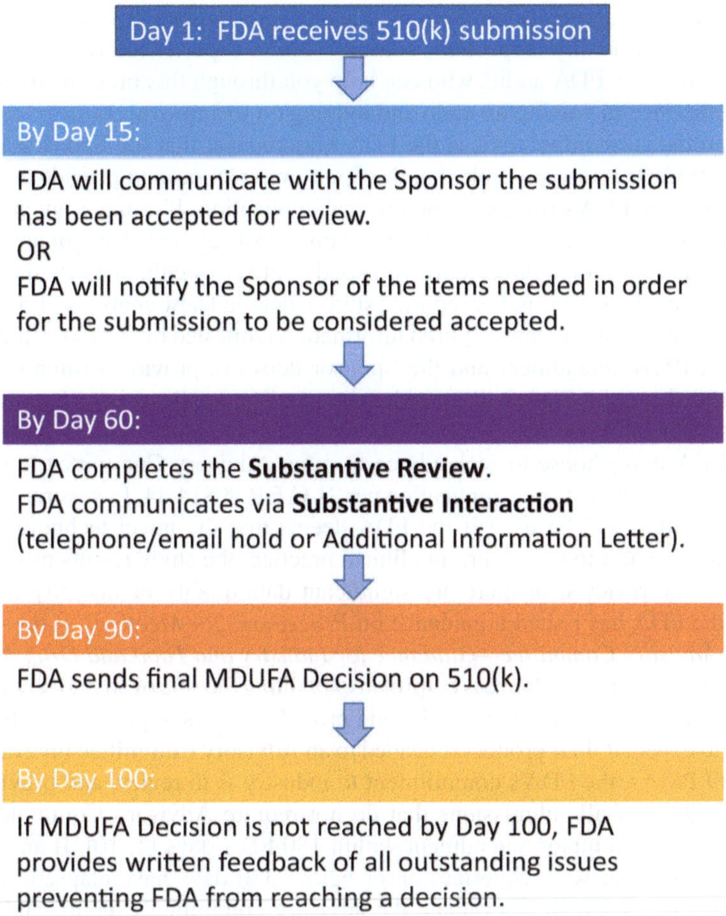

Fig. 5.3 Review process for PMA applications

As with 501(k) applications, the review at Day 15 is *very* preliminary and communication from the FDA will indicate the assigned FDA reviewer and acceptance or rejection of your submission. If your application is not filed (Day 45 review), the FDA will communicate with the Sponsor to identify the specific reasons why the application was not filed and what information is necessary in order to file your application.

Once the FDA has issued acceptance and filing letters, the FDA begins their substantive review and the clock begins on the date of receipt of the last submission or amendment that made the PMA complete. The Sponsor can request a Day 100 Meeting (see Chap. 2 for details regarding this meeting) to discuss the FDA's review and any issues identified to date. During the substantive review period, the FDA will perform inspection(s) on the medical device quality system for all original

PMAs and for supplements for alternative/additional facilities. If you have not experienced an FDA quality inspection, consult with an experienced regulatory expert who specializes in FDA audits who can help you through this process. It can make all the difference in passing an audit and moving on to approval.

During the substantive review, the FDA may request that the Sponsor amend a pending PMA. It is also possible that the Sponsor can choose to provide an amendment during the FDA's review of the original application. If a major amendment is provided, the review period will be extended up to 180 days [8]. A major amendment would include additional information provided related to significant new data (either from a previously unreported or reported study), detailed new analyses of previously submitted data, or significant required information requested by the FDA. If the FDA requests a PMA amendment and the Sponsor does not provide written responses within 180 days, the FDA will consider the pending application to be withdrawn voluntarily by the Sponsor.

The FDA may choose to seek advice from an Advisory Committee (a panel of experts) on a medical device application per 21 C.F.R. § 814.44. In general, an Advisory Committee will be used if the FDA deems that the novel technology could have a significant impact on current clinical practice; the study results provided are unclear to the reviews; or there are significant data quality or integrity concerns. In fact, the FDA has issued a guidance on *Procedures for Meetings of the Medical Devices Advisory Committee—Guidance for Industry and Food and Drug Administration Staff* [9]. This guidance is helpful in explaining the role of advisory committee meetings in the overall review process and provides excellent guidance to Sponsors on what to expect if their product is headed to an Advisory Committee for evaluation.

For all PMAs, the FDA's commitment to industry is to review and provide final action on 90% of all submissions that do not require Advisory Committee input and did not have a major amendment within 180 FDA days [1, 10]. If an advisory committee meeting was requested, on or before 320 days have elapsed, the FDA will complete their review, evaluate the advisory committee's report, and issue a decision. Possible FDA decisions include an approval letter, an approvable letter, a not approvable letter, or a deny approval letter [11, 12].

5.1.3.1 Summary of Safety and Effectiveness Data (SSED)

Per regulation (520(h)(1)(A) of the FD&C Act; 21 C.F.R. § 814.9, the FDA is required to prepare and publicly post a detailed Summary of the Safety and Effectiveness Data (SSED) on which they based their approval or denial decision for an application. The SSED is required to be made publicly available at the time the PMA is either approved or denied approval [11].

The FDA drafts the public SSED based on the SSED written by the Sponsor and included in the original PMA (21 C.F.R. § 814.20, see our discussion in Chap. 3). Unfortunately, SSEDs written by Sponsors are not deemed by the FDA as objective, balanced, and/or appropriate for public dissemination, and often the FDA will require the Sponsor to rewrite their SSED.

The FDA has provided a template for a SSED (*FDA Summary of Safety and Effectiveness Data (SSED) Office of Device Evaluation Template* [13]). The template provides not only the required outline, but also provides (in red italic font) guidance from the FDA for each section. Be sure to consult this helpful guidance when creating the SSED.

The SSED should be written at a level that a college graduate can understand. Not everyone who reviews the SSED is a science graduate, a professional, or an engineer. Technical terms should be used as needed, but our recommendation is to follow a keep-it-simple approach. SSED rewrites cause delays, but can be avoided if Sponsors review and follow previously FDA-released SSEDs (preferably recent ones for similar devices). Following a successful model can go a long way toward ensuring that your SSED is useful to the FDA when posting publicly or in the Federal Register (which is done quarterly).

5.1.4 De Novo Review Process

Historically, De Novo Classification Requests had to first be submitted as a 510(k), and an evaluation of not substantially equivalent (NSE) had to be issued by FDA before a Sponsor could submit a De Novo application. Now Sponsors have the ability to submit direct De Novo applications and forego the initial 510(k) step. This direct De Novo pathway significantly shortens the review cycle and is most efficient for both the Sponsor and the FDA (see *De Novo Classification Process (Evaluation of Automatic Class III Designation)* [14] and *FDA and Industry Actions on De Novo Classification Requests: Effect on FDA Review Clock and Goals* [15]).

If a Sponsor initially submits a De Novo (rather than having received an NSE for initially submitting a 510(k)), the FDA does not begin their review until they have verified that the device in question for the specific indications for use is not the subject of another medical device application (i.e., 510(k) or PMA) from the same Sponsor (Fig. 5.4). If it is, the Sponsor must withdraw the other application(s) prior to starting the De Novo review process.

As of the guidance update in September 2019, similar to 510(k) and PMA reviews, the De Novo review begins with an acceptance review within fifteen (15) calendar days. This first pass check of the application is done to ensure that User Fees have been paid (or were waived) and to ensure that the application is reviewable (an eCopy is provided, per regulation). Additionally, the FDA will determine if, at a high level, the submission contains all the necessary information for a substantive review and evaluates whether the device is appropriately subjected to the De Novo pathway. If it is not, the application is not accepted for filing. The FDA's guidance *Acceptance Review for De Novo Classification Requests* [16] provides details on acceptance and filing, including an easy-to-follow checklist. As always, check the FDA website for an updated, approved version of this guidance.

After the application is accepted and filed, the FDA begins their substantive review. Once a request has been accepted for substantive review, the FDA will only review

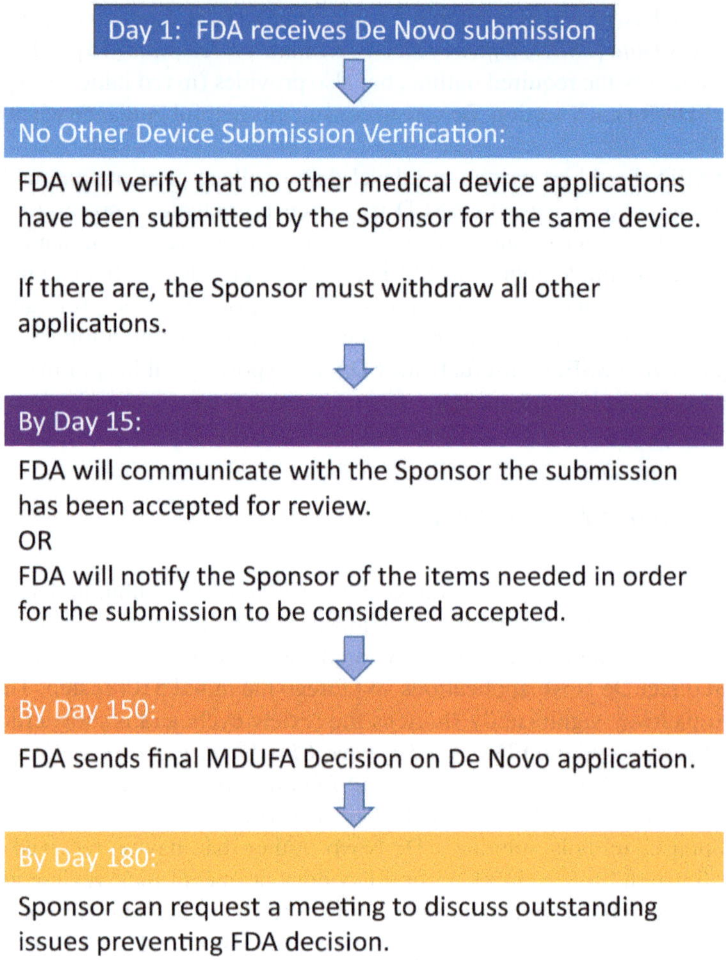

Fig. 5.4 Review process for de novo application

those items they received prior to initiating the substantive review. This means, any supplements or amendments submitted in response to AI letters *after* the FDA starts the substantive review will not be included in their review. As noted before, every effort should be made to respond to the AI letters by the date requested by the FDA.

Per the MDUFA V commitment letter, Congress is holding the FDA to render a decision on De Novo applications within 150 FDA days of receipt of the application for 50% of the applications received. In this context, the 150 days are calculated as the number of calendar days between the date the De Novo request was received and the date of decision, excluding the days a submission was on hold for an AI request. Possible responses include a granted order for classification or a declining order.

If the FDA does not provide a review decision within 180 days and no AIs are issued, the Sponsors can request a meeting with the FDA to discuss all the outstanding issues preventing the FDA from making a decision on the device application.

5.1.5 Combination Products

As indicated in Chap. 4, the FDA will determine if your combination product application is considered a drug with a device component (reviewed as a drug) or a device with a drug component (reviewed as a device). If the primary mode of action is the device, then the review process will follow one of the above timelines.

For drug applications (505(b)(1) or 505(b)(2) applications), a team of physicians, statisticians, chemists, pharmacologists, and other scientists are involved in the review of your application. This is in addition to consult provided for the device component review. Each member of the review team will conduct a full review of their assigned portion of your application. Once each reviewer has completed the review of their section, the entire review team will meet to discuss the application. A Mid-Cycle meeting is held at approximately five (5) months after receipt of the application to discuss issues found to date. As needed, a final meeting, the Wrap Up Meeting, is held at the conclusion of the review, usually about eleven (11) months after receipt of the application.

As with device applications, there is a checklist involved with the review—actually, there are several checklists involved as each discipline has their own checklist. These checklists are available on the FDA website. As with a PMA, the FDA may choose to seek advice from an Advisory Committee on components of an application.

Also, similar to a PMA, the FDA may conduct audits as part of their review. In addition to audits of the manufacturing facilities, the FDA may perform inspections of the clinical study sites. Again, if you have not experienced an FDA quality audit or clinical study site inspection, it is highly recommended that you consult with an experienced regulatory expert. It can make all the difference in passing an audit and moving on to approval.

The FDA will communicate with the Sponsor throughout the review process [17]. The form of communication (telephone call, letter, or a meeting) will depend on the issue. In general, the FDA will make every reasonable effort to promptly communicate any <u>easily correctable</u> (not major) deficiencies discovered during review.

Approximately ninety (90) days after the FDA starts the review [18], a Sponsor may be given an opportunity to meet with the reviewing officials. A 90-day meeting is usually reserved for Sponsors submitting applications related to new chemical entities or major new indications for established (marketed) products. At this meeting, the Sponsor will be informed of the general progress of the review and the overall status of their application, including additional deficiencies known at this time. The meeting agenda may or may not include major deficiencies with the application.

Once the FDA has completed their review, the Sponsor will be given another opportunity to meet with the review team [19]. The purpose of this meeting is to discuss the major deficiencies found during review and the steps the Sponsor needs to complete in order for the FDA to approve the application. All major deficiencies will be communicated via a Complete Response letter.

So, you may be asking yourself why the FDA doesn't discuss major deficiencies "as soon as they are found." The reason for this is simple (but complex at the same time). A major deficiency found by one reviewer may or may not truly be a major deficiency with the overall application. As indicated before, the FDA will host specific internal meetings to discuss your entire application.

Let's consider our CardioColander example. During the clinical review, the FDA reviewer indicates on their checklist that the safety of the device components was insufficiently covered and raised three (3) specific questions that *may* (or may not) be addressed in other sections of the application. During an FDA review meeting, the clinical reviewer brings up their concerns. There are three (3) possible outcomes to this discussion—(1) the information was provided in another section and, upon discussion, these concerns are now non-issues, or (2) the other reviewers (i.e., non-clinical and device team members) may indicate they will review their section again to ensure that adequate information is provided, or (3) other members of the review team had the same basic questions. If the outcome of the meeting is that the other reviewers have to go back and "check" (i.e., during a Mid-Cycle meeting), the Sponsor may not be notified until the issue is confirmed. If the outcome of the meeting is that the other reviewers had the same basic questions, the Sponsor will be notified of the issues in a Complete Response letter.

As one can determine from the example above, these reviews are complex. Timelines can vary depending on the complexity of the application and review. In general, as indicated in 21 C.F.R. § 314.100, the FDA strives to provide a decision within 180 calendar days. However, be aware that it can, and often does, take one (1) year to complete the review of a drug application. Possible responses include an approval letter, an approvable letter, or a deny approval letter.

5.2 Commercial Considerations Post-FDA and Product Launch

Value Proposition Development

A compelling value proposition is essential for the success of any medical device in the market. It is the unique combination of features, benefits, and value (clinical and economic) that helps a device sell to the right market and sets a medical device apart when there are competitors. Differentiation is key in a crowded marketplace, and a strong value proposition helps to communicate why customers should choose your product.

Device companies face a myriad of stakeholders seeking information on the relative benefit and cost, or "value," of the technology, including the care processes in which devices are used. These audiences range from patients, clinicians, specialty societies, and health care facilities to payers. Incentives of such stakeholders frequently vary, if not outright conflict. Manufacturers are challenged to meet everyone's needs and must consider the continuing shift of financial risk from payers to providers, few incentives to consider long-term outcomes, and newer "value-based payment mechanisms," requiring additional resources when crafting an effective value statement.

While the value assessment of drugs is a well-established discipline, principles for evaluating the cost–benefit tradeoffs associated with devices are not as well developed. Characteristics of medical devices hinder direct application of many pharma-focused value frameworks. Evaluating the value of medical devices depends on understanding how devices can substantially alter entire care processes; operator variation; user learning curves; impact of often short product lifecycles; pricing policy and disclosure; and the level of confidence around parameter estimates that are often used in economic evaluations (a major component of the value proposition).

Payers' scrutiny of spiraling health care costs and efforts to control spending has created an increasingly complex additional hurdle (also known as the "post-regulatory hurdle") for medical devices. Inefficiencies and rising costs have been common themes across many countries and striking an acceptable balance between safety, efficacy, and cost has proven to be difficult (but not impossible!).

It is critical to understand that the "value" of a medical device evolves over time. At its core, the value of a health care technology at launch relates to its ability to affect the overall net health outcomes for a particular patient population (Net health outcome is the measure by which all benefits and harms associated with technology are considered). When added together, the net benefit or net harm is one essential element in the evaluation of value.

In developing the initial value proposition of a device, it is essential to address several key strategic questions, including:

- Does this product solve a problem (an identified gap in patient care)?
- Can I effectively show improved clinical outcomes within a specific patient population with the adoption of this technology?
- What is the true cost–benefit for all stakeholders, including patients, providers, consumers, and payer organizations?

The development of a value proposition cannot be considered a "desirable option" to be formed late in product development. Instead, it is a driving factor in a successful launch and continued commercial success. It is an iterative process that needs to start early on, must involve a cross-functional team, and gets continually updated throughout the product's lifecycle.

Economic Considerations

Health economics plays a growing and pivotal role in optimizing healthcare resource allocation, particularly in the medical device sector. As industry faces increasing pressure to reduce costs while maintaining quality, manufacturers are focusing on strategies that will enhance market access and reimbursement rates. Smaller organizations need to leverage health economics to improve the acceptance of their technology, while larger firms aim to maximize value for both devices and patient outcomes. Integrating health economics early in product development can significantly enhance commercialization efforts, enabling companies to disrupt the market effectively and meet evolving healthcare demands.

Health economics is essential in medical device development as it informs decision-making. By evaluating the cost-effectiveness of medical devices, health economics helps manufacturers and policymakers understand the financial value of innovations, guiding reimbursement strategies and market access decisions. Furthermore, health economics addresses challenges such as the fragmented nature of medical device markets and the need for transparent pricing and competition. As regulatory agencies and payer organizations increasingly rely on economic evaluations to support coverage and reimbursement decisions, integrating health economics into the development process becomes critical for the successful commercialization of these products.

The landscape of health economics in the device market is rapidly evolving, driven by increasing scrutiny of cost-effectiveness and value-based care. As healthcare systems face budget constraints and political influence, understanding economic implications in adopting these products has become essential for manufacturers, regulators, and consumers. By understanding and embracing trends in health economics, a shift toward accountability in spending, manufacturers can better position themselves in a competitive market while contributing to improved patient care and system sustainability.

Health economics analysis (HEA) is an essential tool for manufacturers to gain a clear understanding of the direct as well as the indirect costs of medical intervention. Different models are adopted to understand how a particular medical device would get better access to the market as well as meet the manufacturers' reimbursement goals. Health economics analysis provides a clear picture on the sustainability of the business strategy of any medical device company. The analysis does this by examining the following three factors:

Cost-Functionality: By taking technology costs and utility outcomes of the medical devices into consideration, HEA indicates how functional the product is for patients.

Cost-Minimization: Health economics is about finding the best viable alternative medical solution at the least possible cost. For instance, if a low-cost technology can provide the same clinical benefit as an existing high-cost one- it becomes beneficial for both the patient and the manufacturer.

Cost-Effectiveness: Here the clinical outcome of the alternative technologies is examined in a detailed manner. While the cost of technology is measured in monetary terms, an increase in life expectancy and improvement in outcomes and quality of life helps determine the overall benefits of the product.

Early economic assessments should guide R&D decisions to add value and differentiation to the asset, while postmarket studies demonstrate real-world cost-effectiveness to support reimbursement decisions by payers. As healthcare budgets tighten, integrating health economics throughout the product lifecycle is crucial for medical device companies to remain competitive.

Payer Considerations

Prior to launching any new medical device, companies must address crucial payer considerations to ensure successful market access and reimbursement. It is important to recognize the different payer types (Medicare, Medicaid, private insurers) relevant to the target market and patient population. It is also important to perform an initial assessment to identify relevant payors to include in your reimbursement strategy. A second consideration for manufacturers is to engage with payers early in the development process to understand their specific coverage requirements, policies, and evidence requirements (there are often differences between national payer requirements and regional payer policies). In the current healthcare landscape, payers prioritize value-based care and payment models that align with their value-based care VBC goals.

Companies must demonstrate the clinical and economic value of their device beyond just efficacy and safety, aligning with real-world outcomes. Larger payer groups have dedicated technology evaluation committees that conduct independent reviews, while smaller payers may rely on aggregated information and third-party services. It is not enough to have satisfied FDA. Payers often require additional clinical evidence and focus on real-world outcomes to set pricing.

Companies also need to develop economic models and conduct HEOR studies to demonstrate the device's cost-effectiveness and value, as discussed earlier in this chapter. Such models are necessary as payors need to understand the economic impact of device adoption on their monthly patient payment ratios.

The combination of each of these factors, especially data showing cost-effectiveness and improved outcomes, is not only essential for a carefully crafted payer strategy but will become integral to the value proposition and a manufacturer's pricing and reimbursement planning.

Pricing and Reimbursement

Pricing and reimbursement will be one of the first elements a device manufacturer should consider when developing a new product. It is possible to develop a medical device that costs more to manufacture than would be covered by pricing and reimbursement in the U.S. market. Companies may assume that medical need will ensure some level of reimbursement, however, in the age of value-based payment models and risk sharing arrangements, along with a need to reduce health expenditures while simultaneously improving health outcomes, medical need alone is no longer sufficient. A reimbursement strategy is often the first step that companies formulate in determining how to successfully market their product and analysis of the current reimbursement landscape is imperative. In the current regulatory and reimbursement climate, reimbursement studies alone fall short of what is needed to bring in actual sales. Companies need to develop a strategy for how their product will be paid for, which may not always be through government or commercial payors.

Coverage, coding, and payment are the essential pillars of reimbursement, and each stands alone within a reimbursement strategy. Coverage via a National Coverage Decision (NCD) or a Local Coverage Decision will only apply to new medical procedures and technologies that are not currently defined in the regulations. An NCD applies only to a fraction of new devices and requires a significant amount of clinical data, including at least one publication based on the results from a randomized, controlled trial (RCT). An additional challenge is the different emphasis placed on the approval process by the FDA and what is required for reimbursement by private insurance companies that typically follow the Centers for Medicare & Medicaid Services (CMS) guidelines. The FDA stresses safety and efficacy while CMS focuses on superiority of a product relative to the "gold standard" and potential cost savings.

Coding refers to the alphanumeric systems used to classify and identify medical devices, procedures, and related information for various purposes, including regulatory compliance, reimbursement, and tracking. It is the language of CMS, private payers, facilities, and providers. Coding translates into payment. Codes are used to define where a procedure is performed, and what equipment is involved in that procedure. Coding is where a majority of companies focus reimbursement analysis for cleared products. Determining how and where a product fits into the coding landscape can be a complicated and sensitive matter requiring thorough analysis. Creating a new code or obtaining an add-on code for additional payment to an existing procedure is an avenue companies often pursue. This strategy may be successful but requires strong clinical evidence and data, which would ideally be published in a peer- reviewed journal.

Payment

Once coverage determinations and coding hurdles have been crossed, the amount of payment for a new device can vary based on the setting where the device is used (e.g., hospital, ambulatory surgery center, physician office, home use), the type of device, and the payer. Healthcare differs from other industries as the unit of payment is a variable. There is not a single payment amount or method for any device, as payment methodologies and amounts can vary from one payer to another, from one provider to another, and from one patient to another. It is crucial that device manufacturers understand payment models and payment systems and how they may apply to your device. This includes the advancement of newer innovative contracting models with risk sharing whereby manufacturers give providers large discounts or rebates if certain clinical/economic outcomes are not met or gain sharing models where products are provided at a low price, but providers and payers agree to share with the manufacturers a portion of the cost savings or revenue gains from the use of the product.

Let's take an example from the diabetes space. The gain sharing agreement is for people living with Type 1 and Type 2 diabetes who use multiple daily insulin injections. Part of the manufacturer's payment is based on improving clinical outcomes for a national payer's members who transition from multiple daily injections to an insulin pump sold by the company. The agreement considers improvement of the patient experience and clinical outcomes, as well as the total cost of care. While there are a number of barriers to some new payment models, including the need for collecting, storing and sharing data across different entities, CMS and private payers continue to review and implement models in an effort to reduce costs often associated with new technologies. The CMS Innovation Program may be a valuable resource to manufacturers launching new products, through their focus on developing payment models supporting innovative technologies. While CMMI doesn't directly support development, its focus on value-based care within healthcare systems creates an environment where devices that contribute to improved outcomes and lower costs are likely to succeed in these types of programs and initiatives.

Just as no two devices are alike, no two strategies for successful product launch and continued success of commercial distribution are the same. Incorporating pricing and reimbursement strategies early in the development process are necessary to ensure that new, innovative technology is not just available for use but is actually used by patients who can benefit from it. Manufacturers need to understand the true benefits of carefully planned payer and reimbursement strategies to not only successfully launch their product, but also to maximize payment and achieve financial success.

5.3 Advertising and Promoting Your Device

There is considerable regulation and more than a dozen guidances that dictate the format and content of advertising and promotional materials for prescription drugs and biological products. In fact, drugs and biologics Sponsors are required to submit virtually all advertising and promotional materials to the FDA for their evaluation. Jared Lantzy, Operations Analyst at thc FDA, went on record at a 2015 DIA annual meeting stating that the FDA received 46,494 Form FDA 2253 (advertising and promotions) submissions to CDER OPDP in CY2014 (roughly 80,000 materials) [20]. This number has increased dramatically since 2014 with the advent of electronic submissions of advertising and promotional materials. Additionally, the FDA issues warning letters and untitled letters to drug Sponsors for non-compliant advertising and promotional content. The issuance of these letters has increased exponentially in the last ten (10) years, likely reflecting both the ever-greater complexity of related regulation and additional resources for enforcement action by the Agency.

In contrast, there is very little regulation or guidance that specifies the FDA requirements for advertising and promotional materials for medical devices. As mentioned in Chap. 4, there are basic medical device labeling requirements that medical device Sponsors must adhere to or the medical device can be misbranded by the FDA. That being said, medical device Sponsors are required to submit their labeling, advertising, and promotional materials in premarketing applications (510(k), De Novo, PMA) and, if significant changes are planned for their labeling, advertising, and promotional materials in the postmarket setting, the materials may also need to be submitted to FDA for their review (refer to submission-specific guidances regarding changes to a device, which include labeling changes).

We believe the clock is ticking and it is only a matter of time until the FDA requires this information to be submitted. Just like medical device development and registration requirements, the regulatory landscape is changing rapidly for medical device advertisers. The previous Commissioner Scott Gottlieb has indicated that the FDA is working hard on enhancing and modernizing the FDA's approach to medical device safety and innovation [21]. This modernization includes labeling and advertising/promotions. Recently, the FDA has issued several guidances that address common questions and issues that result from advertising and promotion of medical devices. We provide a brief summarization of the topics included in these guidances for your reference.

- *Guidance for Industry Presenting Risk Information in Prescription Drug and Medical Device Promotion* [22]—This guidance provides the FDA's position on the proper way to disclose risk information about a restricted medical device. Specific examples are provided to aid the medical device promoter in ensuring fair balance and an accurate "net impression" of the risks of the product.
- *Guidance for Industry Responding to Unsolicited Requests for Off-Label Information About Prescription Drugs and Medical Devices* [23]—This guidance provides the FDA's recommendations Sponsors on how to respond to unsolicited requests for off-label information from a variety of sources. Specifically, it is the FDA's expectation that medical device Sponsors respond to unsolicited requests for information about marketed medical devices in a way that is "...truthful, balanced, non-misleading, and non-promotional...." Sponsors are permitted to provide additional information related to uncleared/ungranted/unapproved medical device indications or situations of use as long as that information meets the requirements above and the response directly answers a targeted question. The FDA provides further recommendations depending on whether the unsolicited request will be public or non-public.
- *Medical Product Communications That Are Consistent With the FDA-Required Labeling—Questions and Answers Guidance for Industry* [24]—This guidance provides the FDA's thinking on how to ensure a Sponsor's product advertising and promotions are consistent with the product's FDA-required labeling. This guidance also provides a number of helpful examples and scenarios of claims that are inconsistent with FDA-required labeling.
- *Drug and Device Manufacturer Communications With Payors, Formulary Committees, and Similar Entities—Questions and Answers Guidance for Industry and Review Staff* [25]—This guidance provides answers to common questions regarding Sponsor's communication of healthcare economic information for medical devices to payors, formulary committees, or other similar entities. This guidance also provides the FDA's thinking on how to handle dissemination of

medical device information to payors about uncleared/ungranted/unapproved uses of medical devices.

The FDA is making changes rapidly, so be sure to refer back to the FDA's website for updates and newly issued guidances applicable to medical device advertising and promotion.

Although there is currently limited FDA oversight of medical device advertising and promotion (and even more limited enforcement action), this does not mean that the medical device manufacturer(s) should push the "regulatory envelope."

The FDA attends conferences, trade shows, and combs through public sources (via the internet) for misleading advertising and promotions of medical devices. They also receive reports from patients, healthcare providers, hospitals, and competing medical device companies who act as "whistleblowers" for misleading advertising. Recall that the FDA has the authority to misbrand a medical device if it does not comply with all applicable regulations and guidances.

For broad reaching issues that apply to specific devices or classes of devices, the FDA can also issue a "Letter to Industry." These letters are posted on the FDA's website to facilitate awareness at https://www.fda.gov/medical-devices/industry-medical-devices/letters-industry (accessed 20 August 2025).

Example On May 26, 2015, the FDA issued a Letter to Industry for dermal implants requiring a labeling change for all device manufacturers to warn device users of a safety concern of the filler inadvertently being injected into blood vessels which could trigger rare, but serious, adverse events.

If CDRH learns that a medical device is possibly being advertised or promoted in a non-compliant way, FDA may issue an "It Has Come to Our Attention" Letter (IHCTOA Letter). Such a letter is intended for data gathering purposes and typically does not include enforcement action statements. These letters are also posted on the aforementioned FDA website, https://www.fda.gov/medical-devices/industry-medical-devices/letters-industry (accessed 20 August 2025).

5.4 Postmarketing Requirements

Unfortunately, the medical device Sponsor's work is never done. Receiving a positive MDUFA decision is just the beginning. Medical device manufacturers (and some related firms) involved in the marketing and distribution of medical devices are required to follow particular regulations and guidances once they enter into the marketplace [26]. Such requirements include:

- Monitoring for and reporting of device malfunctions, serious injuries, or deaths.
- Conducting and reporting on postmarketing study requirements.
- Annual reporting.
- Registering the establishments where devices are produced or distributed.
- Record keeping.

Don't forget that the FDA has the authority to inspect a medical device manufacturer and audit their manufacturing and record-keeping practices at any time to ensure compliance to keep patients safe. In this section, we will discuss each of the postmarket requirements in turn.

5.4.1 Postmarket Safety Monitoring and Reporting

The safety information that goes into a medical device application prior to a positive MDUFA decision by the FDA is often quite limited in scope. The known safety of a medical device is based on premarket testing under tightly controlled, limited conditions. Once the device hits the open market, the proverbial "flood gates" are opened and more patients are exposed to the device in real-world settings. Sometimes, devices perform exactly as expected based on the data collected during development, but not all do. The purpose of postmarketing surveillance is to systematically collect, analyze, interpret, and subsequently act on medical device use data to ensure the public health and reduce morbidity and mortality.

Medical device postmarketing surveillance in the United States is currently (as of this writing) heavily reliant on mandatory Medical Device Reporting (MDR) of possible device-associated serious injuries, deaths, and malfunctions. These are listed in FDA's MAUDE Database. To supplement this, nearly 300 hospitals and medical centers voluntarily report events via the Medical Product Safety Network (MedSun; listed in the MedSun Database). The FDA has the authority to order postmarket safety studies (postapproval studies and postmarket surveillance studies) for the purpose of collecting additional efficacy and safety assessments, and the FDA conducts their own investigations of medical device safety using a variety of sources. We will delve into each of the current surveillance activities in more detail and then discuss the FDA's safety surveillance activities under development.

Postmarketing safety reporting requirements for combination products are provided in the final version of *Postmarketing Safety Reporting for Combination Products Guidance for Industry and FDA Staff* [27]. The combination product Postmarketing Safety Reporting requirements apply to both Combination Product Applicants and Constituent Part Applicants. A Combination Product Applicant is one that holds the only or all applications for a combination product. An example of such an applicant is a company that holds the approved PMA for a drug-eluting stent. A Constituent Part Applicant holds an application for a constituent part of a combination product marketed under an application held by a different applicant. For example, a company holds the NDA for a light-activated drug which is used in combination with an approved PMA for a laser system held by another entity or company. Both Combination Product Applicants and Constituent Part Applicants must meet the safety reporting requirements consistent with the application type under which their combination product or constituent part received marketing authorization.

Combination Product Applicants are also subject to certain safety reporting requirements related to the constituent parts of the combination product. For a combination product that contains a device as the constituent part, Combination Product Applicants must comply with 5-day, malfunction, and correction reporting requirements as well recordkeeping requirements. A combination product that contains a drug as the constituent part, field alert reporting (FAR) and 15-day reporting requirements must be met. When the constituent part is a biologic, biological product deviation reporting (BPDR) as well as 15-day reporting requirements must be met.

Constituent Product Applicants must share with other Constituent Part Applicants initial information of death, serious injuries, or adverse experience within 5 calendar days from receipt.

5.4.1.1 Mandatory Medical Device Reporting (MDR)

Regulation 21 C.F.R. § 803 supplements the provisions of 21 C.F.R. § 820 and stipulates the mandatory MDR requirements for manufacturers, importers, and device user facilities to report certain device-related adverse events and product problems to the FDA postmarketing of a medical device (see Table 5.1 for details and reporting timelines). These regulations stipulate that Sponsors must have a standardized system in place for capturing postmarketing safety events, and policies that dictate the timely and effective identification and communication of adverse events associated with the device. These regulations also require that Sponsors store and maintain adverse event information in compliance with the FDA's good documentation and record-keeping requirements.

Per regulation, manufacturers must notify the FDA when:

- They learn that their device may have caused or contributed to a death or serious injury, defined as:
  - Being life threatening.
  - Results in permanent impairment or permanent damage.
  - Necessitates medical or surgical intervention to preclude permanent impairment.
- They learn that their device has malfunctioned and may cause or contribute to a death or serious injury if the malfunction occurred again.

This requirement includes events caused by a failure or malfunction of the device; improper or inadequate design of the device; manufacturing issues with the device; inaccurate, misleading, or ambiguous labeling; or user error. For example, the

Table 5.1 FDA mandatory medical device safety reporting requirements

Reporter	What to report	Report form #	To whom	When
Manufacturers	30 day reports of deaths, serious injuries, and malfunctions	FDA 3500A*	FDA	Within 30 calendar days of becoming aware of an event
	5-day reports for an event designated by FDA or an event that requires remedial action to prevent an unreasonable risk of substantial harm to the public health	FDA 3500A*	FDA	Within 5 work days of becoming aware of an event
Importers	Reports of deaths and serious injuries	FDA 3500A*	FDA and the Manufacturer	Within 30 calendar days of becoming aware of an event
	Reports of malfunctions	FDA 3500A*	Manufacturer	Within 30 calendar days of becoming aware of an event

* Or electronic equivalent

fictional Medicus Societas company would be required to notify the FDA if a healthcare provider called and told the company that the CardioColander was implanted in a patient and the patient receiving the device died three (3) days later of a heart attack because the filter stopped working (device malfunction). For this report, Medicus Societas would be required to notify the FDA within thirty (30) days of the date of the call from the healthcare provider.

In addition to the information in Table 5.1, device user facilities (e.g., hospital, ambulatory surgical facility, nursing home, outpatient diagnostic facility, or outpatient treatment facility, but not a healthcare provider's office) must notify the FDA and the manufacturer when they learn that a medical device is suspected to have caused or contributed to a death. User facilities must report a medical device-related serious injury to the manufacturer or directly to the FDA if the medical device manufacturer is unknown. A user facility should notify the manufacturer or the FDA within ten (10) days of learning of the event. A user facility is not required to report a device malfunction, but can voluntarily notify the FDA either electronically, by phone, or by mail. User facilities are welcomed by the FDA to submit eMDRs, but paper MDR is still permitted.

Per the FDA's eMDR Final Rule, medical device manufacturers and importers must submit all MDRs electronically via the FDA's Electronic Submissions Gateway (ESG). Paper reports are no longer permissible (see the FDA's guidance, *Questions and Answers about eMDR—Electronic Medical Device Reporting—Guidance for Industry, User Facilities and FDA Staff*) [28].

> The FDA also requires "supplemental" or "follow-up reports" whenever additional information is obtained after you submit an initial thirty (30)- or five (5)-day report that alters the conclusions drawn in the original MDR (see 21 C.F.R. § 803.56).

Follow-up reports must be submitted to the FDA within thirty (30) calendar days of the receipt of the new information. When you submit a follow-up report, you are required to denote that it is a follow-up report and to provide information in the follow-up that links the follow-up with the initial report. For example, let's return to our previous example where we were required to submit an initial report for the patient who died because the CardioColander malfunctioned. Five (5) days after the submission of the initial MDR, the Medicus Societas company learned that the patient was killed in a tragic skydiving accident. We would want to submit a follow-up report with this information because it changes the interpretation of the cause of death and the association of that death with the CardioColander medical device.

We are commonly asked when the clock starts with respect to the timing of a medical device report. Officially, the FDA's perspective is that the timing for reporting begins when a member of the Sponsor organization becomes aware of the reportable event. This can be any employee from the CEO to the company janitor who picked up the phone because everyone else was out on their lunch break. Regulation 21 C.F.R. § 803.3(b) also stipulates that the clock starts when any supervisory employee becomes aware that an event requires remedial action to protect the public safety. Therefore, it is important that you train all company employees on how to collect safety information and properly report it in accordance with all FDA regulations and guidances. See the FDA's guidance, *Medical Device Reporting for Manufacturers* [29] for an excellent summary of the requirements, important definitions, additional details on what needs to be reported and when, and the very limited number of exceptions to the MDR reporting requirements.

5.4.1.2 Voluntary Medical Device Adverse Event Reporting

Although Sponsors (manufacturers), importers, and user facilities have reporting obligations, adverse events can sometimes fall through the cracks. In an effort to capture as much information as possible, the FDA has recommended voluntary adverse event reporting from healthcare professionals, patients, caregivers, and consumers who observe or experience significant adverse events or product problems. Voluntary reporting can be done by mail, via the MedWatcher mobile app, calling the FDA at 1-800-FDA-1088, or using FDA form 3500B (see the FDA's website: https://www.accessdata.fda.gov/scripts/medwatch/index.cfm?action=consumer.reporting1 (accessed 6 August 2025)).

5.4.1.3 MedSun

Another tool to try to capture medical device safety information is the Medical Product Safety Network (MedSun). MedSun is an adverse event-reporting program intended for collaborative use between the FDA and the clinical community (hospitals, nursing homes, and outpatient treatment and diagnostic centers) to identify, understand, and solve problems with the use of medical devices. The information entered into MedSun is real-time and searchable so that all contributors (including the FDA) can identify emerging safety signals and take appropriate action to ensure patient safety (see https://www.accessdata.fda.gov/scripts/cdrh/cfdocs/Medsun/searchreport.cfm (accessed 6 August 2025) to access the searchable database). Mandatory and voluntary reporting can be done by user facilities using this system.

5.4.2 Postmarketing Study Requirements

The FDA has the authority to require medical device Sponsors to conduct studies of their medical device postmarketing in order to collect additional efficacy and safety information about the device. These studies can take several forms, but typically include studies that must be conducted as either a condition of the positive MDUFA decision (also called postapproval studies; see 21 C.F.R. § 814.82(a)(2)), and/or for continued evaluation of the ongoing safety of the device (also called postmarketing safety surveillance studies).

Sponsors of Class II or Class III medical devices may be required by the FDA to conduct postmarket surveillance studies (also known as "522 studies"). When required, the purpose of these studies is to obtain more detailed information about the safety of the medical device through more comprehensive non-clinical device testing, analyses of publicly available clinical databases, and observational studies. For certain high-risk devices, randomized controlled trials may be required.

Postapproval studies required as a condition of approval for a device approved under a PMA order usually assess the device's safety, effectiveness, and/or reliability. These studies collect valuable data that assess long-term, real-world device performance, and may be clinical or non-clinical in nature. Fortunately, the Sponsor and the FDA usually discuss the postapproval study protocol(s) and timelines during the review of the PMA prior to approval so you, as Sponsor, have some input and influence over the postmarket requirements.

The FDA will indicate in your application decision letter if postmarket studies are required and will provide a binding description of the nature of the requirement(s).

Once issued, Sponsors must undergo a formal process to have postmarket study commitments modified or eliminated (see *Procedures for Handling Post-Approval Studies Imposed by PMA Order*) [30].

In order to have a better understanding of these types of studies, let's look at a recent PMA approval. On September 20, 2017, the FDA issued an approval letter to Abiomed, Inc., for their Impella RP® [31] System PMA [32], which notes that the Impella RP® System is indicated for use as providing temporary right ventricular support for a period of up to fourteen (14) days in patients with a body surface area $\geq$ 1.5 m^2, who have developed acute right heart failure or decompensation following (1) heart transplant, (2) myocardial infarction, (3) left ventricular assist device implantation, or (4) open-heart surgery. In the PMA approval letter, the FDA is requiring the Sponsor to conduct two postapproval studies (PAS):

(1) Impella RP—Real-World Evidence Evaluation and Periodic Reporting: The purpose of this study is to evaluate the long-term, real-world efficacy and safety of the device.
(2) Impella RP Pediatric—Real-World Evidence Evaluation and Periodic Reporting: The purpose of this study is to evaluate the long-term, real-world efficacy and safety of the device in pediatric patients.

Note in the letter that the FDA is quite specific about the study designs, sample sizes, and data to be collected in each of these studies. The FDA also provides the Sponsor with a timeline for reporting the study results to the FDA and updating the labeling once the studies have been completed. The FDA has a guidance that can help Sponsors who are required by the FDA to conduct postapproval studies (see *Procedures for Handling Post-Approval Studies Imposed by PMA Order* [30]). Routine postapproval study reports must be submitted in accordance with the FDA's requirements outlined in the decision letter. Note that the frequency of these reports may differ from the routine (annual) reporting requirements.

The FDA lists all medical device Sponsors who have medical devices with postapproval study commitments or postmarket surveillance study requirements on their website (https://www.accessdata.fda.gov/scripts/cdrh/cfdocs/cfpma/pma_pas.cfm (accessed 6 August 2025)). Under section 502(t)(3) of the Act, 21 U.S.C. 352(t)(3), a device is misbranded if a medical device Sponsor fails to comply with any requirement under section 522 of the act. Failure to conduct postmarketing study requirements or make progress on those requirements can result in misbranding, seizure of medical devices in the marketplace, injunction, prosecution, or monetary penalties. It can also result in some really bad press for your company.

5.4.3 *The Future of Postmarketing Safety Surveillance*

MDRs are a valuable source of information about the safety and performance of medical devices currently being marketed. However, criticism has been raised about the "passive" nature of this reporting methodology. Specifically, the overarching picture of the safety of a medical device may be incomplete or even misleading due to inaccurate reporting of events, a failure to report all events that have occurred, a lack of understanding of the reported events in the context of the total number of uses of the device (denominator for exposure), and a lack of timeliness of reporting. It is important to note that this is not unique to medical device safety reporting, but is a criticism that is also raised with respect to pharmaceutical postmarket safety reporting [33].

In 2012, the FDA provided a white paper entitled, "Strengthening our national system for medical device postmarket surveillance." This document provided a detailed history of medical device postmarketing surveillance activities, current day activities that help the FDA evaluate the safety and efficacy of marketed medical devices (also described above), and information on future directions in safety surveillance.

Fast forward to January 2016, and recall our discussion about unique device identifiers (UDIs) in Chap. 3. UDIs became a new requirement in part because they will enable the use of standardized information across the lifecycle of a medical device to be linked with information from clinical registries, electronic health records, and medical billing claims. Such a database can then be mined for potential safety risks and can help the FDA more proactively balance the safety risks of medical devices with true patient benefit. It is also anticipated that such a database will enable healthcare providers and patients to make more informed and timely treatment decisions.

To this aim, the FDA is working with a variety of partners to create NEST (The National Evaluation System for health Technology), an active surveillance system that works hand-in-hand with more traditional methods of medical device surveillance. NEST became one of CDRH's 2016–2017 strategic priorities and continues to be a primary focus of the Center today. In 2017, the FDA issued the guidance *Use of Real-World Evidence to Support Regulatory Decision-Making for Medical Devices* [34] as further progress toward this initiative.

In October 2012, the FDA also began the CDRH Signal Management Program with a goal of establishing consistency, efficiency, accountability, and transparency in how the FDA reviews, evaluates, and regulates signals identified in the postmarketing medical device space. One hope for this program is that by identifying signals with currently marketed products, the FDA can more proactively shape the development of similar products prior to marketing.

Since the program was put into place in 2012, hundreds of safety signals have been evaluated, resulting in numerous public health and/or regulatory actions taken (letters issued). In 2016, the FDA issued a guidance document on the topic, *Public Notification of Emerging Postmarket Medical Device Signals ("Emerging Signals")* [35]. In 2018, the FDA published their *Medical Device Safety Action Plan* [36] which summarizes the FDA's vision for postmarket safety surveillance. Included in this plan are a summary of the FDA's goals to better capture safety information sooner (and more comprehensively), evaluate safety more effectively in the context of device risk/benefit and patient perspectives, and address cybersecurity issues.

As with almost all areas of device regulation, postmarketing safety surveillance for devices is a hot-topic. Stay on top of updated requirements by monitoring the FDA website on a routine basis.

5.4.4 Periodic (Annual) Reports

Per 21 C.F.R. § 814.84, the FDA requires that the Sponsor of a medical device approved via PMA submit a periodic report on a routine basis (usually annually, but other frequencies (i.e., quarterly) can be required by the FDA). The purpose of the report is to help the FDA see the activities and events that have transpired during the reporting period to routinely assess the risk/benefit profile of the medical device. The periodic report is prepared by the Sponsor and should include:

- The start and end date of the reporting period.
- A summary of any of the following changes (see 21 C.F.R. § 814.39(a) and (b)) that occurred during the reporting period:
  - New indications for use of the device.
  - Use of a different facility or establishment to manufacture, process, or package the device.
  - Changes in sterilization procedures.
  - Changes in packaging.
  - Changes in the performance or design specifications, circuits, components, ingredients, principle of operation, or physical layout of the device.
  - Any extensions of the expiration date of the device based on data obtained under a new or revised stability or sterility testing protocol that has not been approved by the FDA.
- A summary of all MDRs that occurred in the reporting period.
- A summary and all supporting data for the following that was not previously submitted in the application:
  - Unpublished reports of data from any clinical investigations or non-clinical laboratory studies involving the device or related devices and known to the Sponsor.
  - A literature search and review of studies that concern the device published during the reporting period.
- A summary of distribution data (the number of devices sold and distributed during the reporting period).
- A list of all UDIs in circulation during the reporting period, if applicable for your device class.

The FDA will review your periodic report and consider the contents carefully. Signals observed in the literature and in MDRs can result in the FDA requiring labeling changes to improve the proper and safe use of the device. This is another way that the FDA evaluates the overall safety and utility of a medical device in real-world settings.

5.4.5 Medical Device and Establishment Registration

It is important for the FDA to know who is manufacturing and distributing medical devices in the United States, and what specific devices are being marketed. Such requirements enable the FDA to contact Sponsors directly with urgent safety communications related to specific device types or classes. To this end, companies (establishments) that manufacture and distribute medical devices intended for commercial distribution are required to register their company annually with the FDA [37]. Most companies that manufacture and distribute medical devices are also required to list their medical device(s) and the activities associated with the device(s) at that establishment with the FDA (see the FDA's website to determine whether your company is required to register and list). Waivers can be granted under special circumstances.

All device establishments must submit their annual registration between 1 October and 31 December of each year. There is a fee associated with registration. In FY2018, the registration fee was $4624. Registration involves paying the annual registration User Fee and then completing the registration process online. You will receive an email notification from the FDA once the process is complete.

CDRH Learn also has several courses devoted to the topic of registration and listing that are particularly helpful to understanding the "why's" and "how's" to better inform Sponsors on how to adhere to the registration and listing requirements.

5.4.6 Medical Device Tracking

The FDA requires Sponsors (manufacturers) of certain Class II and Class III medical devices to have a defined tracking system for their product [38]. The FDA requires mandatory tracking of certain devices to ensure that manufacturers can pull potentially dangerous or defective devices off the market quickly and/or notify patients of significant device problems. Devices that are subject to mandatory tracking are devices:

- Whose failure would be reasonably likely to have serious, adverse health consequences.
- That are implanted in the human body for more than one (1) year.
- That are life-sustaining or life-supporting and are used outside of a device user facility.

Because our example product, CardioColander, would be implanted and is anticipated to remain in the body for 12 months or longer, we would be subject to mandatory tracking requirements. Similarly, tracking requirements would also apply for our stent example, but would not apply for our blood pressure monitor example.

For our implantable devices, we would be required by the FDA to have a specific written standard operating procedure (SOP) that describes the method we use for tracking our device throughout distribution. This SOP would also have to include a description of our quality assurance program and audit procedures. So that you are aware, the FDA has the authority to audit all SOPs and tracking records at any time. Per regulation, Sponsors are required to maintain records for the useful life of each tracked device they manufacture or distribute. "Useful life" of a device is defined as the time a device is in use or in distribution for use.

In your MDUFA decision letter, the FDA will indicate whether they believe your device requires you to adhere to the mandatory tracking regulations. However, if you are not certain, read the FDA's guidance on the topic, *Medical Device Tracking* [39] and consult with CDRH.

5.4.7 Record-Keeping Requirements

Sponsors must establish and follow quality systems to help ensure that their medical devices consistently meet all applicable requirements and specifications. This is not optional; a mandatory set of requirements must be carefully established and followed. Procedures (SOPs) must be put in place to capture virtually every aspect of company operations, but especially how a company handles:

- MDR activities (21 C.F.R. § 803.17).
- Product complaints (21 C.F.R. § 820.198).
- Corrective and preventive actions (CAPA, 21 C.F.R. § 820.100).
- Situations where the product does not perform to prespecified specifications (21 C.F.R. § 820.90).

Good documentation practices must be a part of the quality system your company establishes (quality systems management is discussed in more detail in Chap. 15).

> As is often said in industry, "If it's not documented, it's not done!" The FDA will certainly take that approach when evaluating a company's postmarket manufacturing, distribution, and safety surveillance activities.

The medical device record-keeping requirements are described as part of the Quality System Regulation (soon to be the "Quality Management System Regulation"), 21 C.F.R. § 820, one of the most important pieces of regulation that a medical device company must follow. The FDA requires that Sponsors (establishments) must maintain all records in a readily accessible facility for FDA auditors to inspect (review/copy). Stored records must be legible, stored securely, and protected from deterioration and loss. Electronic records must be backed up routinely. Any records that contain confidential information must be marked as such to let the FDA know what can and cannot be disclosed to the public [40].

The FDA requires that medical device records are retained for a period of time equivalent to the design and expected life of the device, or for a minimum of two (2) years from the date of release for commercial distribution by the Sponsor, whichever is longer.

5.5 Product Updates

As mentioned before, the medical device Sponsor's work is never done. Your device is on the market and your company is *finally* making a profit. You truly believe everything is under control, including all the postmarketing requirements. It is time for an extended vacation!

Before you pack that suitcase, you may want to consider the pace at which our world is evolving.

> Technology is changing rapidly and unmet medical needs come to light every day. After your device has been on the market for some time, you will receive feedback. In order to maintain or increase your market share, updating a marketed device is inevitable.

There are some updates a Sponsor can make to an approved, granted, or cleared device that do not require a new submission to the FDA prior to marketing. The amount of work required to support an update depends on your initial submission type (510(k) or PMA) as well as feedback from the FDA on your initial submission. Fortunately, the FDA has established guidelines for determining if a submission is required before marketing an updated device.

5.5.1 Updates to Devices Cleared via 510(k)

The FDA guidance *Deciding When to Submit a 510(k) for a Change to an Existing Device* [41] is a very helpful tool for Sponsors who intend to modify a 510(k)-cleared device or other device subject to the 510(k) requirements (e.g., marketing authorization granted via a De Novo). Another helpful guidance document *Deciding when to Submit a 510(k) for a Software Change to an Existing Device* [42] is also available. Both guidance documents provide helpful flow diagrams to walk Sponsors through the process of determining what, if anything, must be provided to the FDA prior to marketing an updated device.

Basically, all changes to a marketed device must be documented [43]. The scope and type of documentation required varies, but should be included as part of the manufacturer's quality system (GMPs under 21 C.F.R. § 820 including design control requirements under 21 C.F.R. § 820.30). It is possible to make a number of changes to your device without the requirement to submit a new 510(k). However, each time a change is made, the updated device should be compared to the originally cleared device to determine if it would be considered substantially equivalent. Although this evaluation of substantial equivalence is not *exactly* the same, it is a helpful tool for Sponsors to determine if a new 510(k) is required.

If a new 510(k) is submitted, all the changes made that triggered the requirement for submission must be fully described. To ensure FDA understanding, any submission should also include any changes previously made that did not require a new 510(k) submission.

Possible changes or updates would include:

- Labeling updates.
- Technology, engineering, or performance changes.
- Software changes or updates.
- Changes in materials.

5.5.1.1 Determining Impact to Safety and Effectiveness of the Device

With any change, the very first thing Sponsors must consider is if the change was made with an intent to significantly affect safety or effectiveness of the device. Even if the change did not originate with that specific intent, Sponsors must consider if any change *could* affect safety or effectiveness. Sponsors must evaluate all changes through this lens, by conducting a risk-based assessment to identify and evaluate all new risks or changes in known risks based on the proposed change. If multiple changes are proposed, each change should be evaluated individually as well as together before a final decision is made regarding whether to submit a 510(k).

Typical risk-based assessments include risk estimations, acceptability, controls, risk/benefit analysis and overall risk evaluations as well as hazards and hazardous situations. However, the risk-based assessment should be very thorough. For example, would there be any unintended consequences of the proposed change? Even something as simple as updating the sterilization process could have unintended consequences on device materials or performance. A change based on feedback from a specific user group may have unintended consequences on another user group that could negatively impact device use (and therefore safety).

Even if you determine a 510(k) submission is not required, the Sponsor should confirm this decision with successful, routine verification and validation testing. If unexpected results are obtained, the decision concerning submission should be reconsidered.

5.5.1.2 Updates to Devices Approved via PMA

Much like a 510(k), if the original device submission was a PMA, then a PMA supplement [44] must be submitted before a change impacting safety or effectiveness is released to the market. As with 510(k)s, the Agency expects you to perform an evaluation to determine the impact of the change(s) on safety and effectiveness (Sect. 5.5.1.1). The PMA holder is responsible for determining if a PMA supplement is required, but, in general, the following changes will require a PMA supplement:

- New instructions for use (IFU).
- Labeling changes.
- Use of a different facility or establishment to manufacture, process, or package the device.
- Change in sterilization process.
- Change in packaging.
- Changes in performance or design specifications, circuits, components, ingredients, principle of operation, or physical layout of the device.
- Extension of an expiry date under an unapproved (by the FDA) new or revised stability protocol.

There are caveats and special situations related to the above list. The FDA guidance *Modifications to Devices Subject to Premarket Approval (PMA)—The PMA Supplement Decision-Making Process* [45] provides a table of the types of PMA supplements and a flow diagram to help Sponsors determine the pathway for submitting information supporting the device change. If you are unsure what you should do in a specific instance, consult with a regulatory medical device expert or directly with the FDA.

As with 510(k)s, a PMA holder can make changes to a device without submitting a PMA supplement if the change does not affect safety or effectiveness. Again, all changes to a marketed device must be documented [43] and reported to the FDA in postapproval periodic reports (Sect. 5.4.4). The scope and type of documentation required varies, but should be included as part of the manufacturer's quality system (GMPs under 21 C.F.R. § 820 including design control requirements under 21 C.F.R. § 820.30).

5.5.1.3 Updates to Devices Granted via De Novo

Changes or updates to De Novo-granted devices are evaluated similarly to those of 510(k)-cleared devices and may result in the need for a new 510(k) or even De Novo submitted to FDA, depending on the type of change(s).

5.6 Recalls, Corrections, and Removal of Devices

We all know it happens sometimes—an issue has come up or a device in the market requires repairs. As you have probably deduced, the FDA has rules around different actions a Sponsor can take and the reporting requirements related to each action. As defined in 21 C.F.R. § 810:

- A **device recall** means a correction or removal of a device for human use where there is a reasonably probability that the device would cause serious, adverse health consequences or death.
- A **device correction** is considered as any repair, modification, adjustment, relabeling, destruction, or inspection (including patient monitoring) without the physical removal of the device from the point of use.
- A **device removal** means the physical removal of a device from the point of use to some other location for repair, modification, adjustment, relabeling, destruction, or inspection.

The FDA has provided a helpful summary of the actions and associated reporting on their website (https://www.fda.gov/medicaldevices/deviceregulationandguidance/postmarketrequirements/recallscorrectionsandremovals/default.htm (accessed 6 August 2025)). You can search recalls in FDA's Medical Device Recalls Database (this is one of the many databases listed on CDRH's Medical Device Databases webpage, referenced in Sect. 3.1.1).

5.6.1 Voluntary Device Recall

A voluntary recall (21 C.F.R. § 7) is an action taken because the manufacturer and/or distributor has carried out their responsibility to protect the well-being and health of the public from products that do or could present a risk of injury or gross deception or are otherwise defective. For radiation emitting electronic products, recalls are covered under 21 C.F.R. § 1003 and 1004.

If a voluntary recall is being considered or initiated, the FDA must be notified. A recall strategy should be included with the notification. The FDA will conduct a health hazard evaluation. This evaluation will include elements such as if a disease or injury has already occurred, whether conditions could contribute to a health hazard for humans (or animals) and what segment of the population is involved, the likelihood that a hazard will occur and an assessment of the consequences, both immediate and long-term, associated with the hazard. Based on the outcome of the evaluation, the FDA will assign a recall classification based on the degree of health hazard:

- **Class I** indicates a reasonable probability that the use of or exposure to the product *will cause* serious adverse health consequences or death.

- **Class II** indicates the use of or exposure to the product *may cause* temporary or medically reversible adverse health consequences; however, the probability of serious consequence is remote.
- **Class III** indicates the use of or exposure to the product is not likely to cause adverse health consequences.

The FDA will also evaluate the recall strategy and recommend any changes needed to ensure public health. During the recall, the FDA will want periodic recall status reports. It is important to note that the initiation of a recall should not be delayed awaiting the FDA's review of the proposed strategy.

5.6.2 Mandatory Device Recalls

In rare instances, the FDA under 21 C.F.R. § 810 may issue a recall order. If such an order is issued, it will be served in person or via certified/registered mail with a return receipt. The order will indicate the extent of the recall (wholesale, retail, or user level), the timeline for the completion of the recall, and the requirement for submitting a recall strategy and periodic status reports. The person named on the order will have the opportunity to consult with the FDA.

If, in the FDA's opinion, there is a reasonable probability that the device would cause serious, adverse health consequences or death, the FDA will issue a Cease Distribution and Notification order. This order will indicate that distribution should stop immediately, and that healthcare professionals and user facilities should be notified and instructed to cease use of the device. The person named on the order will have the opportunity either to provide a written request asking for the order be modified, vacated, or amended, or have a regulatory hearing.

5.6.3 Device Corrections and Removals

As indicated in 21 C.F.R. § 806 and in Section 519(f) of the FD&C Act, all device manufacturers and importers are to:

- Promptly report (within ten (10) working days) certain actions concerning device corrections and removal.
- Maintain records of <u>all</u> corrections and removals *regardless* of whether such actions are required to be reported.

If a device correction or removal was initiated in order to reduce a health risk by the device, or to remedy a violation or error caused by the device that may present a health risk, it must be reported. It is important to note that even if the error was cause by "user" or "use error," it still must be reported. If the correction or removal has already been provided to the FDA via mandatory medical device reporting (see 21

C.F.R. § 803; Sect. 5.4.1.1 of this chapter), you do not need to submit an additional report. Additionally, if the correction or removal was initiated by an order from the FDA, you do not need to submit an additional report.

There are many caveats and special situations related to the correction or removal of devices. For example, there are specific obligations related to repurchase, repairs or replacement of electronic products [46]. If you are unsure what you should do in a specific instance, consult with a regulatory medical device expert or directly with the FDA.

References

1. *MDUFA [Medical Device User Fee Amendments (MDUFA)] Performance Goals and Procedures, Fiscal Years 2023 Through 2027*. Available online at https://www.fda.gov/media/158308/download. Accessed 5 August 2025.
2. FDA's guidance for Industry and Food and Drug Administration Staff: *eCopy Program for Medical Device Submissions, Guidance for Industry and Food and Drug Administration Staff* – FINAL GUIDANCE (April 2020). Available online at https://www.fda.gov/regulatory-information/search-fda-guidance-documents/ecopy-program-medical-device-submissions. Accessed 5 August 2025.
3. FDA's guidance for Industry and Food and Drug Administration Staff: *The 510(k) Program: Evaluating Substantial Equivalence in Premarket Notifications [510(k)]* – FINAL GUIDANCE (July 2014). Available online at https://www.fda.gov/regulatory-information/search-fda-guidance-documents/510k-program-evaluating-substantial-equivalence-premarket-notifications-510k. Accessed 20 July 2025.
4. FDA's guidance for Industry and FDA Staff: *Types of Communication During the Review of Medical Device Submissions* – FINAL GUIDANCE (April 2014). Available online at https://www.fda.gov/regulatory-information/search-fda-guidance-documents/types-communication-during-review-medical-device-submissions. Accessed 5 August 2025.
5. FDA's guidance for Industry and Food and Drug Administration Staff: *Refuse to Accept Policy for 510(k)s)* – FINAL GUIDANCE (April 2022). Available online at https://www.fda.gov/regulatory-information/search-fda-guidance-documents/refuse-accept-policy-510ks. Accessed 5 August 2025.
6. FDA's guidance for Industry and Food and Drug Administration Staff: *FDA and Industry Actions on Premarket Notification (510(k)) Submissions: Effect on FDA Review Clock and Goals* – FINAL GUIDANCE (October 2022). Available online at https://www.fda.gov/regulatory-information/search-fda-guidance-documents/fda-and-industry-actions-premarket-notification-510k-submissions-effect-fda-review-clock-and-goals. Accessed 5 August 2025.
7. FDA's guidance for Industry and Food and Drug Administration Staff: *Acceptance and Filing Reviews for Premarket Approval Applications (PMAs)* – FINAL GUIDANCE (October 2022). Available online at https://www.fda.gov/regulatory-information/search-fda-guidance-documents/acceptance-and-filing-reviews-premarket-approval-applications-pmas. Accessed 5 August 2025.
8. 21 C.F.R. § 814.37.
9. FDA's guidance for Industry and Food and Drug Administration Staff: *Procedures for Meetings of the Medical Devices Advisory Committee - Guidance for Industry and Food and Drug Administration Staff)* – FINAL GUIDANCE (September 2017). Available online at https://www.fda.gov/regulatory-information/search-fda-guidance-documents/procedures-meetings-medical-devices-advisory-committee. Accessed 5 August 2025.
10. 21 C.F.R. § 814.40.

11. 21 C.F.R. § 814.44.
12. 21 C.F.R. § 814.45.
13. *FDA Summary of Safety and Effectiveness Data (SSED) Office of Device Evaluation Template.* Available online at https://fda.report/media/113810/SSED-Template.pdf. Accessed 5 August 2025.
14. FDA's guidance for Industry and Food and Drug Administration Staff: *De Novo Classification Process (Evaluation of Automatic Class III Designation)* – FINAL GUIDANCE (October 2021). Available online at https://www.fda.gov/regulatory-information/search-fda-guidance-documents/de-novo-classification-process-evaluation-automatic-class-iii-designation. Accessed 5 August 2025.
15. FDA's guidance for Industry and Food and Drug Administration Staff: *FDA and Industry Actions on De Novo Classification Requests: Effect on FDA Review Clock and Goals* – FINAL GUIDANCE (October 2022). Available online at https://www.fda.gov/regulatory-information/search-fda-guidance-documents/fda-and-industry-actions-de-novo-classification-requests-effect-fda-review-clock-and-goals. Accessed 5 August 2025.
16. FDA's guidance for Industry and Food and Drug Administration Staff: *Acceptance Review for De Novo Classification Requests* – FINAL GUIDANCE (October 2021). Available online at https://www.fda.gov/regulatory-information/search-fda-guidance-documents/acceptance-review-de-novo-classification-requests. Accessed 5 August 2025.
17. 21 C.F.R. § 314.102.
18. 21 C.F.R. § 314.102(c).
19. 21 C.F.R. § 314.102(d).
20. Lantzy, J. (2015). FDA's Module 1 Update: From Ideas to Implementation. DIA eRegulatory and Intelligence Annual Conference. Philadelphia, PA.
21. Gottlieb, S. FDA Statement: Statement from FDA Commissioner Scott Gottlieb, M.D., on new efforts to enhance and modernize the FDA's approach to medical device safety and innovation, April 17, 2018, available online at https://www.fda.gov/news-events/fda-brief/fda-brief-fda-brings-additional-efficiency-and-modernization-regulation-digital-health-part-digital#:~:text=In%20this%20case%2C%20the%20FDA,and%20for%20regulating%20tobacco%20products (accessed June 23, 2025).
22. FDA's draft guidance for Industry and Food and Drug Administration Staff: *Guidance for Industry Presenting Risk Information in Prescription Drug and Medical Device Promotion* – DRAFT GUIDANCE (May 2009). Available online at https://www.fda.gov/downloads/drugs/guidances/ucm155480.pdf. Accessed 5 August 2025.
23. FDA's draft guidance for Industry and Food and Drug Administration Staff: *Guidance for Industry Responding to Unsolicited Requests for Off-Label Information About Prescription Drugs and Medical Devices* – DRAFT GUIDANCE (December 2011). Available online at https://www.fda.gov/regulatory-information/search-fda-guidance-documents/responding-unsolicited-requests-label-information-about-prescription-drugs-and-medical-devices. Accessed 5 August 2025.
24. FDA guidance for Industry: *Medical Product Communications That Are Consistent With the FDA-Required Labeling – Questions and Answer* – FINAL GUIDANCE (June 2018). Available online at https://www.fda.gov/regulatory-information/search-fda-guidance-documents/medical-product-communications-are-consistent-fda-required-labeling-questions-and-answers. Accessed July 1, 2025.
25. FDA guidance for Industry and Review Staff: *Drug and Device Manufacturer Communications With Payors, Formulary Committees, and Similar Entities – Questions and Answers* – FINAL GUIDANCE (June 2018). Available online at https://www.fda.gov/regulatory-information/search-fda-guidance-documents/drug-and-device-manufacturer-communications-payors-formulary-committees-and-similar-entities. Accessed July 1, 2025.
26. 21 C.F.R. § 814.82.
27. FDA guidance for Industry and FDA Staff: *Postmarketing Safety Reporting for Combination Products* – FINAL GUIDANCE (July 2019). Available online at https://www.fda.gov/media/111788/download. Accessed 6 August 2025.

28. FDA guidance for Industry, User Facilities, and FDA Staff: *Questions and Answers about eMDR - Electronic Medical Device Reporting - Guidance for Industry, User Facilities and FDA Staff* – FINAL GUIDANCE (February 2014). Available online at https://www.fda.gov/regulatory-information/search-fda-guidance-documents/questions-and-answers-about-emdr-electronic-medical-device-reporting-guidance-industry-user. Accessed 6 August 2025.
29. FDA guidance for Industry and Food and Drug Administration Staff: *Medical Device Reporting for Manufacturers* – FINAL GUIDANCE (November 2016). Available online at https://www.fda.gov/regulatory-information/search-fda-guidance-documents/medical-device-reporting-manufacturers. Accessed 6 August 2025.
30. FDA guidance for Industry and Food and Drug Administration Staff: *Procedures for Handling Post-Approval Studies Imposed by PMA Order* – FINAL GUIDANCE (October 2022). Available online at https://www.fda.gov/regulatory-information/search-fda-guidance-documents/procedures-handling-post-approval-studies-imposed-pma-order. Accessed 6 August 2025.
31. U.S. Registration No. 4,765,728. Owner: Abiomed, Inc.
32. FDA PMA Approval Letter to Abiomed, Inc. Available online at https://www.accessdata.fda.gov/cdrh_docs/pdf17/P170011A.pdf. Accessed 6 August 2025.
33. Sharrar RG and Dieck GS, *Monitoring product safety in the postmarketing environment.* Ther Adv Drug Saf. 2013:4(5) 211–219.
34. FDA guidance for Industry and Food and Drug Administration Staff: *Use of Real-World Evidence to Support Regulatory Decision-Making for Medical Devices* – FINAL GUIDANCE (August 2017). Available online at https://www.fda.gov/regulatory-information/search-fda-guidance-documents/use-real-world-evidence-support-regulatory-decision-making-medical-devices. Accessed 6 August 2025.
35. FDA guidance for Industry and Food and Drug Administration Staff: *Public Notification of Emerging Postmarket Medical Device Signals ("Emerging Signals")* – FINAL GUIDANCE (December 2016). Available online at https://www.fda.gov/regulatory-information/search-fda-guidance-documents/public-notification-emerging-postmarket-medical-device-signals-emerging-signals. Accessed 6 August 2025.
36. FDA guidance for Industry and Food and Drug Administration Staff: *Medical Device Safety Action Plan: Protecting Patients, Promoting Public Health.* Available online at https://www.fda.gov/about-fda/cdrh-reports/medical-device-safety-action-plan-protecting-patients-promoting-public-health. Accessed 6 August 2025.
37. 21 C.F.R. § 807.
38. 21 C.F.R. § 821.
39. FDA guidance for Industry and FDA Staff: *Medical Device Tracking* – FINAL GUIDANCE (March 2014). Available online at https://www.fda.gov/regulatory-information/search-fda-guidance-documents/medical-device-tracking. Accessed 6 August 2025.
40. 21 C.F.R. § 820.20.
41. FDA's guidance for Industry and Food and Drug Administration Staff: *Deciding When to Submit a 510(k) for a Change to an Existing Device* – FINAL GUIDANCE (October 2017). Available online at https://www.fda.gov/regulatory-information/search-fda-guidance-documents/deciding-when-submit-510k-change-existing-device. Accessed 26 July 2025.
42. FDA's guidance for Industry and Food and Drug Administration Staff: *Deciding When to Submit a 510(k) for a Software Change to an Existing Device* – FINAL GUIDANCE (October 2017). Available online at https://www.fda.gov/regulatory-information/search-fda-guidance-documents/deciding-when-submit-510k-change-existing-device. Accessed 26 July 2025.
43. 21 C.F.R. § 820.
44. 21 C.F.R. § 814.39.
45. FDA's guidance for Industry and FDA Staff: *Modifications to Devices Subject to Premarket Approval (PMA) - The PMA Supplement Decision-Making Process* – FINAL GUIDANCE (October 2017). Available online at https://www.fda.gov/regulatory-information/search-fda-guidance-documents/modifications-devices-subject-premarket-approval-pma-pma-supplement-decision-making-process. Accessed 6 August 2025.
46. 21 C.F.R. § 1004.

Part II
Overview of Intellectual Property Rights for Medical Device Technology

Chapter 6
Patents

6.1 Patent Protection as an Intellectual Property Strategy for Medical Device Technology

> Congress shall have power .... To promote the progress of science and useful arts, by securing for limited times to authors and inventors the exclusive right to their respective writings and discoveries [1].

A U.S. Patent is a contract between the United States and the inventor(s) and/or assignee(s) in which the owner is granted a limited monopoly to exclude others from making, using, selling, offering for sale, or importing a patented invention into the United States during the term of the patent. In exchange for these exclusive rights, which is granted by the federal government through the United States Patent and Trademark Office (USPTO), the inventor and/or assignee is required to disclose the full and complete details of the invention to the public. The theory behind the patent system is that if the public has access to complete inventive disclosures, it will develop new and better ways of solving the same problems.

The patent monopoly has some limitations. A patent does not give an owner the right to make, use, or sell an invention. For example, a patent owner can be prevented from selling its patented invention if a competitor's earlier patent covers some part of the patented invention. Furthermore, a U.S. Patent is only enforceable within the United States; it is not enforceable outside the United States, as each country offers its own patent protections within its borders.

The right to exclude others from making, using, selling, offering for sale, or importing a patented invention creates barriers for competitors from entering the market. Such barriers can either keep competitors out of the market or may be a source of revenue to a company that is willing to license its patents.

Patent licenses can take many different forms, and a patent owner can license all or only a portion of the patent rights. For example, a medical device company can grant a license to a subsidiary company to sell a patented invention in one specific

G. B. Halt Jr. et al., *FDA and Intellectual Property Strategies for Medical Device Technologies*, https://doi.org/10.1007/978-3-032-12649-8_6

market (e.g., the field of cardiovascular surgery), while granting a second license to a second company to sell the patented invention in another, separate market (e.g., the field of dental and oral surgery). Developing a strong portfolio of patent rights (i.e., a portfolio of various barriers to market entry) can have numerous benefits for a medical device company, including creating new business opportunities and reducing the risk of competition.

Example Medicus Societas holds a patent on the CardioColander LDL blood filter device. However, the centerpiece for the CardioColander is the process by which the filter is made. Medicus Societas licensed the patented manufacturing technique used to form the tightly coiled stainless steel filter that is the CardioColander from a company called Fiberfacture. Fiberfacture holds several patents on techniques for producing densely coiled products from thin fibers. The CardioColander filter is formed by a specific manufacturing technique perfected by Fiberfacture where thin stainless steel fibers are coiled in a particular way to produce unique properties in the overall filter structure that are well-suited for use as an LDL blood filter.

While Medicus Societas has the right to exclude others from making, using, selling, offering for sale, or importing the CardioColander within the scope of Medicus Societas' medical device patent rights, Medicus Societas in turn needs a license from Fiberfacture to use Medicus Societas' manufacturing technique to create Medicus Societas' CardioColander filter products.

A medical device company can use patents either offensively (i.e., to keep others out of the marketplace), or defensively (i.e., as a counter-threat to competitors that have threatened patent infringement litigation). Patents can also be a valuable corporate asset, and may be monetized in a number of ways (for example, patents can be licensed as discussed in Chap. 12).

6.1.1 Research Agreements

There are many medical device innovations that are the result of collaborative research efforts. Collaborative research occurs whenever two or more entities join forces in the furtherance of technological innovation or advancement of a science. In the medical device space, it is common for research to be sponsored by the government (e.g., through a government grant), or by a university.

Many medical device startups have their beginnings formed from collaborative research efforts. The startup (or its original founders before the startup company was officially created) may have been awarded a government grant or a university research grant that enabled them to pursue research in a particular medical field. Through the course of the sponsored research, it is not uncommon for intellectual property to be created by the researcher. However, government and university sponsored research is usually contingent upon the researchers entering into a research agreement.

Research agreements lay out the terms of the collaborative research effort. The agreement usually addresses terms such as the scope of the research, how long sponsorship will last, how much money is to be given to the researchers and how ownership of any intellectual property will be shared. Usually, the sponsor and the researcher/inventor of intellectual property subject to a research agreement are considered joint owner of the intellectual property that is produced as a result of the joint research effort.

When applying for patent protection on any invention stemming from a joint research agreement, patent applicants are required to disclose the joint research agreement and the joint ownership in the patent application.

6.1.2 Making an Invention Disclosure Statement to Your Patent Attorney

At the outset of the patent application process, inventors generally must first make an invention disclosure statement to their patent attorney. An invention disclosure statement is usually a form that contains a number of questions for the inventor or patent applicant to answer concerning the invention. The purpose of an invention disclosure statement is to help the patent attorney understand the invention, as well as to learn about any disclosures that the inventor or patent applicant has made to the public concerning the invention. The answers to the questions posed in the invention disclosure statement inform the attorney as to whether the invention may be patent eligible, and whether any potential time bars may exist that could prevent the inventor or patent applicant from seeking patent protection.

> ***Example*** An example invention disclosure statement for the CardioColander LDL cholesterol blood filter is included in Appendix A of this book.

6.2 Patentability Requirements in the United States

In order for an invention to be patentable, it must be:

- Patentable subject matter,
- Useful,
- Novel, and
- Non-obvious.

6.2.1 Patentable Subject Matter

Pursuant to the patent statute, "[w]hoever invents or discovers any new and useful process, machine, manufacture, or composition of matter, or any new and useful improvement thereof, may obtain a patent therefore, subject to the conditions and requirements of this title [2]."

(1) A "process" is a way to produce a result. For example, a process may consist of mixing certain ceramic elements at a particular pressure and temperature to create a new ceramic composite for dental fillings or to create a novel form of hydroxyapatite/PMMA composition for bone cement applications. Not all processes are patentable. For example, a pure mathematical algorithm is not patentable. However, a mathematical algorithm included in a process used to determine a useful, concrete, and tangible result will in most circumstances be considered patentable subject matter. An example of a patentable process involving a mathematical algorithm might be a unique algorithm applied to sensed glucose data generated by a glucose sensing device that is worn by a patient.
(2) A "machine" is a device with assembled parts that move to perform a desired operation. For the most part, medical devices are considered machines for patentability purposes.
(3) A "manufacture" or "article of manufacture" is typically regarded as a man-made, tangible object that is not naturally occurring. An example of a manufacture could include an artificial hip and socket joint formed by a specific process—the process could be a manufacturing process or a chemical process.
(4) A "composition of matter" is any compound, substance, mixture, etc., that is the result of combining two or more ingredients. New and non-obvious biomaterials could qualify as compositions of matter.

Based on the above definitions, it is no surprise that patentable subject matter has been said to "include anything under the sun that is made by man [3]." Patentable subject matter can even include things that might not immediately strike you as being one of the four types of patent-eligible subject matter discussed above. For instance, genetically engineered life forms, such as bacteria, can be patent-eligible, and intangibles like software may also be eligible for patent protection. Although there have been many software patent cases that have thrust patents into the limelight in more recent years, software, including computer programs, applications, and internet-related inventions, are patentable subject matter.

There are, however, some recognized exceptions to what is patent-eligible subject matter, including:

- Laws of nature,
- Natural phenomena, and
- Abstract ideas.

Inventions may often encompass more than one category of patentable subject matter. Accordingly, patents will often have more than one type of claim.

Example During product testing of the CardioColander LDL cholesterol filter, a senior engineer discovered that adding a drug eluting polymer coating of a specific thickness to the stainless steel filter resulted in a longer useful life of the filter. The engineer decides to patent the invention.

In this example, the engineer may pursue patent protection for both the CardioColander LDL cholesterol filter and the method of making the filter with a polymer drug eluting coating of a particular thickness. A patent with product claims may give the inventor broader protection because the claims would give the inventor the right to exclude competitors from making the product regardless of any method they use to make the product (e.g., the drug eluting polymer coating could be added to the stainless-steel filter by a dipping process, a chemical vapor deposition process, an extruding process, etc.). Method claims are often desirable because even if the specific product claims are not held to be novel, the method of making the product may still be novel.

6.2.1.1 Eligible Subject Matter Under 35 U.S.C. § 101

35 U.S.C. § 101 sets the threshold for what kinds of things may be patented in the United States. The statute permits patents for "any new and useful process, machine, manufacture, or composition of matter," but the Supreme Court has long recognized exceptions for laws of nature, natural phenomena, and abstract ideas. The modern test comes from two decisions that work together. In *Mayo Collaborative Services v. Prometheus Laboratories, Inc.*, the Court explained that a claim that merely recites a natural law with instructions to "apply it" is not patentable unless additional features integrate the law into a specific application that meaningfully limits preemption [4, pp. 72–80]. In *Alice Corp. v. CLS Bank International*, the Court formalized a two-step inquiry. Step one asks whether the claim is "directed to" one of the judicial exceptions [5, pp. 217–222]. If it is, step two asks whether the elements, individually and as an ordered combination, add an "inventive concept" that transforms the ineligible idea into a patent-eligible application.

Two implications follow from this framework. First, at step one, courts look to the claim language in light of the specification to identify the focus of the claimed advance. Claims that capture a technological improvement in a device or software architecture tend to clear step one, while claims that state a goal or result implemented on generic computers tend to be characterized as abstract, and thus patent-ineligible. Second, at step two, conventional components and high-level functional verbs do not supply an inventive concept. The added elements must do more than merely automate a longstanding practice or invoke off-the-shelf computation. [4, pp. 77–80; 5, pp. 218–226].

The Federal Circuit's decision in *CardioNet, LLC v. InfoBionic, Inc.* shows what it looks like when a medical-device claim meets step one [6, pp. 1368–1371]. The asserted claims recited a cardiac monitoring device that measures beat-to-beat timing, identifies premature ventricular beats, evaluates the effect of those beats on timing variability, and generates clinically meaningful events when specified conditions are met. Reading the claims in light of the specification, the court held that they were directed to a specific improvement in cardiac-monitoring technology, not to an abstract idea of "detecting arrhythmias." Because the claims described a defined device architecture and processing logic tied to improved performance under real-world constraints, eligibility was resolved at step one, and there was no need to search for an inventive concept at step two.

Contour IP Holding LLC v. GoPro, Inc. reinforces the same principle in a consumer-hardware setting with obvious lessons for connected medical devices and software as a medical device (SaMD) [7, pp. 1379–1387]. The asserted POV camera claims require generating two video streams from image data in parallel, recording the higher-quality stream locally, and wirelessly transmitting a lower-quality stream to a remote device that allows real-time viewing and control of settings. The district court characterized the claims as "creating and transmitting video" and found them abstract. The Federal Circuit reversed, holding at step one that the claims are directed to a technological solution to a technological problem: a specific camera architecture that simultaneously generates streams and uses wireless transfer for live preview while preserving high-quality capture. The court emphasized that the claims, as construed, were limited to parallel generation and that the written description provided a basis for the improvement in concrete device operations.

In contrast, *Recentive Analytics, Inc. v. Fox Corp.* addresses patents that use established machine-learning techniques to produce business outcomes [8, pp. 1209–1217]. The claims collected data, iteratively trained a model, generated event schedules and network maps, and updated outputs over time. The specifications repeatedly allowed "any suitable" machine learning technique and invoked only generic computing. The Federal Circuit affirmed dismissal under § 101, holding at step one that the claims were directed to abstract ideas of generating schedules and network maps with generic machine learning, and at step two that there was no inventive concept because the patents recited only conventional machine learning (ML) and generic hardware. The panel framed a question of first impression and answered it directly: claims that do no more than apply established methods of machine learning to a new data environment are not patent-eligible absent a specific technological improvement to the computing or learning technology itself.

For medical devices and SaMD, these court decisions chart a workable path through § 101. Claims that tie signal acquisition, data conditioning, feature extraction, decision logic, and control actions to a particular device or software architecture, and that show how that architecture improves performance under clinical constraints, fit comfortably within *CardioNet* [6, pp. 1368–1371] and *Contour* [7, pp. 1379–1387]. A cardiac monitor that specifies how it distinguishes arrhythmias in noisy ambulatory signals, or an infusion system that claims a defined control loop with measured stability properties and safety interlocks, presents a concrete technological implementation that can satisfy step one. By contrast, a SaMD claim that states "train a model on clinical data, then classify or schedule" on generic hardware, while permitting "any suitable" algorithm, risks a similar fate to *Recentive's* [8, pp. 1209–1217]. Eligibility is not about whether machine learning is mentioned, but about whether the claimed advance improves the technology that performs the work rather than claiming a result achieved with conventional tools.

The practical drafting lesson involves putting the "how" into the claims and anchoring it in the specification with measurable technical gains. For software-heavy devices, that usually means reciting the sensor regimen, preprocessing, feature vectors, decision thresholds, and the device actions that follow, together with a description of why that pipeline improves positive predictivity, latency, battery life, bandwidth use, or safety. For connected hardware, it means memorializing the architecture that solves a device-level problem, as seen in Contour's parallel streaming, which enables real-time preview without sacrificing high-quality capture. For artificial intelligence (AI)-enabled SaMD, this means claiming concrete data representations, training protocols, or deployment architectures that alter how the model or system operates, rather than relying on off-the-shelf training to achieve a clinical or operational result. Courts applying *Mayo* and *Alice* will continue to police the boundary; however, these cases demonstrate that careful claim design, paired with a specification that documents real technical improvements, often determines the outcome.

6.2.2 Invention Has Utility

A patent application must also demonstrate that the claimed invention is "useful" for some purpose to meet the legal requirement of utility. In most technical fields, this utility requirement has a very low threshold, which is easily satisfied by demonstrating any useful result. For a patented invention to fail to satisfy the utility requirement, it must be "totally incapable of achieving a useful result," which is rare in applications for processes, machines, articles of manufacture, and software methods.

While rare in those instances, failure to satisfy the utility requirement is more common in biotechnology and chemical applications. In the biotechnology and chemical fields, the USPTO typically requires that applications disclose a practical or real-world benefit that can be derived from the invention. In other words, the invention must have a specific, substantial and credible utility.

(1) **Specific Utility**. Specific utility requires that the patent applicant have knowledge of what the invention does.
(2) **Substantial Utility**. Substantial utility requires that the claimed invention has a real-world benefit (e.g., a treatment for a disease).
(3) **Credible Utility**. Credible utility requires that the claimed invention be believable based on the current state-of-the-art.

In the chemical field, claims may be rejected for lack of utility if a compound or reaction creates a reasonable doubt as to whether there is a credible utility.

6.2.3 *Invention is Novel*

In order for an invention to be patentable, it must be new or "novel" (i.e., not in the prior art). If the prior art shows every element of a claim, the claim is unpatentable as "anticipated" by the prior art. In the U.S., prior art is "everything" in the public domain that existed before the filing date of a patent application. In order for a patented invention to be rejected over a prior art reference, the reference must have been public somewhere in the world. Secret or non-public materials cannot act as prior art. Rules regarding prior art differ around the world. For most foreign countries, prior art is "everything" prior to the priority filing date of a patent application (i.e., most countries do not recognize a "one-year" grace period that is recognized in the U.S.).

The AIA Prior Invention—35 U.S.C. § 102(a)(1) and § 102(a)(2)

Under § 102(a)(1), if an invention was known, used, on sale, or was disclosed in a printed publication anywhere in the world before the effective filing date of the invention, it is not patentable. This section of the America Invents Act (AIA) expands prior art to include uses and sales anywhere in the world, not just in the United States. However, there are a few exceptions regarding disclosures to the public, which are codified in § 102(b). Section 102(a)(1) also considers prior art to be public use, on sale or "otherwise available to the public before the effective filing date of the claimed invention." For example, displaying a product at a trade show anywhere in the world is likely to be considered a use that bars patentability.

Prior to the enactment of the AIA on March 15, 2013, the United States had a first-to-invent patent system, meaning that if an inventor could prove that he or she was the first person to invent an invention, he or she would be entitled to patent protection. Section 102(a)(2) of the AIA, in essence, switched the U.S. patent application system from a "first-to-invent" to a "first-to-file" regime by indicating that a patent is not available if there is an earlier-filed application describing the claimed invention that "names another inventor."

Statutory Bars—35 U.S.C. § 102 (B)

Under the AIA, § 102(b) offers several exceptions to § 102(a). Any public disclosure, use, offer for sale, or sale of the invention, made by the inventor, joint inventor, or another who obtained the subject matter of the invention directly or indirectly from the inventor is not considered prior art under this section so long as the disclosure is within one (1) year of the effective filing date of the patent application. Public disclosure of an incomplete invention may not rise to a statutory bar.

Under the AIA, the inventor's own actions will not result in a § 102(b) statutory bar so long as an application is filed within one (1) year of the inventor's act of disclosure. For example, an inventor's public disclosure of the invention at a trade show or offer to sell the invention to anyone is a permissible public disclosure only if an application is filed within a year of that public disclosure. More than one (1) year prior to the application filing date can be a statutory bar. An exception to the public use statutory bar is if the invention is being publicly used for bona fide testing or evaluation.

Example Two co-inventors, Mary Fey Ellen and Estrella Lauder, equally contribute to making a new LED light-emitting mask for facial dermatological treatments. The light intensity, wavelength and temperature can be controlled through a smart device mobile application. Mary Fey Ellen is interested in publishing a press release for their company's web-based blog to disclose their new breakthrough LED light mask and mobile app. If the two inventors intend to pursue patent protection for their invention, Mary Fey Ellen should wait until after the patent application is filed to publish her press release. If Mary Fey Ellen publishes her press release prior to the date the patent application is filed, however, the two co-inventors will have one year from the first date of the publication on their company's webpage to file a U.S. patent application on the LED light mask and mobile app. By publishing the release prior to the patent application's filing date, the inventors may be prohibited from filing foreign patent applications later down the road.

Example During the research and development phase of a new touch screen design for a wearable infusion pump, a company wants to test the sensitivity of the touch screen to pressure and touch to determine whether adjustments need to be made. In doing so, the company goes to the local college and allows students to test the product under the condition that the students agree to complete an evaluation related to the product.

After the product testing is complete and the company makes a determination that the product's sensitivity range is acceptable, the company also makes a competitor's product available to the students at the college for a one-day-only "demo" in order to evaluate their preference between the company's product, and the competitor's product. One year and one day after this preference-testing, the company files a patent application for the wearable infusion pump with the optimized touch screen design.

In this example, the company will likely be able to argue that the sensitivity testing is not a statutory bar because it was conducted for bona fide experimental purposes in order to determine whether the touch screen's sensitivity range is acceptable for use by patients.

6.2.4 Invention is Non-obvious

An invention is obvious if the differences between the subject matter sought to be patented and the prior art are such that the subject matter as a whole would have been obvious at the effective filing date of the invention to a person having ordinary skill in the art (also commonly referred to as a POSITA, or PHOSITA) to which said subject matter pertains. For example, merely substituting a screw for a pin in a bone fracture repair procedure would normally not be a patentable distinction to the method of performing the fracture repair, since both are commonly used fasteners in bone fracture repair and since it would have been obvious to one of ordinary skill in the art at the effective filing date of the invention to have swapped one type of well-known fastener for the other.

In conducting an obviousness analysis, a patent examiner may combine multiple prior art references. The patent examiner, however, cannot combine references arbitrarily. The non-obviousness requirement requires that a patent examiner step into the shoes of a person of ordinary skill at the effective filing date of the invention and determine whether the claimed invention would have been obvious without using hindsight obtained by reviewing the patent application.

Every obviousness determination considers four (4) factual inquiries:

(1) The scope and content of the prior art;
(2) The differences between the prior art and the claimed invention;
(3) The level of ordinary skill in the pertinent art field at the effective filing date of the invention; and
(4) Objective evidence of obviousness or non-obviousness (i.e., "secondary considerations").

The scope and content of the prior art includes art that is directed to the same field of invention as claimed in a patent application, and any other art that is logically relied upon. The prior art used in determining whether an invention is obvious is the same material defined as "prior art" under 35 U.S.C. § 102. For example, if an invention is directed to wearable infusion pump that has a touch screen lock feature with fingerprint recognition unlock capability, an examiner might look to the relevant wearable infusion pump art, touch screen art, smart device art, fingerprint recognition software art, and any other art concerned with combining touch screens with wearable infusion pump devices or fingerprint scanning and reading technology.

Determining the differences between the prior art and the claimed invention is a useful starting point to determine whether the claimed invention would have been obvious in view of the prior art. If the differences between the prior art and claims are trivial, the claimed invention will likely be unpatentable as obvious in view of the prior art.

The level of skill required of a hypothetical person having ordinary skill in the art is more than that of an ordinary layperson but less than that of an expert in the field of the invention. Determining the level of skill in the art is a factual question that is often open to debate. Factors that are often considered in such a determination can include the level of sophistication in the technology, the education of ordinary people in the field, and prior art attempts to solve related problems.

When patent cases go to court over whether a patent was issued in error because it was obvious in view of the prior art, the courts refer to objective evidence of obviousness or non-obviousness as "secondary considerations." Such secondary considerations include: long-felt need for the invention, failure of others to create the invention, commercial success of the invention, unexpected results, and copying by others. For example, if there was a long-felt need for the claimed solution to a problem, or if the invention is commercially successful, the claimed invention is likely not obvious. Also, showing the prior art teaches away from the claimed invention can be used to support a finding of non-obviousness.

6.3 Public Disclosures

There are many situations in which inventors make public disclosures about their inventions prior to or during the patent application process. Public disclosures can lead to a bar against obtaining patent protection if the disclosure is made to the public more than one year before the patent application for the invention is filed. Once a disclosure is made to the public, it is considered "prior art,"—art or knowledge that was known by the public before the filing date of the invention's patent application. The following chart summarizes the types of materials and acts that are considered to be "prior art" when disclosed to the public:

WHAT	WHO	WHERE	WHEN
The invention is **publicly known**	By another	Anywhere	Before the applicant's effective filing date
The invention is **publicly used**	By another	Anywhere	Before the applicant's effective filing date
The invention is **described in a patent**	By another	Anywhere	Before the applicant's effective filing date
The invention is **on sale**	By anyone	Anywhere	Before the applicant's effective filing date
The invention is described in a **publicly available printed publication**	By another	Anywhere	Before the applicant's effective filing date

(continued)

(continued)

WHAT	WHO	WHERE	WHEN
The invention is otherwise **made available to the public**	By anyone	Anywhere	Before the applicant's effective filing date
The invention is **publicly used**	By inventor, joint inventor, or another who obtained the subject matter of the invention directly or indirectly from the inventor	Anywhere	More than 1 year prior to the application effective filing date
The invention is **on sale**	By the inventor, joint inventor, or another who obtained the subject matter directly or indirectly from the inventor	Anywhere	More than 1 year prior to the application effective filing date
The invention is otherwise **made available to the public**	By the inventor, joint inventor, or another who obtained the subject matter directly or indirectly from the inventor	Anywhere	More than 1 year prior to the application effective filing date

6.4 Provisional Applications

In advance of filing a non-provisional or regular patent application, a provisional patent application may be filed to preserve an early filing date for one (1) year. The requirements for filing a provisional application include:

- Providing a specification,
- Providing drawing figures (if necessary to an understanding of the invention),
- Payment of the official filing fee, and
- Providing the name and home residence of each inventor.

A provisional patent application is not examined by an examiner at the USPTO. A provisional application may be converted into a non-provisional patent application at any time during the twelve (12) month period after filing. Once the application is converted into a non-provisional patent application, then the application will be reviewed by a patent examiner.

In addition, an applicant has one (1) year from its provisional patent application filing date to file any foreign patent applications claiming priority to the provisional patent application filing date.

Medical device companies often consider the benefits of filing provisional patent applications because they offer several advantages. Provisional patent applications have lower costs, provide the rights to an earlier effective filing date, and have minimal filing requirements. Provisional patent applications remain confidential (and a potential trade secret) if the twelve (12) month period lapses and the applicant decides not to pursue a non-provisional patent application.

> ***Example*** Returning to the Mary Fey Ellen and Estrella Lauder example discussed earlier, assume that Mary Fey Ellen inadvertently forgot to inform the company she works for that she submitted a description of the LED light-emitting mask, including some preliminary drawings of the LED light-emitting mask, its circuitry and a brief description of how the mobile application works, to the *Journal of Dermatological Discovery* for publication. The *Journal of Dermatological Discovery* will publish tomorrow, and the company wants to file a patent application to preserve its rights to file a foreign patent application before the publication.
>
> Because the *Journal of Dermatological Discovery* publishes tomorrow, it is unlikely that the company and its patent attorney will have sufficient time to prepare a thorough non-provisional patent application for the LED light-emitting mask and mobile app developed by Mary Fey Ellen and Estrella Lauder. Under this scenario, the company should file a provisional application with as much data as it can possibly gather and submit before the *Journal of Dermatological Discovery* publishes Mary Fey Ellen's article. The company will then have one year to file a non-provisional patent application of the LED light-emitting mask and mobile app, or any foreign patent applications claiming priority to its provisional application's filing date.

6.5 Non-provisional Applications

Once the decision is made to pursue patent protection for an invention, a non-provisional patent application should be filed with the USPTO. The USPTO assigns a filing date and an application serial number to the application upon its submission to the USPTO. The filing date is important because it sets a date that "prior art" references must predate in order to reject the claims of the application.

6.5.1 The Anatomy of a Patent Application

A regular, or non-provisional, patent application must provide a specification that:

(1) Describes the invention in sufficient detail to show one skilled in the art that the inventor possessed the claimed invention at the time of filing ("written description requirement"); and

(2) Describes the invention in a manner that would allow one skilled in the art to make and use the claimed invention without undue experimentation (e.g., "enablement requirement").

The specification must conclude with one or more claims that particularly point out and distinctly claim the novel subject matter of the invention. Drawings may be required if they are necessary for the understanding of the subject matter sought to be patented, according to 37 C.F.R. § 1.81(a). Finally, the application must include an Application Data Sheet that names the true and correct inventors, and must include the requisite filing fee.

6.5.1.1 Claims

A patent application's claims are critical to defining the scope of protection sought in a patent. The claim's scope has been described as defining the "metes and bounds" of the patented invention. These "metes and bounds" define a patent holder's rights to exclude others from making, using, selling, offering to sell, or importing an accused invention. Thus, if an accused invention falls within a patent claim's scope, it infringes the patent's scope. Before reaching infringement, however, the claims must meet certain requirements.

The claims must be supported by the specification. If the specification describes parts of an invention that are not defined in the claims, it is possible that such disclosure will be dedicated to the public. For this reason, the claims must particularly point out and distinctly claim the novel subject matter of the invention, and should describe the invention as broadly as possible based on the specification.

The claims in a patent application are typically structured to include independent claims that broadly define the claimed invention, and dependent claims that further limit the scope of the independent claims. A dependent claim includes all of the limitations of an independent claim, but includes additional elements that further limit the independent claim. For example, a dependent claim may read, "The apparatus of claim 1, further comprising [additional elements]."

Example An example of an Independent Method Claim from U.S. Publication No. 2011/0125110:

25. A method of negative pressure wound therapy, which method comprises:

    Packing a wound cavity with one or more wound packing elements, and. applying a reduced pressure to the wound cavity,
    wherein the wound packing element is in the form of an airlaid nonwoven material.

Example An example of a Dependent Method Claim from U.S. Publication No. 2011/0125110:

36. A method as claimed in claim 25, wherein the wound packing element is supplied in the form of an oversized sheet which may be cut to the appropriate dimensions immediately prior to use, to fit the wound cavity.

6.5.1.2 Specification

The specification of a patent is the written description of the invention. It serves as a disclosure of the invention to the public. If the patent issues and becomes involved in litigation and a term in the claims is ambiguous, the court will look to the specification for guidance on how to construe the claim (i.e., how the court will interpret the claim). There are several required sections that must be included in the specification. These sections include:

(1) ***The title of the invention***. The essence of the invention should be captured in as few words as possible. The title is limited to 500 characters.
(2) ***Statement regarding federally sponsored research (if applicable)***. If the invention is the result of federally sponsored research funding, the U.S. requires that this be disclosed in the patent application. If the invention is not the result of federally funded research, this section can be excluded from the specification.
(3) ***Abstract***. Each application must include an abstract of the invention. The abstract may not exceed 150 words.
(4) ***Background of the invention***. This section of the application discusses select, relevant art in the field and emphasizes the major differences between the art and the invention being claimed in the application. The purpose is to point out precisely any and all improvements and challenges that are overcome by the invention.
(5) ***Summary of the invention***. The summary is separate and distinct from the abstract of the invention. The summary focuses on the claimed invention. The summary is sometimes used to discuss problems that exist with the prior art and is often used to highlight the advantages that the present invention has over the prior art.
(6) ***Description of the drawings***. This section provides a general overview and roadmap for navigating the drawings included in the application. This section identifies each figure, and provides a brief description of each.
(7) ***Detailed description of the invention***. This section is the heart of the patent application. An applicant provides a detailed description of the invention; its characteristics, preferred embodiments, definitions of terms, and specific examples of how to practice the invention are provided. The examples illustrate how the invention is meant to operate, and do not limit the claims of the invention in any way.

(8) ***Sequence listing (if applicable).*** A sequence listing is not very common because it is only a requirement for those applications that include nucleic acid or amino acid sequences as part of the invention. If sequences are disclosed in the specification, they must be included in a listing for easy reference.

Aside from these required sections, the specification must also satisfy three requirements set forth in the first paragraph of 35 U.S.C. § 112: the specification must contain a *written description* of the invention, and of the manner and process of making and using it (i.e., the *best mode* for making and using the invention), in such full, clear, concise, and *exact terms as to enable* any person skilled in the art to which it pertains, or with which it is most nearly connected, to make and use the same.

Written Description Requirement

The patent specification must clearly describe the invention so that someone skilled in the relevant field can tell that the inventor actually had the idea and understood how it worked at the time of filing. In other words, the description should show that the inventor was truly in possession of the invention, not just claiming an idea after the fact. It must also explain the invention in enough detail so that a knowledgeable person could make and use it without needing to guess or experiment excessively.

Example A patent application specification that describes high-definition Wi-Fi video camera technology and control system for use in an endoscopy medical device is filed by a medical device company. After filing the patent application, the company learns that healthcare professionals prefer a Wi-Fi video camera with recording capabilities, programmability, and mobile device application controls, and so the company would like to pursue patent protection for these other embodiments of the invention. Because the company's original patent application did not disclose video recording capabilities, programmability and mobile device application controls, any claims to such embodiments will likely be rejected as not being supported by the written description of the specification. Instead, the company can pursue these other embodiments via the filing of a continuation-in-part patent application (which is discussed in Sect. 6.5.6 of this chapter).

Enablement Requirement

The enablement requirement requires that the specification, at the time the application is filed, describe the invention in such a manner that a person of ordinary skill in the art could *make* and *use* the claimed invention without undue experimentation. The fact that a person of ordinary skill in the art is required to perform *some* experimentation when carrying out the claimed invention does not mean that such experimentation is "undue" [9, p. 785]. However, the quality or quantity of any such experimentation must not be unreasonable or unduly burdensome.

Courts apply a factual inquiry to determine whether such experimentation is undue or unreasonable. This determination is guided by the eight non-exclusive Wands factors:

(1) The quantity of experimentation necessary,
(2) the amount of direction or guidance provided,
(3) the presence or absence of working examples,
(4) the nature of the invention,
(5) the state of the prior art,
(6) the relative skill or those in the art,
(7) the predictability of unpredictability of the art, and
(8) the breadth of the claims [10, p. 737].

The conclusion of whether experimentation is undue is not drawn from a single factual consideration; it requires balancing of the factors and analyzing the evidence as a whole. [10, p. 737]. What is considered reasonable experimentation depends on the nature of the invention and the prior art [11, p. 612]. The level of detail needed to meet the enablement requirement depends on how much is already known in the field and how predictable the art is [12, p. 1254]. Predictability in the art refers to whether one skilled in the art can reasonably deduce the outcome of the changes to the invention. If they cannot, then the art is unpredictable and therefore requires greater enablement [13].

Example A doctor has developed a new surgical glue composition, which has a unique chemical composition that results in an even heat distribution and high-strength adhesive properties—the glue warms evenly on the patient's skin during use and is noticeably more durable once the glue sets. The doctor found that mixing the composition in a low-temperature environment in a low-pressurize atmosphere gave the new surgical glue enhanced qualities.

In a patent application for the new surgical glue, the doctor openly discloses the specifics of the chemical formulation in the specification of the patent application. The doctor also discloses the use of a low-pressurized atmosphere in the production of the glue, but the doctor guards the details of the pressurization process as if they were a trade secret.

Since the doctor failed to disclose the details of how to use the low-pressurized atmosphere in the production process of the surgical glue, a person of ordinary skill in the relevant art would not be able to determine the exact temperature and pressure to use without undue experimentation. The doctor has not provided sufficient description because the disclosure is not sufficient to enable others to practice the invention.

Best Mode Requirement

The specification must identify the "best" way to make and use the invention, as determined by the inventor. That is not to say that the disclosed best mode in the specification is the absolute best way to use or make the invention—there may be ways to make or use the invention that the inventor never thought of. Rather, in order to satisfy the best mode requirement, the specification must explain at least the best mode for making and using the invention that is known to them at the time of filing.

Example A healthcare professional has invented a method for treating the common cold by administering a combination of Vitamin C, Zinc and Echinacea to a patient in specific quantities according to a specific dosing schedule. In a patent application for the method of treating a patient having the common cold, the inventor must describe the best mode for administering the combination of Vitamin C, Zinc and Echinacea to the patient, and must explain the dosages that should be administered and provide details concerning the dosing schedule.

6.5.2 The "Duty to Disclose"

Under U.S. patent law, those involved in the patent application process have certain duties and obligations that they are required to adhere to as they interact with the USPTO. During the patent application process, inventors, patent applicants (assignees) and attorneys or patent agents must conduct themselves in an honest way. Those who interact with the USPTO during the patent application process have an obligation to act with candor when working with the USPTO and must deal with the USPTO in good faith. This includes a duty to disclose to the USPTO any and all relevant prior art that the inventor, patent applicant or attorney or patent agent may have knowledge of. Any prior art that is material to the patentability of a patent application that is pending before the USPTO must be brought to the attention of the USPTO in a timely manner for review and consideration.

Prior art subject to disclosure to the USPTO generally takes the form of a printed publication, such as a printed patent or application. Sometimes the prior art can be a relevant journal article or product brochure for a competitor's product. Additionally, any prior art that is identified by a foreign Patent Office or searching authority (e.g., any prior art cited in a foreign patent application examination action or cited in an international search report) should also be disclosed to the USPTO.

Failure to disclose known, relevant prior art to the USPTO can serve as grounds for inequitable conduct, i.e., a breach of one's duty of candor and good faith while dealing with the USPTO. When there is evidence of inequitable conduct, an issued patent can be held unenforceable by the courts because the patent was obtained with "unclean hands," or through flagrant misconduct. When a patent is held to be unenforceable due to inequitable conduct, all of the claims in the patent are unenforceable [14]. Inequitable conduct can also be used as a defense to patent infringement allegations.

6.5.3 *Assignment of Patent Rights*

A patent conveys to the patent holder a bundle of rights: the right to exclude others from making, using, selling, offering for sale, or importing a patented invention into the United States. Inventors may hold the rights to a patent, or the inventors may assign their rights in the patent or patent application to another. For instance, there are many inventors whose job is to create new, patentable inventions, and as part of their job, these inventors are often required to assign any and all patent rights to their employer.

A patent assignment is the legal transfer of ownership of a patent or patent application to another. Assignment is often part of an employment agreement, or other contract. A sample assignment can be found in Appendix C.

Example A sample assignment can be found in Appendix C of this book.

6.5.4 *Inventorship Versus Ownership*

6.5.4.1 Inventorship

The actual inventors of the subject matter that is being patented must file a U.S. patent application. Determination of inventorship can be a difficult task that requires a thorough legal analysis. "Conception" of the invention is typically considered the key to determining inventorship. Conception is the mental formulation and disclosure by the inventor or inventors of a complete idea for a product or process. Mere contributions of labor, such as carrying out experiments that are designed for others, or general supervision are typically insufficient to vest inventorship rights in the invention. In the academic setting, it is often discretionary to name contributors of a research project in published articles; however, naming inventors in a patent application is not discretionary. If the inventorship on an issued patent is incorrect, a court may invalidate the patent.

Example Two entrepreneurs, Mike and Richard, equally contributed to the conception of a new app that tracks when a user last took a dose of specific medicine. Users can input the medications that they are prescribed to take, dosing schedules, and confirmation that the appropriate medication was taken on schedule. The app also provides users with an alarm reminding them to take their medications as appropriate.

Two interns, Kevin and Nicole, initially tested the user friendliness of the app under the direction of Mike. Kevin and Nicole together decided it would be easier for users to use the app if there was a swipe vertically up feature that the user could perform to record special notes about the medication, e.g., "take medication with food" or "avoid sun exposure," and swipe vertically down feature that would allow the user quickly update that the medication was taken on time.

The startup company that employs Mike, Richard, Kevin, and Nicole, decides to file a patent application for this invention and needs to determine the inventor(s).

In this example, if the startup company decides to pursue claims in a patent application directed to the app itself, Mike and Richard should be considered the inventors. If the company decides to pursue claims to a method of using the app that includes the swipe vertically up or down feature, then Mike, Richard, Kevin, and Nicole should be identified as the inventors.

6.5.4.2 Ownership

Ownership is a different matter. A patent will always need to list the inventors, but ownership can be held by someone other than the inventors, such as the employer of the inventors (i.e., the assignee). Whoever owns the rights to a patent or patent application has the ability to sell the patent, or to do anything else they would like with the patent rights that have been assigned to them. This could include:

- Licensing the right to sell the invention,
- Licensing the right to use the invention,
- Making a new assignment of the rights to the patent or patent application, or
- Selling the rights to the patent or patent applications.

6.5.5 What to Expect: The Patent Application Examination Process

After filing a patent application, the USPTO assigns the application to a patent examiner for examination. During examination, the examiner ensures that the application satisfies all formal requirements for the specification, claims, and drawings. The examiner also conducts a search of available prior art references using search databases, including the internet. Following the examiner's initial examination and search, the examiner will usually issue an objection to the application for failing to satisfy a formal requirement, or reject the claims as anticipated or obvious in view of the prior art discovered during the examiner's search in an Office Action.

In response to an Official Action, the applicant, typically through his or her patent attorney, can submit a formal response to address the rejections noted by the examiner and distinguish the claimed invention over the prior art. By distinguishing the claimed invention over the prior art, the applicant may amend the claims. Claim amendments are not required and may be particularly unnecessary when an examiner misinterprets a reference or improperly combines references to support a rejection.

After filing a response to an Official Action, the examiner considers the arguments or amendments and determines whether to issue a subsequent Official Action or to allow the application. If the examiner issues another Official Action, the Applicant will be given another opportunity to respond. There is no limit on the number of Official Actions that can be issued in the patent application process, although after a first action, examiners will usually issue a "final" office action, which can have the effect of closing prosecution. If prosecution is closed in an application, an applicant can file a Request for Continued Examination (RCE) along with a response and the USPTO official fee. A timely filed RCE will place the application in the queue again for an additional round of examination and prosecution. On the other hand, if the examiner decides to allow the application, the applicant will receive a Notice of Allowance, which will have a set period of time for the applicant to pay a fee in order to have the application officially issued as a U.S. patent.

6.5.6 Continuing Applications

United States patent law allows applicants to file continuing patent applications claiming the benefit of the disclosure and filing date of an earlier-filed co-pending application ("parent application"). The parent application does not have to be the first or earliest filed application in a chain of continuing applications; it simply must be a related application that is co-pending at the time of filing. Continuing applications must share at least one common inventor with the parent application, make a specific claim of priority to the parent application, and be filed while the parent application is co-pending. Although continuing applications claim the benefit of the earlier-filed "parent" application, they are newly filed applications that restart the examination process.

There are three (3) types of continuing patent applications recognized in U.S. patent practice:

(1) Continuation applications;
(2) Continuation-in-part applications; and
(3) Divisional applications.

Continuation applications have the same specification as the parent application, but with different claims. Continuation applications are useful to:

(1) Claim subject matter that was disclosed but not fully claimed in the parent application;

(2) Seek broader, narrower, or different claim coverage;
(3) Present new arguments in support of allowance of the application after a final rejection is received or prosecution is closed[1]: or
(4) Keep an application pending to capture developments not specifically addressed by any of the issued claims.

Divisional applications can be filed in response to an Office Action from the USPTO, which states that the claims of the parent application are directed to two or more distinct inventions (e.g., claims to a product and claims to a method of making a product can be considered distinct inventions).

A continuation-in-part application ("CIP") is a later-filed application that repeats some substantial portion, if not all, of the parent application's disclosure, and adds new subject matter not disclosed in the parent application. Claimed subject matter that is supported by the parent application is entitled to the effective filing date of the parent application. Claimed subject matter that is not supported by the parent application has the filing date of the CIP. Generally, CIPs claim new or related embodiments of an invention not disclosed in the parent application while effectively maintaining the filing date of the parent application for all originally disclosed subject matter. Although the applicant always has the option of filing a new application for this new subject matter, the priority claim for a CIP application may prevent the parent application itself from being cited to reject any original subject matter from the parent application that is claimed in the CIP.

Example Medicus Societas files a patent application describing the CardioColander. After filing the patent application, Medicus Societas learns that coating the CardioColander filter with a drug eluting polymer could be immensely beneficial to patients. Medicus Societas may be able to file a continuation-in-part (CIP) application to describe and claim the new embodiments of the CardioColander that have drug eluting capabilities. By filing a CIP application, the priority date for all subject matter in the CIP that overlaps with the original application disclosure for the CardioColander will have the original application's filing date. All new subject matter related to the drug eluting coating for the CardioColander will be entitled to the CIP application's filing date. Such a continuation-in-part application must be filed while Medicus Societas' original patent application or an application claiming priority thereto is still pending (i.e., not an issued patent).

[1] A Request for Continued Examination (RCE) under 37 C.F.R. § 1.114 can be filed upon payment of the requisite fee to present new arguments or claims in an application after a final rejection as an alternative to filing a continuation application.

6.6 Going Global: International Patent Considerations

The rules for obtaining a patent vary from country to country. Patent protection in other countries typically requires international filings, which are submitted to each country's patent office. Most countries permit applicants a non-extendible period of one (1) year from the date of filing a U.S. patent application in which to file their patent application. In most countries, if a foreign patent application is filed within one (1) year of the U.S. patent application filing date and claims priority to it, the U.S. patent application filing date is the applicable priority date of the application.

6.6.1 Patent Cooperation Treaty (PCT)

The U.S. and over one hundred and fifty (150) other countries are signatories to the Patent Cooperation Treaty (PCT) that permits patent applicants to file international patent applications, also known as PCT applications. A PCT application is similar to a U.S. provisional application in that it preserves priority and does not issue as a patent. Within thirty (30) months from the PCT priority date, the applicant must file individual patent applications in all countries in which examination is desired (i.e., PCT applications provide an additional eighteen (18) months of time to file foreign application beyond the typical one (1) year period for filing priority foreign applications). Filing a PCT application can be advantageous in the following respects:

(1) If an applicant is interested in filing a patent application in numerous countries, a PCT application permits the applicant to have the benefit of a PCT patent examiner's prior art search and results before incurring the expense of filing numerous patent applications,
(2) A PCT application gives an applicant additional time (30 months from the PCT filing date) to delay the expenses associated with applying for patent protection in individual countries, and
(3) Many countries give deference to a PCT examiner's examination search and opinion on patentability, which can reduce the costs of prosecuting a patent application in individual countries.

Example The inventor of the CardioColander assigns his patent to his employer, Medicus Societas. Medicus Societas wants to pursue U.S. and foreign patent protection for the CardioColander LDL blood filter. The company would like to file its patent applications as soon as possible, but is unsure as to how successful the product will be and is hesitant to spend too much on international patent protection.

If Medicus Societas files a PCT application, it will have up to 30 months to determine which countries it will need to pursue protection in for the Cardio-Colander. This will provide the company with additional time to evaluate the commercial success of the CardioColander product and target select foreign markets. A PCT application will also give the company the benefit of a single examination, which can assist its determination of how much to invest in both U.S. and international patent protection.

6.7 Special Considerations for Patents

6.7.1 Meaning of "a" in a Claim

In medical devices and SaMD, architecture is rarely monolithic. A cardiac pacemaker may include a sensing Application-Specific Integrated Circuit (ASIC), a therapy microcontroller unit, and firmware services. A hospital platform might split ingestion, classification, and alerting across edge hardware and cloud microservices. This engineering reality makes tiny claim words in a patent application do outsized legal work.

In a patent claim, the indefinite article "a" typically permits plurality in open-ended claims that use "comprising," yet when the claim later ties functionality to "said" element, courts require that at least one instance of that element itself perform each recited function. For device and software teams that intentionally distribute tasks across components or packets, this is where infringement and validity can pivot: a claim that recites "a processor" and "said processor" that acquires electrocardiography data, derives heart-rate variability, and drives therapy cannot be met by three different processors that split the verbs among them unless the claim text expressly allows collective performance. The same risk appears with "a message" that must carry multiple fields, or "a catheter" that must both sense and ablate.

The indefinite article "a" usually means "one or more" in open-ended claims that use "comprising." That background rule, however, does not license claim drafters or litigants to distribute later-recited functions across multiple different instances of that element when the claim ties those functions back to the "said" element. The Federal Circuit has been consistent: when a claim introduces "a [thing]" and then uses "said [thing]" to recite additional capabilities, at least one instance of that thing must be capable of performing each of the functions attributed to the "said" element [15, pp. 1315–1317]. A few exemplary Federal Circuit cases that shed light on this topic, include:

- *Salazar v. AT&T Mobility, LLC*, affirming claim construction that "requires a singular element—'a microprocessor'—to be capable of performing all of the recited functionality," while recognizing the claim could still include more than one microprocessor so long as at least one performs all the recited functions) [15, pp. 1315–1317].

- *Convolve, Inc. v. Compaq Comput. Corp.*, 812 F.3d 1313, 1321 (Fed. Cir. 2016) reading "a user interface ... with a processor ... comprising" to require a single processor having the subsequently recited characteristics [16, p. 1321].
- *In re Varma*, rejecting a construction that "would allow multiple different requests to collectively satisfy a limitation reciting 'a request' that itself corresponds to two or more investments," and explaining with the now-familiar illustration: "For a dog owner to have 'a dog that rolls over and fetches sticks,' it does not suffice that he have two dogs, each able to perform just one of the tasks." [17, pp. 1362–1363]
- *Harari v. Lee*, concluding that "a bit line" and "said bit line" require one bit line that activates multiple cells [18, pp. 1341–1342].

Those decisions live comfortably with the general "one or more" principle [19, p. 1342; 20, p. 1356]. The point is that "a" may allow plurality in the claim as a whole, but the back-reference to "said [element]" anchors the required functionality in at least one singular instance.

The practical effect is straightforward. If a claim recites "a control message ... wherein the control message has power control bits for a plurality of UEs," the limitation is not met by splitting the work across two different control messages, each carrying bits for only one UE. By the same logic as *Salazar* and *Varma*, there can be many control messages in the system, but at least one of them must itself have power control bits for the plurality of UEs that the claim requires. [15, p. 1317]. The same reading applies when the claim repeats the pattern in a dependent claim.

Two drafting lessons follow. First, if the intent is to distribute functionality across multiple instances, say so in the claims. Language such as "one or more processors collectively configured to" or "a plurality of control messages collectively carrying" makes the collective nature explicit and avoids *Salazar* problems. Second, a patentee can act as its own lexicographer, but only with "reasonable clarity, deliberateness, and precision" [21, p. 1366; 22, p. 1249]. If your technology truly uses several microprocessors or several messages that split the claimed work, define the claim term in the specification to capture that collective implementation and then draft the claim text to match. Absent that clarity, courts will not rely on a generic "one or more" presumption to dissolve what the "said [element]" language requires.

Litigation strategy mirrors the drafting playbook. An accused infringer should map each "said [element]" back-reference to a single accused component and show that no one component performs all the recited functions. A patentee should resist attempts to reframe the claim at too high a level of abstraction by walking the court through the element-by-element linkages and, where appropriate, by pointing to claim language that makes collective performance explicit. Either way, the inquiry is textual. Open-ended "comprising" language does not override the syntactic reality that "said [element]" gathers later limitations onto at least one instance of the thing first introduced with "a."

For medical devices and SaMD, the "a/said" pattern in claim drafting appears frequently, and it matters because architectures are often distributed across edge hardware and cloud services. If a claim recites "a processor" and later "said processor acquires ECG data, derives heart-rate variability, and drives pacing-mode transitions," you cannot satisfy the claim by pointing to a digital signal processor that acquires and derives, while a separate microcontroller unit drives pacing; at least one processor must itself perform the full set of recited functions, as in the microprocessor line of court cases discussed above.

The same applies to messaging and telemetry. If a claim recites "a message" and later "said message encodes patient ID and therapy lockout flags," two different packets, one with ID and another with flags, will not do; there must be a single message that carries both fields. Likewise, "a catheter" followed by "said catheter senses electrograms and delivers ablation" requires one catheter with both capabilities, not separate sensing and ablation catheters. In cloud-connected SaMD, phrases like "a server" and "said server" can be trapdoors: a microservices stack where one service ingests signals, another classifies, and a third issues clinical prompts that will not satisfy a claim that anchors all three verbs to "said server," unless the claim text makes collective performance explicit. The drafting and litigation takeaway is simple for regulated platforms that evolve through iterations: either claim the collective expressly—for example, "one or more processors collectively configured to"—or tether the verbs in the claim to a single component or message that really does all the work.

6.7.2 Nonce Words

Medical devices and SaMD are built from modular hardware and layered software. Teams naturally reach for shorthand labels like "module," "handler," "unit," or "component" that are "configured to" do things. These shorthand labels are known as nonce terms in patent law. A nonce term is a generic placeholder that lacks a definite structure but also serves to describe a function. 35 U.S.C. § 112(f) is where those labels within the claim language can generate issues for patent applicants. If a claim term is just a placeholder for a function, a court may treat the term like a means-plus-function limitation, or as a nonce term, and confine it to the structure disclosed in the specification. The touchstone is what a skilled artisan would understand: does the term, in context, connote sufficiently definite structure, or does it merely recite what the thing does [23, pp. 1348–1351]. *Williamson v. Citrix Online, LLC*, explains that generic words such as "module," "mechanism," or "device" often function as nonce terms when divorced from structural understanding [23, pp. 1348–1351]. *Zeroclick, LLC v. Apple Inc.*, cautions against invoking § 112(f) where the record shows the term names a recognized class of structures [24, pp. 1007–1010].

A recent decision underscores the point. In *Fintiv, Inc. v. PayPal Holdings, Inc.*, the Federal Circuit affirmed an indefiniteness ruling because the claims used "payment handler" and "payment handler service" to recite functions like "using APIs of different payment processors" and "exposing a common API," yet the term and the specifications did not provide sufficiently definite structure or an implementing algorithm [25, p. 1377]. The court analogized "handler" to the nonce term "module," rejected the argument that connectors like "configured to" imparted structure, and found the specifications merely parroted the claim language without disclosing how the functions were performed. The figures did not cure the defect because they did not show "how" the payment handler used the different application programming interfaces (APIs) or exposed the common API. With § 112(f) invoked and no corresponding algorithmic structure, the claims were indefinite.

For software-heavy devices, structure often lives in interfaces, data flows, and named protocols. The Federal Circuit's "Noah" framework distinguishes the total absence of structure from disclosures a skilled team can implement without undue creativity [26, pp. 1311–1318]. References to well-known protocols can, in some settings, supply the needed structure. In *AllVoice Computing PLC v. Nuance Commc'ns, Inc.*, the Federal Circuit noted pointing to the Windows DDE protocol sufficed because experts explained how a person having ordinary skill in the art would implement it [27, pp. 1245–1246]. And in *Sisvel Int'l S.A. v. Sierra Wireless, Inc.*, naming a known communications protocol helped move the analysis out of "no structure at all" [28, pp. 1379–1384]. These cases do not create a safe harbor; they show what it takes to properly claim an invention. Naming a protocol is helpful only if, in context, it conveys concrete structure and the specification ties that structure to the claimed function. *Media Rights Techs., Inc. v. Capital One Fin. Corp.*, (treating "compliance mechanism" as a black box) [29, pp. 1372–1376] and *Advanced Ground Info. Sys., Inc. v. Life360, Inc.*, [30, pp. 1347–1348] remind that generic software nouns, even when they sound technical, can still be nonce words. When § 112(f) applies, algorithm disclosure matters: saying what result occurs is not enough [31, pp. 1007–1008; 32, pp. 1333–1338; 33, pp. 1134–1135].

For drafting in medical device and SaMD claims, applicants should name recognized structures and tie functions to the interfaces and algorithms a skilled team would expect. If a "telemetry handler" integrates with Health Level Seven (HL7), Fast healthcare Interoperability Resource (FHIR), Digital Imaging and Communication in Medicine (DICOM), Bluetooth Low Energy, or a hospital gateway application programming interface (API), explicitly state it as such, and explain the role it plays in the data path. If a "controller" executes a control law, identify inputs, outputs, and state updates. Doing so preserves the presumption against § 112(f) even without the word "means," and if § 112(f) applies, it supplies the corresponding structure that bounds the claim. Conversely, avoid bare containers such as "module configured to" without architectural anchors; *Williamson* shows how quickly they can collapse into § 112(f). Where you must rely on protocol names, ensure the specification makes clear that a POSITA would recognize the protocol's structures, sequence of operations, and role in accomplishing the claimed function, and consider expert declarations that explain why implementation would have been straightforward at the priority date, as in *AllVoice* and *Sisvel.*

The litigation strategy playbook tracks the same themes. A patent owner should emphasize ordinary technical meaning, protocol context, and concrete architectural roles, and use expert testimony to connect the claim term to recognized structures and to the disclosed algorithm. A petitioner should show that the term is a black-box label with no accepted structural meaning in the field, then walk through the specification to demonstrate the absence of an algorithm tied to the claimed function. In the wake of *Fintiv*, courts will discount connectors like "configured to" and "operable to" when the claimed "handler" or "service" simply describes what the software should accomplish rather than how it does so.

The bottom line for regulated health technology is to be practical in claiming the invention. Section 112(f) is not hostile to software or modular device claims; it is hostile to black boxes in the claim language. In medical devices and SaMD, where safety cases and interoperability often ride on well-known APIs and workflows, claims that name recognized components and interfaces, and that are backed by a specification showing ordinary meaning and concrete implementation to a skilled artisan, will avoid "nonce" traps and keep claim scope aligned with the technology actually built.

6.7.3 Clearly Defining over Crowded Prior Art

Both the specifications and the claims can be utilized to distinguish an invention from prior art references in a crowded technological space. A prior art search conducted prior to filing a patent application is highly recommended. The prior art search will reveal documents (e.g., patent applications, publications, and journal articles) that are similar to the invention. Based on the findings of the prior art search, the specification and the claims can be tailored to clearly identify how the medical device is patentable in light of the prior art and can be written to provide specific support for the features needed to distinguish over the cited references.

For instance, the specification can be enhanced with additional technical details to avoid or overcome a potential novelty rejection under 35 U.S.C. § 102 by clearly identifying how the medical device differs from the prior art. A novelty rejection from the USPTO may be avoided by utilizing the background section of the specification to expressly state how the particular prior art teaches a device having a particular technical feature. The specification can then elaborate on how this particular technical feature differs from the technical features of the claimed medical device. Care needs to be exercised when discussing the safety and efficacy of the claimed medical device (see more on this topic in Chap. 15). Yet, when appropriately used, the specification can be used to suitably locate the invented medical device in the proper context of the prior art and clearly identify the novelty.

Novelty is not the only requirement. An invention must also be non-obvious under 35 U.S.C. § 103. Particularly, in crowded art areas, it is likely that a patent application will face an obviousness-type rejection under 35 U.S.C. § 103. This type of rejection can be rebutted by identifying secondary considerations (as discussed above in Sect. 6.2.4).[2]

For instance, the specification may include particular information to address "unexpected results" by showing that the particular claimed range is critical to the invention by generally showing that the range as recited in the claims achieves unexpected results when compared to the prior art range [34]. Alternatively, the specification may be used to distinguish a prior art reference where the range disclosed by the prior art "is so broad as to encompass a very large number of possible distinct compositions [35, p. 1330]." Accordingly, the specification may be used to argue in favor of the patentability of an invention by showing that the "claimed range produces new and unexpected results when compared to the prior art [36, p. 1322]."

In addition, the specification may be used to evidence the "long-felt but unsolved need" in the market and the failure of others to create the invention. The specification can be particularly useful in this case because "absent a showing of long-felt need or the failure of others, the mere passage of time without the claimed invention is not evidence of nonobviousness [37, p. 990]." As a result, the specification can provide explicit evidence of the well-known need and demonstrate how others have failed to satisfy this need. This "[e]vidence is particularly probative of obviousness when it demonstrates both that a demand existed for the patented invention, and that others tried but failed to satisfy that demand [38, p. 1082]." As a result, the specification provides an opportunity for the applicant to expressly leverage the findings of the prior art search, demonstrating how others have attempted and failed to solve the particular problem addressed by the medical device.

The specification may also be used to demonstrate how the invention is more than a matter of design choice. For example, in *In Gardner v. TEC Syst., Inc.* [39] the Federal Circuit held that, where the only difference between the prior art and the claims was a recitation of relative dimensions of the claimed device and a device having the claimed relative dimensions would not perform differently than the prior art device, the claimed device was not patentably distinct from the prior art device [40]. As a result, the specification can be used to expressly describe how the particular dimensions are critically important to the performance of the medical device. It may be possible to obtain patent protection for a medical device even if the prior art is similar in shape if the specification includes sufficient detail.

Including as much information as possible in the specification is critically important because the scope of the claims is limited to what is supported in the specification, i.e., claims may be no broader than the supporting disclosure. As such, a narrow disclosure will limit the breadth of the claims [41, p. 1480]. With a broad specification that discusses technical innovation in the context of the prior art, the claims can be tailored to avoid an obviousness rejection under 35 U.S.C. § 103. For instance, the claims can be tailored to specifically include a particular technical feature included in the specification that is not disclosed by the prior art. Alternatively, the claims can

[2] As a reminder, the secondary considerations include: long-felt but unsolved need for the invention, failure of others to create the invention, commercial success of the invention, unexpected results, and copying by others.

be tailored to recite a specific range of values and rely upon the unexpected results described in the specification. By properly utilizing the specification, a patent may be obtained on a medical device even in a crowded technological space.

6.7.3.1 Avoiding Design-Arounds with Broad Claims

The goal of the claims in a patent application is to have the broadest possible scope while meeting the requirements for patentability. It is the claims that define the scope and boundaries of the patent owner's exclusive monopoly rights once a patent is granted. Therefore, broader claims are generally preferred because broad claims include fewer limitations than narrow claims and provide a larger monopoly right. Specifically, the monopoly right is larger, because in order "[t]o establish infringement, every limitation set forth in a patent claim must be found in an accused product or process exactly or by a substantial equivalent [42, p. 1535]." However, the broader the claim is, the more difficult it potentially is to satisfy the novelty requirements under 35 U.S.C. § 102 and the non-obviousness requirement under 35 U.S.C. § 103.

A broader claim allows for a broader range of prior art to be applied to the claim during examination by the USPTO, which may prevent a patent application from issuing during prosecution of the patent application, or could invalidate a patent during a postgrant processing (see Sect. 12.7.2). As a result, narrower claims that are specific to one particular element or product and contain more details than a broader claim are often required to distinguish an invention from the prior art.

Yet, "[o]ne of the benefits of a patent system is its so-called 'negative incentive' to 'design around' a competitor's products, even when they are patented, thus bringing a steady flow of innovations to the marketplace [43, p. 745]." Narrower claims are easier to "design around" because a competitor can avoid infringing on a patent with a narrow scope by either not implementing a particular limitation of the claim, or implementing the limitation in a substantially different way. Therefore, in determining the proper scope of the claims, the ease with which competitors can design around the claims must be balanced against the range of available prior art.

An ideal manner to balance the scope of the patent is to focus the independent claims on the key technology innovation of the medical device. Then the dependent claims can be drafted to add additional limitations and detail in order to generate narrow claims that are focused on exactly how the medical device operates, e.g., how the various components of the medical device interact with one another. By obtaining a patent that includes a mix of broad independent claims and narrower dependent claims, the patent is more likely to survive a validity challenge and is better suited to protect against potential design around.

In addition, a manufacturer may also attempt to guard against "design arounds" by including method claims in a patent application that are directed to what the device does and are not bound to a particular apparatus. However, the inclusion of method claims that are not bound to a particular apparatus claims that are contained in the same application as claims directed to a particular apparatus, will often be met with a restriction requirement in the United States or a unity of invention rejection in Europe. This rejection can be overcome by filing a divisional application and prosecuting the method and apparatus claims in two separate patent applications.

A good strategy for avoiding design-arounds is to seek patents directed to a mix of apparatus and method claims, so that the ability of competitors to design around patents covering a medical device can be minimized.

6.7.4 Analogous Prior Art

In medical devices and SaMD, obviousness lives or dies on what counts as references that are "in the technical neighborhood." A single platform often blends hardware, firmware, and cloud analytics, so petitioners routinely reach across domains: automotive fasteners for injector housings, aerospace redundancy for infusion safety, and web-scale ranking papers for arrhythmia detection, to assemble § 103 combinations. The law does not let patent examiners apply prior art references from any technical field. Before any reference can be part of a motivation-to-combine theory, it must be analogous technology to the patent, either from the same field of endeavor as the claimed device or reasonably pertinent to the specific problem the inventors set out to solve. That gatekeeping step is uniquely consequential in regulated health tech, where superficially similar components may address very different constraints (sterility, biocompatibility, clinician workflow, cybersecurity, latency, and auditability).

Analogous prior art is the gatekeeper for 35 U.S.C § 103. Before references can be marshaled into a motivation-to-combine theory, each one must first qualify as "analogous" to the challenged patent. The Federal Circuit applies two separate tests. First, is the reference from the same field of endeavor, regardless of the problem addressed. Second, if it is not in the same field, is it nevertheless reasonably pertinent to the particular problem with which the inventor was involved [44, p. 1325; 45, p. 1377]. The inquiry is factual and turns on what a person of ordinary skill would have considered relevant at the time, informed by the invention's subject matter, function, and structure.

The field-of-endeavor test looks to the patent's explanations and embodiments to determine the appropriate field. In *In re Bigio*, for example, the Board treated toothbrush art as analogous to a claimed hairbrush because, viewed with common sense, both fell within a broader brush field that shared structure and function; the Federal Circuit affirmed on substantial evidence [44, pp. 1323–1326]. The touchstone is reality of the circumstances rather than labels.

If the art is outside the field of endeavor, the reasonably pertinent test asks whether the reference would logically have commended itself to a skilled artisan's attention in seeking a solution to the inventor's problem [46, p. 1382; 47, pp. 1359–1360]. Identifying that "problem" is not free-floating; it must be defined from the perspective of a POSITA by reference to the patent itself, and then compared to the reference's problem [48, p. 1360]. The analogous art inquiry therefore hinges on the similarity of problems and the closeness of subject matter, both as seen through the eyes of the skilled artisan.

Sanofi-Aventis v. Mylan is the cautionary case for patent petitioners who try to stitch together references without first tying them to the patent's problem [45]. The patent claimed a drug-delivery device with a spring washer secured by multiple fixing elements. Mylan combined two device references and added a third reference, to supply snap-fit grips; it argued the third reference was pertinent to a problem addressed in another prior art reference rather than the patent itself. The Patent Trial and Appeal Board (PTAB) accepted that framing. The Federal Circuit reversed because the challenger never carried its burden to show that the third reference was analogous to the*'614 patent*, as opposed to being analogous to the another of the references cited; even if a reference is pertinent to one problem in another reference, it does not follow that it is pertinent to the challenged patent's problem [45, pp. 1377–1383]. That holding also marks a quiet divergence from the EPO's problem–solution approach, under which the "objective technical problem" is formulated with reference to the closest prior art and the differences between it and the claims. U.S. law asks a different question: is the reference analogous to the patent's field or problem, not to another reference's problem?

These principles have practical effect in Inter Partes Reviews (IPRs). It is the petitioner's burden to prove that each reference is analogous to the challenged patent, and the burden is not met by arguing only that the references are analogous to each other. Patent owners should press that threshold, especially at the Patent Owner's Preliminary Response stage, by defining the problem solved by the patent in terms that are grounded in the specification yet meaningfully different from the problems faced by the cited art. If the Board adopts the patentee's problem framing and the petitioner's references fall outside it, the petition fails at the gate. Petitioners, for their part, should build the record from the outset around the patent's problem and field; they should avoid relying on cross-reference analogies that leave the patent's own problem underdeveloped.

Prosecution records also matter. *Elekta Ltd. v. Zap Surgical Systems, Inc.* shows how an applicant's silence can expand the universe of "relevant" art in later proceedings [49]. There, the patentee's Information Disclosure Statement (IDS) included radiation imaging patents while pursuing claims directed to radiation therapy systems; because the applicant had not clarified the limited relevance of those imaging references, the PTAB and the Federal Circuit treated both imaging and therapy art as belonging to the pertinent field and found a motivation to combine accordingly [49]. The lesson is not to under-disclose; the duty of candor remains. It is to consider whether, when submitting an IDS, the applicant should include a neutral statement that the submission satisfies the duty to disclose information material to patentability, while making no admission about the relevance of any reference or whether any combination would supply a motivation to combine. Practitioners should weigh the risks of more specific statements, since admissions can be weaponized; nevertheless, where an examiner's characterization of field or problem is trending overbroad, a contemporaneous prosecution paper that anchors the field or problem in the specification can preserve arguments for later.

For medical devices and SaMD, analogous art disputes often turn on whether a reference from an adjacent engineering domain would have commended itself to a POSITA working on the claimed device's problem. A mechanical fastener reference from the automotive area may not be reasonably pertinent to a safety critical drug injector, unless the patent's problem is truly mechanical retention under vibration rather than, say, aseptic cartridge loading. A machine learning paper on content ranking may be outside the field of an arrhythmia detection SaMD, unless the paper addresses the same detection problem under comparable signal and constraint regimes. The analysis is very fact-intensive, but the constants are the two tests and the vantage point of the POSITA.

Strategies naturally follow. Petitioners should define the patent's field and problem with expert support, then map each reference to one or both tests with specific, technical bridges; generalized "optimize as needed" or "common design goals" sound bites rarely carry the day. Patent owners should resist attempts to broaden the field of endeavor beyond what the specification supports and should require the petitioner to prove reasonable pertinence with more than superficial analogies; expert declarations that define the problem with the fidelity of real-world design constraints can be decisive. When the record is thin or ambiguous on the patent's problem to be solved, *Sanofi-Aventis* [45] and *Donner* [48] provide the framework for holding petitioners to their burden; when the record is overbroad due to prosecution silence, *Elekta* [49] warns that tribunals may fill the gap.

The bottom line is straightforward. Analogous prior art is not a formality. It is the factual foundation that determines which references are even in play for obviousness under 35 U.S.C. § 103. Keep the inquiry tethered to the patent's field and problem to be solved, build a record from the POSITA's perspective, and do not let combinations proceed without first passing that gate.

6.7.5 *AI as a POSITA*

The "person of ordinary skill in the art" remains the hinge for almost every major patent doctrine, yet who that person is and how they reason has never been static. After *KSR International Co. v. Teleflex Inc.*, the skilled artisan is not a rote technician but "a person of ordinary creativity," capable of applying common sense to combine familiar elements when a reason exists to do so [50, p. 421]. That framing matters in obviousness, where tribunals ask what a skilled artisan knew, why they would have combined references, and whether success would have been reasonably expected, and it reverberates in enablement and written description, which turn on what such an artisan can practice from the specification without undue experimentation, and in claim construction questions such as whether a software term connotes structure under 35 U.S.C. § 112(f).

Recent advances in generative AI invite a fresh look at these touchpoints because large language models now process technical corpora, synthesize disparate teachings, and propose plausible modifications in ways that can resemble the evaluative moves courts attribute to the POSITA. The question is not whether AI replaces the legal construct but how it can inform the record about what a skilled artisan would have known, tried, or understood at the priority date.

Obviousness offers the clearest on-ramp. Section 103 asks whether the claim as a whole would have been obvious to a person having ordinary skill in the art, and Federal Circuit decisions emphasize three fact questions: disclosure of each limitation, motivation to modify or combine, and reasonable expectation of success, all evaluated from the POSITA's perspective and reviewed for substantial evidence on appeal [50, pp. 415–422]. In *Fleming v. Cirrus Design Corp.*, the court affirmed an obviousness determination where the Board credited a motivation to combine two references even though neither alone taught all features, emphasizing that obviousness does not demand a perfect match so long as a skilled artisan could reasonably combine the teachings using ordinary creativity [51, pp. 1221–1222]. In other words, the law expects more than mechanical recall; it expects reasoned synthesis. Contemporary AI can help surface that synthesis by scanning contemporaneous literature, identifying connections, and articulating rationales a skilled artisan might find persuasive, with the critical caveat that any analysis must be anchored to pre-priority evidence rather than hindsight.

Disclosure doctrines pull the same thread from the other end. Enablement and written description require that the specification equip a skilled artisan to make and use the full claim scope and demonstrate that the inventor possessed what is claimed. *In re Wands* remains the lodestar for what counts as "undue" experimentation, listing factors such as the quantity of experimentation, guidance in the specification, working examples, the nature and predictability of the art, and the breadth of the claims [10, p. 737]. Written description asks whether the four corners of the document show possession as of the filing date [52, p. 1351]. Used carefully, AI can pressure-test a specification against those standards by flagging missing steps, non-portable examples, and unsupported generalizations, and by checking whether representative embodiments plus generalizable rules carry a practitioner across the full breadth. The point is not to substitute a model's output for expert testimony, but to use it as a disciplined way to ask what a skilled team can actually implement from the document.

Claim construction, including § 112(f), also runs through the POSITA. Whether a term like "application," "code," or "module" conveys sufficiently definite structure turns on what a skilled artisan would understand in context. *Dyfan, LLC v. Target Corp.* is a recent reminder that software words are not per se nonce terms; in that record, "code" and "application" had recognized structural meaning to practitioners, especially given evidence that off-the-shelf implementations existed [53, pp. 1366–1371]. The more the intrinsic record ties a label to known interfaces, state machines, or protocol roles, the more a tribunal will credit the term's structural connotation. AI can assist by testing how consistently a technical corpus uses a term to denote a class of components rather than a black box, which can inform, but not replace, the human expert's explanation of ordinary meaning.

In practice, defining the POSITA is a matter of careful, field-specific detail. Courts weigh factors such as education and years of experience, the types of problems and solutions in the field, the sophistication and predictability of the technology, and the pace of innovation [54, p. 1579; 56, pp. 1256–1257]. In complex areas like medical devices and SaMD, the realistic artisan is often multidisciplinary, combining embedded systems, control theory, and signal processing with an appreciation for clinical constraints such as safety margins, alarm burden, cybersecurity, interoperability, and regulatory validation. That richer definition matters because it conditions what combinations are truly predictable, what constitutes "routine" software changes, and what a disclosure must teach across a wide design space. It also disciplines the use of AI in building the record. If a model is used to probe motivation to combine or to sanity-check enablement, the workflow should be time-boxed to pre-priority sources and documented so that a human expert can explain why the prompts, corpus, and outputs fairly reflect what the skilled artisan would have done and understood at the relevant time. The model then becomes a lens for finding and organizing contemporaneous evidence, rather than a black-box oracle.

The litigation and prosecution strategy playbooks naturally follow. On § 103, it is important to frame motivation and reasonable expectation of success at the level of the defined POSITA, not a caricatured generalist or an omniscient expert. Also, using contemporaneous technical sources to show why the path was predictable or, conversely, why it was not, and let AI help surface those sources without letting it smuggle in hindsight is necessary. On § 112(a), one should draft and argue with commensurability in mind, pair representative embodiments with portable rules that a skilled team can apply across the full scope, and use AI to stress-test the journey from examples to generalization. On § 112(f), software terms should be tied to recognized structures with protocol names, data-flow roles, and state machines that a practitioner would treat as architecture rather than aspiration, and, if needed, use corpus-based analyses to corroborate ordinary meaning while expert testimony does the legal work. Across all of it, the constant remains: the law asks what a person of ordinary skill would have known, tried, expected, and understood at the priority date.

6.8 Unique Patent Challenges for Medical Devices

Medical devices face their own unique hurdles when it comes to obtaining patent protection. For starters, patent seekers must first determine what types of claims will be needed in order to protect the invention. With medical devices, inventions could take the form of innovative new devices, new methods of use, new diagnostic procedures, and more. Patent seekers must hone the claims for each patent application to specifically focus on the invention.

On a related note, some medical devices contain software, which could be its own patentable invention aside from the physical medical device and method of using the medical device. There are specific challenges when it comes to patenting software before the USPTO, and the use of software that implements artificial intelligence introduces additional hurdles. Further, the FDA may consider the software itself, and software implementing AI, a medical device (i.e., software as a medical device or SaMD).

Finally, when medical devices are subject to regulatory review, such as review by the FDA, situations could arise where a patent issues for the medical device, but the patent owner cannot yet sell the device because FDA approval is pending. In such cases, the patent owner could lose out on some of the life (e.g., patent term) of the patent because they are unable to commercialize the product until FDA approval is granted. In such situations, patent owners may seek an extension of their patent term, known as patent term extension (PTE) (see Sect. 6.9.2).

These patenting challenges are unique to medical devices. While each of these challenges may be time-consuming, they are surmountable.

6.8.1 Medical Device Claims Versus Medical Device Use Claims

When it comes to patenting a medical device, the physical device itself may be eligible for a patent as a machine or article of manufacture under 35 U.S.C. § 101. Additionally, the use of the device may also be eligible for a patent as a process, such as a method/process for using the medical device.

There is no clear distinction made between an "article of manufacture" and a "machine" for purposes of patentability. An "article of manufacture" is often claimed as the common name of the object (e.g., a thermometer, a stent, a sphygmomanometer (blood pressure monitor), or a Computer Readable Storage Medium) whereas a machine is often claimed as an apparatus, device, or system. Regardless whether directed to an "article of manufacture" or a "machine," these patent claims describe the physical object that is the invention. Infringement occurs when the physical object is made, sold, used, or imported into the United States.

Process claims, on the other hand, recite a sequence of steps to be performed in order to accomplish a specific goal. Process claims are most often identified as a method and will recite a sequence of gerund forms of verbs. By way of example, a method of using a medical device would follow the general format of:

> A method comprising the steps of:
>
> **providing** a medical device,
> **inserting** the medical device into a patient,
> **using** the medical device on the patient, and
> **removing** the medical device from the patient.

Infringement occurs when any person, machine, or computer performs the steps recited in the claims. However, a process claim may require a particular apparatus to perform the process.

Yet a claim that recites both a medical device and the method steps of using the medical in the same claim is considered indefinite under AIA U.S.C. § 112(b) or pre-AIA 35 U.S.C. § 112, second paragraph. The Federal Circuit in *IPXL Holdings, L.L.C. v. Amazon.com, Inc.* held that when a claim "recites both a system and the method for using that system, it does not apprise a person of ordinary skill in the art of its scope, and it is invalid under section 112, paragraph 2 [56, p. 1384]."In medical device cases, it is critically important to identify in the claims whether the invention relates to the device itself or to the use of the device.

6.8.2 Diagnostic Methods

U.S. Patent law does not define a "diagnostic method" for the purposes of patent eligibility. However, the U.S. Supreme Court has stated that diagnostic methods are not patent-eligible because the claims do "nothing significantly more than instruction to doctors to apply the applicable laws when treating their patients [57, p. 79]."

Regarding diagnostic methods, for example, in *Mayo Collaborative Servs. v. Prometheus Labs,* two patents, U.S. Patent No. 6,355,623 and U.S. Patent No. 6,680,302, directed to diagnostic tests using thiopurine for the detection of autoimmune diseases, such as Crohn's disease and ulcerative colitis, were at issue. Prometheus was the exclusive licensee of the two patents in question and commercially sold kits for performing the diagnostic test to *Mayo. Mayo* subsequently began selling a similar test with a higher metabolite level. Based on the sale of the similar test, Prometheus sued mayo for patent infringement.

The infringed claims were directed to a method of measuring metabolite levels in the blood and determining the relationship between toxicity and efficacy. A representative claim from U.S. Patent No. 6,355,623 is claim 1, which recites:

1. A method of optimizing therapeutic efficacy for treatment of an immune-mediated gastrointestinal disorder, comprising:
   (a) administering a drug providing 6-thioguanine to a subject having said immune-mediated gastrointestinal disorder; and
   (b) determining the level of 6-thioguanine in said subject having said immune-mediated gastrointestinal disorder,

wherein the level of 6-thioguanine less than about 230 pmol per 8×10^8 red blood cells indicates a need to increase the amount of said drug subsequently administered to said subject and

> wherein the level of 6-thioguanine greater than about 400 pmol per 8×10^8 red blood cells indicates a need to decrease the amount of said drug subsequently administered to said subject.

In *Mayo*, the Supreme Court held that the claims were directed to "relationships between concentrations of certain metabolites in the blood and the likelihood that a dosage of a thiopurine drug will prove ineffective or cause harm" and that this "relation is a consequence of the ways in which thiopurine compounds are metabolized by the body—[an] entirely natural processes [57, p. 79]". The Court reasoned that allowing a patent on this naturally occurring phenomenon would "inhibit the development of more refined treatment recommendations (like that embodied in *Mayo*'s test) that combine *Prometheus*' correlations with later discovered features of metabolites, human physiology, or individual patient characteristics [57, p. 79]." Accordingly, the Court concluded that the claims simply "inform a relevant audience about certain laws of nature and add nothing significant beyond the sum of their parts taken separately [57, pp. 79–80]." The identification of naturally occurring relationships to determine a medical condition is a patent-ineligible diagnostic method.

Similarly, in *Ariosa Diagnostics, Inc. v. Sequenom, Inc.* [58], a diagnostic test that used levels of cell-free fetal DNA (cffDNA) in maternal plasma and serum to diagnosis certain fetal characteristics was determined to be a patent-ineligible diagnostic method. Specifically, at issue in *Ariosa* were the claims of U.S. Patent No. 6,258,540. By way of example, claim 1 recites:

> 1. A method for detecting a paternally inherited nucleic acid of fetal origin performed on a maternal serum or plasma sample from a pregnant female, which method comprises
>
>   amplifying a paternally inherited nucleic acid from the serum or plasma sample and
>   detecting the presence of a paternally inherited nucleic acid of fetal origin in the sample.

The Federal Circuit, applying *Mayo*, held that the claims at issue in *Ariosa* were directed to a patent ineligible diagnostic method because "[t]he only subject matter new and useful as of the date of the application was the discovery of the presence of cffDNA in maternal plasma or serum [58, p. 1377]." The Court further reasoned that the claimed "method amounts to a general instruction to doctors to apply routine, conventional techniques when seeking to detect cffDNA [58, p. 1377]." Accordingly, since the existence of cffDNA is a natural phenomenon, identifying its presence is merely claiming the natural phenomena itself. As a result, the claimed invention "begins and ends with a natural phenomenon [58, p. 1376]." As such, the Court held that a diagnostic method was again patent ineligible because it sought to combine naturally occurring relationships with what would be considered the common knowledge of a physician.

Likewise, in another Federal Circuit case, *Genetic Techs., Ltd. v. Merial LLC* [59], a method for detecting a coding region of genomic deoxyribonucleic acid (DNA) was found to be a diagnostic method and therefore patent ineligible under *Mayo*. At issue in *Genetic Techs.* was U.S. Patent No. 5,612,179, which claims a method where alleles of a particular gene may be detected by amplifying and analyzing non-coding regions known to be linked to the coding region. Specifically, the representative claim states:

1. A method for detection of at least one coding region allele of a multi-allelic genetic locus comprising:
   (a) amplifying genomic DNA with a primer pair that spans a non-coding region sequence, said primer pair defining a DNA sequence which is in genetic linkage with said genetic locus and contains a sufficient number of non-coding region sequence nucleotides to produce an amplified DNA sequence characteristic of said allele; and
   (b) analyzing the amplified DNA sequence to detect the allele.

The Federal Circuit found these claims to be a patent ineligible diagnostic method because "[t]he product of the method of claim 1 is information about a patient's natural genetic makeup—at least one coding region allele" and "[t]he claim relies on the existence of linkage disequilibrium between the non-coding and coding regions [59, p. 1375]." Accordingly, the court held that "the claim is directed to a natural law—the principle that certain non-coding and coding sequences are in linkage disequilibrium with one another [59, p. 1375]," and therefore is a patent ineligible diagnostic method.

As shown in *Genetic Techs.*, *Ariosa* and *Mayo*, a method of diagnosing a condition based upon a natural law or naturally occurring relationship is not patent-eligible. However, a device that is used to perform a diagnostic test may be patent-eligible. For example, in *Exergen Corp. v. Kaz USA, Inc.* [60], a sensor that determined a person's core temperature by detecting the temperature of the forehead directly above the superficial temporal artery was found to be a patent-eligible medical device. The Federal Circuit held that, unlike in *Mayo*, *Ariosa*, and *Genetic Techs*, U.S. Patent Nos. 6,292,685 and 7,787,938 at issue in *Exergen* is directed to "the measurement of a natural phenomenon (core body temperature) [60, p. 11]." Specifically, a representative claim for U.S. Patent No. 7,787,938 is claim 48, which recites:

48. A body temperature detector comprising:

    a radiation detector; and

> electronics that measure radiation from at least three readings per second of the radiation detector as a target skin surface over an artery is viewed, the artery having a relatively constant blood flow, and that process the measured radiation to provide a body temperature approximation, distinct from skin surface temperature, based on detected radiation.

The Court concluded the claims recited more than "the coefficient representing the relationship between temporal-arterial temperature and core body temperature [60, p. 7]," and instead "incorporated that discovery into an unconventional method of temperature measurement [60, p. 12]" and thus "transformed the process into an inventive application of the [relationship] [60, p. 12]."

Claims that are directed to identifying a naturally occurring relationship to determine the presence of (i.e., diagnose) a disease or condition are patent-ineligible method claims. However, a device that measures a naturally occurring relationship may be patent-eligible if the other requirements for patentability are satisfied (e.g., the medical device is novelty and non-obviousness under U.S.C § 102 and U.S.C § 103, respectively).

6.8.3 *Methods of Treatment*

Methods of treating a patient are patent-eligible subject matter. Unlike diagnostic methods, which the courts have found to be patent ineligible because the claims "amount[] to nothing more than observing or identifying the ineligible concept itself [61, p. 1048]," methods of treatment are "directed to a new and useful method of [treating a condition] [61, p. 1048]." For example, in *Rapid Litig. Mgmt. v. CellzDirect, Inc.*, claims directed to "a new and useful laboratory technique for preserving hepatocytes [61, p. 1048]" were patent-eligible because the claims recited steps that achieved a better way of preserving the hepatocytes. Specifically, the representative claim of U.S. Patent No. 7,604,929 at issue in *CellzDirect* recites:

> 1. A method of producing a desired preparation of multi-cryopreserved hepatocytes, said hepatocytes being capable of being frozen and thawed at least two times, and in which greater than 70% of the hepatocytes of said preparation are viable after the final thaw, said method comprising:
>
>   (A) subjecting hepatocytes that have been frozen and thawed to density gradient fractionation to separate viable hepatocytes from nonviable hepatocytes,
>
>   (B) recovering the separated viable hepatocytes, and

> (C) cryopreserving the recovered viable hepatocytes to thereby form said desired preparation of hepatocytes without requiring a density gradient step after thawing the hepatocytes for the second time, wherein the hepatocytes are not plated between the first and second cryopreservations, and wherein greater than 70% of the hepatocytes of said preparation are viable after the final thaw [61, p. 1046].

The Federal Circuit held that these claims were "not simply an observation or detection of the ability of hepatocytes to survive multiple freeze–thaw cycles [61, p. 1048]" which would be a patent-ineligible diagnostic method. Instead, the court determined that the claims were a patent-eligible treatment method because the claims were "directed to a new and useful method of preserving hepatocyte cells [61, p. 1048]." The court concluded that the claimed "new and improved technique, for producing a tangible and useful result, falls squarely outside those categories of inventions that are "directed to" patent-ineligible concepts [61, p. 1050]."

Similarly, in *Vanda Pharm. Inc. v. W.-Ward Pharm. Int'l Ltd.* [62], claims for a method of treating schizophrenia with iloperidone based on the patient's genotype were found to be patent-eligible. Specifically, at issue in *Vanda* was U.S. Reissue Patent 39,198 and U.S. Patent No. 8,586,610, each of which is for an atypical antipsychotic drug approved by the FDA as Fanapt® (iloperidone). The representative claim of the U.S. Patent No. 8,586,610 recites:

> 1. A method for treating a patient with iloperidone, wherein the patient is suffering from schizophrenia, the method comprising the steps of:
>
> determining whether the patient is a CYP2D6 poor metabolizer by:
>
> obtaining or having obtained a biological sample from the patient; and performing or having performed a genotyping assay on the biological sample to determine if the patient has a CYP2D6 poor metabolizer genotype; and
>
> if the patient has a CYP2D6 poor metabolizer genotype, then internally administering iloperidone to the patient in an amount of 12 mg/day or less, and
> if the patient does not have a CYP2D6 poor metabolizer genotype, then internally administering iloperidone to the patient in an amount that is greater than 12 mg/day, up to 24 mg/day,
>
> wherein a risk of QTc prolongation for a patient having a CYP2D6 poor metabolizer genotype is lower following the internal administration of 12 mg/day or less than it would be if the iloperidone were administered in an amount of greater than 12 mg/day, up to 24 mg/day [62, pp. 2–3].

The Court held that the claims recited "more than the natural relationship between CYP2D6 metabolizer genotype and the risk of QTc prolongation [62, p. 32]" and instead "recite[d] a method of treating patients based on this relationship that makes iloperidone safer by lowering the risk of QTc prolongation [62, p. 32]." The Court distinguished the claims in *Vanda* from the claims that were at issue in the U.S. Supreme Court Case *Mayo Collaboration Services v. Prometheus Laboratories* [57], by concluding that the claims in *Mayo* "were not directed to a novel method of treating a disease [62, p. 5]," "[i]nstead, the claims [in Mayo] were directed to a diagnostic method [62, p. 29]." Accordingly, the court held the claims were "directed to a specific method of treatment for specific patients using a specific compound at specific doses to achieve a specific outcome [62, p. 32]" and therefore patent-eligible.

In summary, an invention that is directed to the treatment of a medical condition, as compared to the diagnosis of a medical condition, is patent-eligible.

6.8.4 Medical Device Software

Software is quickly becoming recognized as an integral part of many new medical devices that are being introduced to the market. Many wearable medical devices and health trackers interface with mobile apps and other software platforms. In fact, the inclusion of software in medical device technologies is now so prevalent that the FDA issued Guidance [63] on software-as-a-medical-device (SaMD) in late 2017.

The FDA Guidance notes that software that provides "significant" information or "independent review" regarding diagnosis or treatment of a health or medical condition should be considered SaMD. Conversely, software with the sole purpose of driving a particular piece of medical device hardware, or for retrieving information, organizing data, or optimizing performance, or an associated medical device, is not considered SaMD.

A medical device having software that provides "significant" information would be software that:

(1) Assesses the seriousness of a medical condition, such as classifying a condition as non-serious, serious, or critical.
(2) Assesses a medical condition for identification purposes, e.g., identifies the medical condition.
(3) Provides information concerning the treatment or diagnosis of a medical condition.
(4) Provides information that informs or drives clinical management.

The FDA Guidance also indicates that SaMD has a certain level of "independence," meaning that the software should be capable of performing its intended medical function by being run on any general-purpose computer, e.g., the SaMD cannot be the driver of a medical device hardware and should be able to be implemented on devices other than its "normal" housing.

While certain software may be considered a medical device in the eyes of the FDA, this designation has no bearing on the patentability of the medical device, process of using the software associated with the medical device or any other patentable aspect of the SaMD. The mere fact that a medical device may use software in its functionality is not dispositive of patentability of the medical device. The Federal Circuit has held that "[s]oftware can make non-abstract improvements to … technology just as hardware improvements can, and sometimes the improvements can be accomplished through either route [64, p. 1335]."

However, inventions that are entirely embodied in software are potentially problematic from a patenting standpoint. For example, the Federal Circuit has held "the concept of (1) collecting data, (2) recognizing certain data within the collected data set, and (3) storing that recognized data in a memory [is] abstract [65, p. 1340]" and therefore ineligible for patent protection. Accordingly, inventions that rely upon "big data" and artificial intelligence to mine large datasets must involve more than "[m]erely requiring the selection and manipulation of information—to provide a 'humanly comprehensible' amount of information useful for users [66, p. 1355]".

Additionally, simply adding software to a patent-ineligible concept does not transform the claimed invention into patent-eligible subject matter. The Supreme Court has ruled against "interpreting [patent eligibility under 35 U.S.C. § 101] in ways that make patent eligibility depend simply on the draftsman's art [5, p. 2360]." Consequentially, adding "a handful of generic computer components configured to implement the [abstract] idea" does not make the claimed abstract idea any less abstract [5, p. 2360].

Therefore, the utilization of software in an invention does not necessarily hurt or improve the patentability of a medical device. However, the inclusion of software does allow for additional protection under copyright.

6.8.5 Printed Matter Doctrine

The Printed Matter Doctrine states that "printed matter," meaning information, instructions, or content for human or machine interpretation, is not patentable unless it has a new and non-obvious functional relationship with the physical substrate (such as paper, a label, or a computer-readable medium) on which it is placed. The printed matter doctrine is relevant in medical device practice because the industry relies heavily on information. Labels, instructions for use, radiographic markers, on-screen alerts, barcodes, and machine-readable tags are everywhere, often because FDA rules require them. Patent law draws a line between protecting functional technology and claiming the content of a message. In its modern form, the Federal Circuit applies a two-step test. First, a limitation is "printed matter" only 'if it claims to contain information. Second, if and only if the limitation is printed matter, the tribunal asks whether the information is functionally or structurally related to the claimed "substrate" so that it should still receive patentable weight [67, pp. 848–851; 68, p. 1381]. The Federal Circuit recently reiterated this framework, emphasizing that "printed matter" turns on communicative content, not on the medium or the mere fact of communication [69, pp. 1404–1412].

A core distinction runs through the cases: information for human cognition versus information that a machine must process to make the system work. In the classic Court of Customs and Patent Appeals (C.C.P.A.) (the predecessor court to the Federal Circuit) decision *In re Bernhart*, the court explained that the "printed matter" cases addressed arrangements of characters "useful and intelligible only to the human mind," and it held the doctrine inapplicable where the claimed invention requires a computer to process the information and a plotting machine to draw the output [70, pp. 1399–1400]. Building on that reasoning, *In re Lowry* held that software data structures are not printed matter because they are processed by a machine, impose a physical organization on memory, and improve computer operation [71, pp. 1583–1584]. Together, *Bernhart* and *Lowry* combine the doctrine to human-facing content and keep machine-processed structures in play.

The 2024 *IOENGINE* decision underscores this point [69]. The PTAB had discounted limitations reciting "encrypted communications" and "program code" as printed matter, but the Federal Circuit reversed because the claims did not purport to own any particular informational content. The court emphasized that encryption is a form, not a message, and that "program code" was not claimed for its content. Because the limitations were not printed matter at step one, the inquiry ended [69, pp. 1408–1412]. The court also collected examples of true printed matter in life sciences settings, including dosage instructions on an FDA label, a label instructing a patient to take a drug with food, instructions for performing a DNA test, and numbers printed on a wristband. Those are human-facing communications; they differ from machine-parsed signals that change device behavior [69, pp. 1409–1410].

For medical device patents that necessarily touch FDA labeling, *C.R Bard* shows both the risk and the limit of the doctrine [68]. Bard's vascular access ports used radiographic "CT" markers, shapes, and other identifiers so clinicians could confirm power-injection capability after implantation. The district court held the claims invalid as "printed matter," but the Federal Circuit reversed, explaining that the claims were not *solely* directed to printed matter and were patent-eligible under § 101; factual disputes on anticipation remained as well [68, pp. 1375–1384]. The takeaway for prosecutors is not that labels can carry a claim, but that structural or operational features tied to those labels can keep a claim alive even when some recited content is disregarded.

Procedurally, *Praxair* confirms that printed matter questions can and often should be resolved at claim construction, and it reiterates that the doctrine can bear on § 102 and § 103 as well as eligibility [72, pp. 1032–1035]. On the merits, courts have repeatedly refused to credit claim language that merely instructs a person to do something with a known product or method. In *In re Ngai*, a kit claim reciting instructions did not gain patentable weight because the instructions lacked a functional relationship to the kit components [73, pp. 1338–1339]. And in *King Pharmaceuticals, Inc. v. Eon Labs, Inc.*, an "informing a patient to take metaxalone with food" step could not salvage novelty where the method's result flowed inherently, making the "informing" content printed matter lacking patentable weight [74, pp. 1278–1280].

Against that backdrop, the regulatory landscape supplies important context but not, by itself, patentable substance. FDA's Unique Device Identification (UDI) rules require that device labels and packages bear a UDI and, for certain reusable devices, that the UDI also appears as a direct mark on the device; the UDI must be presented both as easily readable text and as automatic identification and data capture (AIDC) (for example, a barcode) [75–78]. Those mandates explain why so much information surrounds medical devices, yet the content of those labels is still "printed matter" unless the claim ties the data to device behavior. The claims should therefore connect the machine-readable elements to concrete device actions like enabling therapy modes, altering parameter sets, or initiating interlocks.

These principles map neatly onto wireless and software-driven devices. Consider a hospital wristband if your claim distinguishes over the art by reciting that the band "bears a patient number," which is printed matter. If, instead, the band encodes a machine-readable token that the infusion pump must scan before enabling a high-risk therapy mode, and the claim ties scanning and mode-enablement to that token, it becomes a functional relationship. The first claim likely fails at step two; the second is squarely functional. Now consider a ventilator that displays "ALERT: RESP RATE HIGH," which is a message. However, if the ventilator claims a control loop that uses the alert condition to automatically adjust trigger sensitivity and logs clinician overrides for audit, the novelty lies in machine behavior, not human-directed text.

Similarly, a radio-frame indication that toggles a therapy device between modes is not a human instruction; it is a machine-parsed control signal. Step one fails for the challenger because the limitation is not claimed for its content. Even at step two, the indication has a functional relation to the device's operation because it changes what the device does.

Therefore, for claim drafting, avoid staking novelty on what a label says to a human. "Wherein the package states ...," "the IFU indicates ...," or "displaying a warning that ...," without more, invites step-one and step-two headwinds. Instead, claim machine-readable structures and the machine reaction. A wristband that "bears a number" is printed matter; a wristband whose encoded token must be scanned to unlock a high-risk infusion mode, where the claim expressly ties scanning to the device's interlock state, sets up the functional relationship courts credit. This is the practical implementation of *Lowry* [71, pp. 1583–1584] and *Bernhart* [70, 1399–1401] in the SaMD era: data that devices must parse to change state are part of the technology, not mere messages.

For litigation, an accused infringer should isolate human-facing content and argue either that the limitation is not to be given patentable weight at step-one or, at step-two, that no functional relationship exists. *Ngai* [73, pp. 1338–1339] and *King* [74, pp. 1278–1280] are workhorses for those points; *Praxair* [72, pp. 1032–1035] supports addressing them at claim construction. A patentee, by contrast, should map each supposedly "informational" field to specific device behavior, emphasize *IOENGINE* [69, pp. 1404–1412] when the claim does not recite content at all, and, when necessary, prove the *In Re Gulack* functional tie between message and substrate [79, pp. 1386–1387].

The bottom line for the med-device–regulatory intersection is straightforward. FDA rules will continue to require a great deal of labeling. Patent law will continue to discount the communicative content of that labeling. The way through this dichotomy is to draft and argue claims so that information is part of what the system *does*. When a claimed limitation is processed by a machine and changes how the device operates, the printed matter doctrine should not stand in the way [69, pp. 1408–1412; 68, pp. 1375–1384; 71, pp. 1583–1584; 79, pp. 1386–1387].

6.8.6 Routine Optimization

In medical devices and SaMD, small parameter choices often drive big clinical consequences. Contact force and irrigation flow shape lesion quality in ablation, filter windows, and decision thresholds govern alarm fatigue in monitoring, and control gains determine safety margins in infusion. When the prior art itself teaches that a parameter affects the outcome of interest, a tribunal will view tuning that parameter as routine optimization, which strengthens a § 103 case. When the record does not tie the parameter to the clinical or performance result, motivation evaporates and obviousness weakens.

Result-effective variables sit at the center of many obviousness disputes because they provide the bridge between what the prior art teaches in principle and what a person of ordinary skill in the art would actually try in practice and try to optimize. A "result-effective" variable is a parameter that the prior art itself recognizes as affecting a property or outcome of interest. When that recognition exists, tribunals often treat changes to the parameter as routine optimization. When it does not, the motivation to adjust the parameter evaporates, and a prima facie case of obviousness weakens. The Federal Circuit has made this framework explicit. In *Genentech, Inc. v. Hospira, Inc.*, the court explained that a patentee can rebut a presumption of obviousness by showing either that a claimed range is special or critical, or that the process parameter at issue was not recognized in the art as result-effective [80, p. 1341]. *Genentech* thus ties two core showings together: lack of recognition that the variable drives the result, and evidence that the particular claimed range produces an effect that matters.

The prima facie inquiry begins with the record. It is not enough for an examiner or challenger to identify a parameter that could be adjusted in theory. The proponent of obviousness must ground the motivation to optimize in the art by pointing to disclosures that link the variable to the property at issue. The Federal Circuit captured this point in *In re Applied Materials, Inc.*, noting that the absence of disclosure regarding the relationship between a variable and the affected property may preclude a finding that the variable is result-effective [81, p. 1297]. Conversely, where the art does recognize that a parameter affects the relevant property, the parameter is result-effective and optimization may be obvious [82, p. 1009]. The motivation requirement and the result-effective variable concept, therefore, rise and fall together. If the art tells the skilled artisan that changing temperature, pressure, electrode spacing, signal threshold, or drug concentration will predictably influence safety or performance, the law presumes that the artisan would try reasonable adjustments that fall within the art's teachings.

Two C.C.P.A. decisions show the other side of the coin. In *In re Antonie*, the court confronted claims involving a specific ratio of variables that the prior art neither disclosed nor suggested was meaningful. The court emphasized that because the art did not recognize the ratio as affecting the desired property, there was no motivation to explore or optimize that relationship; the claims were not rendered obvious on that record [83, p. 618]. *In re Yates* is to the same effect. There, the court rejected an obviousness theory because the record failed to establish any teaching in the prior art that the variable at issue influenced the property of interest; without evidence of that relationship, there was no basis to deem the variable result-effective [84, p. 1056]. These decisions continue to matter because they anchor the evidentiary burden. Identification of a parameter is not enough; the prior art must also link that parameter to the result that matters.

Range disclosures add a second layer to the analysis. When the art discloses a narrow range for a variable already known to drive results, the law often treats movement within that range as routine. When the art discloses an extremely broad field of possibilities, however, the breadth itself undercuts the inference that routine optimization would predictably yield the claimed invention. In *Genetics Institute, LLC v. Novartis Vaccines & Diagnostics, Inc.*, the Federal Circuit explained that a prior-art range so broad that it encompasses a very large number of distinct compositions may not render a specific claimed composition obvious because arriving at that composition requires more than routine optimization [85, p. 1306]. This principle complements the result-effective variable doctrine. A challenger who relies on broad disclosures should expect to meet rebuttal evidence of criticality, nonlinearity, or unexpected results within the claimed sub-range.

These doctrines have practical impact in medical device work, where small shifts in parameters often have disproportionate effects on safety and performance. Consider ablation power and contact force in electrophysiology. If the art recognizes that contact force directly influences lesion depth and steam-pop risk, then contact force appears to be a result-effective variable, and predictable optimization arguments have traction. If the record shows that a newly claimed interaction between saline irrigation flow and a particular pulsed-power waveform produces a non-intuitive lesion geometry, and the art did not link irrigation flow to that waveform's result, then irrigation flow in that context may not be a result-effective variable. Similar patterns recur in imaging systems that rely on machine learning thresholds, in infusion pumps that adjust delivery based on sensor noise metrics, and in implant manufacturing, where surface roughness parameters interact with sterilization temperature and time. The question is always the same: does the art teach that the specific parameter influences the specific property, so that a skilled artisan would have a reason to tune it?

Regulatory frameworks provide additional context without resolving the patent question by themselves. Design controls and process validation regimes require manufacturers to identify critical process parameters and critical quality attributes, document how those variables influence outcomes, and show that processes remain in control. That documentation can help or hurt in litigation and prosecution. When regulatory records, white papers, or standards show that a variable was widely understood to affect a property, they can bolster a result-effective finding for the challenger. When the records instead show that the claimed parameter was not previously tracked as critical, or that the relationship is non-linear and emergent only under a newly claimed combination of conditions, they support a patentee's argument that the variable was not recognized as result-effective in the prior art. The legal test still looks to what the art taught at the priority date, but the regulated environment often supplies a rich evidentiary record.

These principles translate into concrete strategies. On the prosecution side, when facing a § 103 rejection premised on "routine optimization," the examiner should be pushed to identify where the art actually recognizes the parameter as result-effective for the property that matters, and insist on record evidence that ties the parameter to the outcome. If the art is silent or speculative, *Applied Materials* [81, p. 1297] and *E.I. du Pont de Nemours* [82, p. 1009] should be cited to show that the motivation premise fails. If the art discloses very broad ranges, *Genetics Institute* should be used to argue that the path to the claimed sub-range is not routine and then reinforce that argument with data showing criticality, non-monotonic behavior, or unexpected results [85, p. 1306]. Where appropriate, the record should be framed in *Genentech*'s terms by showing that the claimed range is special or critical, or that the parameter was not recognized as result-effective at all [80, p. 1341].

When drafting a patent application, parameters should be connected to performance in the specification with care. If a parameter will become the fulcrum of patentability, the relationship with the property of interest should be described, including comparative data where feasible, and explaining why the observed behavior is not a matter of routine optimization of a variable already recognized in the art. In medical devices, that often means documenting how a control parameter interacts with physiological or environmental conditions rather than acting in isolation. If the invention resides in a previously unappreciated interaction among variables, it should be expressly stated, and shown why prior teachings would not have motivated the skilled artisan to look there. When ranges are claimed, bare recitation should be avoided. Technical rationale and exemplification should be provided that illuminate why the claimed bounds matter to the result. Doing so strengthens both the non-obviousness story and the enablement and written-description record.

Litigators can use the same map in reverse. A challenger should assemble art that does not merely list variables, but that articulates or demonstrates the relationship between the specific variable and the specific property, and should avoid overreliance on omnidirectional "optimize as needed" statements. Expert testimony should translate those teachings into a concrete motivation story that explains why the skilled artisan would have tuned the parameter and what predictable result would have been expected. A patentee should probe for gaps in that linkage, highlight where the art is broad or agnostic, and press for evidence that the claimed range or combination yields an effect that was not foreseeable from routine optimization. When the record is ambiguous, *Antonie* [83, p. 618] and *Yates* [84, p. 1056] remain powerful reminders that speculation cannot supply the missing recognition.

The through line is simple. Result-effective variables supply the evidentiary and motivational backbone of many obviousness theories. When the art identifies a parameter as affecting the outcome, routine optimization may well follow; when it does not, the case for obviousness falters. In a medical device ecosystem defined by complex interactions and stringent regulatory oversight, being precise about what the art actually taught, and about which variable truly drives which outcome, often decides the issue under 35 U.S.C. § 103.

6.8.7 Full-Scope Enablement

In medical devices and SaMD, breadth is tempting and dangerous in equal measure. A single platform often spans sensors, embedded control, edge analytics, and cloud services, and teams want claims that read on evolving algorithms, parameter ranges, and deployment modes. Section 112(a) sets the price of that breadth. If you claim wide envelopes for ablation duty cycles, contact force, irrigation flow, signal thresholds, or model performance, the specification must teach a skilled team how to make and use the invention across that entire scope without undue experimentation. Courts measure this at the patent application filing date and through the lens of predictability, guidance, examples, and the size of the design space. In practice, that means pairing representative embodiments with reproducible rules a practitioner can apply to reach the rest of the claimed waterfront, not leaving clinicians and engineers to "screen until it works."

The stakes are higher in regulated health tech because safety and effectiveness turn on parameter interactions. A claim that promises stable catheter tracking across a stated metric range, closed-loop infusion performance across patient conditions, or arrhythmia detection at a target F1 score must be matched by teaching that gets a practitioner to those results throughout the range and across contemplated model families. Sparse examples and generic exhortations to tune or try alternatives will not be sufficient.

Full-scope enablement under 35 U.S.C. § 112(a) is the core of the patent bargain: in exchange for exclusivity, the specification must teach a person of ordinary skill in the art to make and use the claimed invention across its full scope without undue experimentation. The broader the claim, the heavier the disclosure lift. Enablement is assessed as of the filing date, and the Federal Circuit's *In re Wands* factors remain the lodestar for what counts as "undue"—the quantity of experimentation required, the amount of guidance provided, the presence of working examples, the nature of the invention, the state of the prior art, the skill level, the predictability of the art, and the breadth of the claims [10, p. 737]. Although distinct from written description, enablement certainly interacts with it: enablement measures how to practice the claim; written description confirms that the inventor possessed what is claimed [52, p. 1351].

The Supreme Court's decision in *Amgen Inc. v. Sanofi* reanchored the analysis around full-scope teaching. Amgen sought genus claims to potentially millions of anti-PCSK9 antibodies largely defined by function; the Court held the disclosure failed because it offered a research "hunting license" rather than instructions that enable the full genus without undue trial-and-error [11, 603–610]. The takeaway transcends biologics: breadth must be matched by commensurate, reproducible guidance or by representative embodiments plus reliable rules that carry a POSITA to the rest of the scope. Claims cannot reserve outcomes while relegating the public to screening its way to working embodiments after the fact.

Federal Circuit guardrails reinforce this calibration of scope to teaching. Ranges and open-ended claims fail when the specification teaches only a subset of the claimed performance envelope; *MagSil Corp. v. Hitachi Glob. Storage Techs., Inc.* invalidated claims where only small resistance-change values were taught, but the claim covered far larger changes [86, pp. 1384–1385] Genus breadth cannot outrun disclosure; limited exemplars rarely enable vast functional families without rules for making the rest, as in *Wyeth & Cordis Corp. v. Abbott Labs* [87, pp. 1384–1389], and *Genetics Inst., LLC v. Novartis Vaccines & Diagnostics, Inc.* [88, p. 1306]. One way to achieve a result does not enable claims to all ways of achieving it [89, pp. 1344–1346. And even when enablement is the focal requirement, courts expect "blazemarks" or comparable signposts that guide the artisan through the breadth being claimed [90, p. 1570].

These principles are critical in medical device and SaMD claims, which often rely on parametric breadth (e.g., ablation duty cycles, contact force, irrigation flow, filter coefficients, sensor thresholds) or functional genus (families of control algorithms, classes of signal-processing pipelines, machine learning models defined by outcome). After *Amgen* [11], two questions dominate: does the specification teach the "whole waterfront," or just a "few islands;" and are there reproducible rules of application? If you claim "signal-quality metrics in the range X–Y that yield stable catheter tracking," you must teach how to achieve stable tracking across the range, not merely at a couple of points. For closed-loop infusion pumps, a genus claim to "control algorithms" defined by a performance outcome should supply design constraints, control laws, and stability criteria that a POSITA can apply to instantiate alternatives within the claim. For imaging and diagnostics with ML, claims to "models configured to detect arrhythmic episodes with F1 $\geq$ Z" should disclose the data regimen, feature engineering, training/validation procedures, and hyperparameter regimes that predictably attain that floor across contemplated model families.

On the prosecution side, breadth should be calibrated to provide a clear "teaching" of the invention. If broad claims are desired, representative species should be provided that span the design space and generalizable rules, algorithms, constraints, scaling laws, that map the rest. It is advisable to provide a clear and complete record with working examples, quantify the predictability of the art, and explain why only routine tuning is needed (or, if not, narrow the scope). Comparative data that shows commensurate results across the claimed range should be provided, not just at sweet spots, nonlinearities or failure modes should be addressed so a POSITA can avoid undue trial-and-error. The lessons of the *Amgen* case should be kept in mind: "screen different models/parameters until desired performance is achieved" should be replaced with stepwise protocols and acceptance criteria [11]. And because § 103 and § 112 often converge in practice, ensure that any "routine optimization" story the examiner advances is consistent with your enablement record: if optimization is truly routine, your specification should already teach how to hit the claimed results across the range [11].

At the PTAB and in court, challengers should frame the mismatch—claim breadth outstrips teaching—and use expert testing to show that practicing the edges or representative corners of the scope demands inventive, not routine, effort. *Amgen* is the fulcrum for genus claims [11]; *MagSil* speaks to expansive ranges [86]; *Wands* quantifies undue experimentation [10]. Patentees should map each zone of the claim to disclosure plus rules that carry a POSITA there, and present extrapolation logic, such as control-theory bounds, scaling laws, ablation studies, thereby showing that the teachings are commensurate with scope. In short, the whole claim should be supported or the claim should be narrowed.

The modern "center of gravity" for § 112(a) is unmistakable. Courts are policing the bargain: full-scope enablement demands commensurate, reproducible instruction that allows the public to practice the invention once the patent expires. In medical devices and SaMD, where parameters interact and performance is safety–critical, the winning path is a specification that pairs representative embodiments with portable rules a POSITA can apply to reach the remainder of the scope without undue experimentation.

6.8.8 Result-Oriented Claims

In medical devices and SaMD, it is easy to write claims that sound like clinical goals and hard to write claims that capture the engineering that achieves those goals. Regulators, clinicians, and product teams express themselves in outcomes: fewer false alarms, earlier arrhythmia detection, tighter glucose control, reduced over-infusion. Section 101 does not grant exclusivity for outcomes. It asks what the device or software actually does to reach them. Under the *Alice* [5] and *Mayo* [4] framework, claims that promise the result without reciting the concrete signal processing, control logic, data handling, or device actions that implement it read as abstract. That risk is highest in architectures that spread functionality across sensors, firmware, and cloud services, where verbs like "analyze", "optimize", and "control" can float free of specific mechanisms, unless the claim text ties them to defined inputs, transformations, and outputs.

Result-oriented claims fail because they claim the goal and skip the details that make the claim work. 35 U.S.C. § 101 tolerates functional language, but not when it merely states a desired outcome and leaves the *how* to later ingenuity or routine, postgrant trial-and-error. The *Alice/Mayo* framework makes that divide explicit. At step one, courts ask whether the claim is directed to a patent-ineligible concept, most often an abstract idea or a law of nature that is cloaked in result-oriented phrasing. At step two, they look for an "inventive concept" that concretely implements the idea; bare recitals that something "analyzes," "optimizes," "controls," or "reduces" without saying how to not supply that transformation. The doctrine is ultimately about preemption: a claim to the goal of "achieving X" risks monopolizing *all* ways of achieving X, particularly where X reflects an abstract method of organizing information or a natural law [4, 5].

Federal Circuit decisions map the legal framework. In *Interval Licensing LLC v. AOL, Inc.*, claims to an "attention manager" that displayed content on unused screen space were ineligible because they promised a result, namely providing information to a user, without the concrete rules that accomplish it [91, pp. 1343–1345]. *American Axle & Mfg., Inc. v. Neapco Holdings LLC* held claims ineligible where the supposed advance was "tuning" a driveshaft to attenuate vibration, but the claims gave no specific teaching beyond invoking the natural law that certain structural parameters affect modes and frequencies; reciting the result of improved damping without claiming a particular tuning process could not clear § 101 [92, pp. 1292–1299]. In the software arena, *Free Stream Media Corp. v. Alphonso Inc.* rejected claims that broadly sought targeted content delivery across devices; the claims articulated functional aspirations but not a technological implementation that transformed the abstract idea [93, pp. 1363–1369]. And *In re Killian* reiterated that demands for a result, their determining benefits eligibility by comparing data from two repositories, remain abstract absent claimed steps that do more than instruct a practitioner to apply generic analysis [94, pp. 1380–1387]. Two companion guideposts keep patentees from back-filling missing "how" with prose in the specification: *ChargePoint, Inc. v. SemaConnect, Inc.*, [95, pp. 769–775] and *Accenture Glob. Servs., GmbH v. Guidewire Software, Inc.* [96, pp. 1345–1347], both emphasize that eligibility turns on the *claims*; unclaimed implementation details cannot rescue result-only claiming.

The flip side is equally instructive: when claims pair a functional objective with a concrete, technological way of achieving it, they can pass § 101. In *Contour IP Holding LLC v. GoPro, Inc.*, the court credited claims that did not just aspire to "improved remote viewing," but recited a specific camera architecture that captures high- and low-quality streams in parallel and transmits the lower-bandwidth stream in real time while recording the higher-quality stream locally, thereby solving a bandwidth/latency problem at the device level [7]. The difference is not wordsmithing; it is that *Contour's* claims teach the network and device operations that *do the work*.

These lines matter acutely for medical devices and SaMD, if one is tempting to claim the clinical aim, for example reduce false alarms, detect arrhythmias sooner, maintain glucose within a target, or prevent over-infusion, rather than the engineering that gets you there. A claim to adjust an infusion pump to maintain a patient's parameter within a therapeutic window is result-oriented if it omits the control law, sensed inputs, sampling strategy, and actuator behavior that implements the adjustment. A claim to identify likely atrial fibrillation is abstract if it lists generic operations, for example, receive signals, analyze, classify, without specifying the signal conditioning, feature extraction, model architecture, and decision thresholds that yield the detection under the device's constraints. By contrast, claims that tie detection to a particular signal-processing pipeline, for example multi-lead segmentation, noise gating under defined signal-to-noise ratio ranges, feature vectors with bounded variability, that specify a trained model class with defined inputs and outputs and a training regimen, and that bind the classifier's output to concrete device actions, for example lockouts or pacing-mode transitions, supply the how that § 101 demands.

For prosecutors, the strategy is to convert clinical ambitions into specific and detailed technical recipes. Terminology such as "reduce false alarms" should be replaced with the pipeline that reduces them; "optimize dosage" should be replaced with the estimator and controller that do so under noise, delay, and safety constraints; "improve image quality" should be replaced with the reconstruction algorithm, sampling pattern, and artifact-suppression routine that achieve the improvement on the claimed hardware. Claim language should be drafted in such a way that memorializes those mechanisms, (i.e., state machines, control gains, feature sets, training protocols, buffer structures, communication frames), so that the inventive concept appears on the face of the claims. Preemption should be treated as a drafting smell test: if the claim would block *any* way of achieving the result, narrow it to the way you actually taught. For litigators, the same heuristics run in reverse. A challenger should chart the gap between result and method, and show that the claims aggregate generic verbs without the structures that accomplish them. A patentee should map each functional phrase to specific claimed mechanisms, resisting attempts to characterize the claims at too high a level of abstraction and, where appropriate, invoking cases like *Contour* to demonstrate that the "how" is baked into the claimed architecture.

The goal is simple. Eligibility tolerates functions, but it requires implementations. Claims that promise outcomes without specifying means fail *Alice* [5] step one and contribute no inventive concept at step two; claims that teach how a particular device or software system achieves the outcome, and that recite how in the claims, clear section 101. That is as true for targeted advertising as it is for closed-loop infusion, arrhythmia detection, or MRI-safe pacemaker modes.

6.8.9 Obvious-Type Double Patenting and Terminal Disclaimers

Medical devices and SaMD portfolios are built in layers. A single product line may spawn hardware claims on sensors and actuators, method claims on control and signal-processing pipelines, computer-readable-medium claims on software logic, and follow-on continuations that capture incremental improvements as clinical feedback and regulatory testing refine the design. That staggered maturation creates uneven expiration dates across a family, often because some applications accrue Patent Term Adjustment during protracted examination while others do not, and because one keystone patent may later receive Patent Term Extension after regulatory review. In this environment, obviousness-type double patenting is not a formality. It determines whether a later-expiring family member can stand on its own or must live and die with an earlier patent through a terminal disclaimer, and it can decide whether years of exclusivity tied to software features, cloud workflows, or therapy algorithms survive into the revenue tail of a device platform.

The business stakes amplify the doctrine's legal mechanics. SaMD updates and postmarket modifications frequently lead to continuations that issue late in the life-cycle, which increases exposure under the adjusted-expiration rule and invites challenges that a later patent is not patentably distinct from an earlier sibling. Terminal disclaimers can resolve the timing issue by aligning terms; however, they condition enforceability on common ownership, which directly impacts co-development arrangements, field-of-use licenses, and future corporate transactions. On the other side of the ledger, a well-planned Patent Term Extension on a single, regulatory reviewed device patent can carry the flagship asset through commercialization while related family members are synchronized with terminal disclaimers or distinguished by real claim-scope differences.

Obviousness-type double patenting prevents an unjust extension of patent rights by blocking multiple patents that claim patentably indistinct subject matter. The modern analysis is practical: do staggered expirations in a family allow the patentee to enjoy extra term for essentially the same invention; if so, a terminal disclaimer is often the cure. The Federal Circuit's decision in *In re Cellect, LLC* reanchored that inquiry in the term that actually matters. The court held that when a patent has received Patent Term Adjustment, obviousness-type double patenting must be assessed against the patent's expiration after that adjustment is added and that earlier-expiring family members can serve as obviousness-type double patenting references for later-expiring relatives [97, pp. 1224–1231].

Cellect sharpened portfolio risk. Because the analysis uses the adjusted expiration, a continuation that benefits from substantial Patent Term Adjustment can be vulnerable to earlier-expiring siblings unless the claims are patentably distinct or a terminal disclaimer aligns term and enforceability. The court also made clear that post-grant challengers are not foreclosed simply because an examiner did not raise obviousness-type double patenting during prosecution [97, pp. 1229–1231].

Allergan USA, Inc. v. MSN Laboratories Private Ltd. clarifies timing and posture. There, the district court read *Cellect* to invalidate an earlier, first-filed and first-issued patent based on later-filed, later-issued relatives that expired sooner [98, pp. 1364–1370]. The Federal Circuit reversed, explaining that obviousness-type double patenting should not be used to cut short the full, lawfully adjusted term of the original patent simply because later family members have earlier expirations [98, pp. 1364–1370].

The doctrinal roots for using earlier-expiring family members as references remain *Gilead Sciences, Inc. v. Natco Pharma Ltd.*, which approved using a later-issued, earlier-expiring patent as an obviousness-type double patenting reference against an earlier-issued, later-expiring patent [99, pp. 1212–1218]. And the line between Patent Term Adjustment and Patent Term Extension is critical. The court has insulated a validly obtained Patent Term Extension from obviousness-type double patenting challenges [100, pp. 1371–1376; 101, pp. 1321–1326]. A terminal disclaimer is the practical tool that resolves many obviousness-type double patenting problems. In substance, it is a statement by which the applicant disclaims any portion of the later patent's term that would extend beyond the term of a reference patent and agrees that the later patent is enforceable only while it and the reference patent are commonly owned [102, 103]. The regulation authorizes disclaiming the terminal part of a patent term, and the MPEP explains the required common-ownership provision for nonstatutory double patenting. Filing the disclaimer aligns expiration dates and prevents separate enforcement after a split in ownership.

Two practical consequences follow. First, Patent Term Adjustment should be treated as a double-edged sword. It can add years to a patent, but it also becomes the yardstick for obviousness-type double patenting, which means a later-expiring continuation with Patent Term Adjustment can be vulnerable to earlier-expiring siblings unless the claims are genuinely distinct or a terminal disclaimer is in place. Second, the instinct to hold terminal disclaimers in reserve should be revisited. *Cellect* expressly contemplates filing terminal disclaimers proactively, even without an examiner's rejection, where the family and claim scope suggest that an obviousness-type overlap exists [97, 1230–1231].

Strategy should reflect posture and value of each patent asset. During prosecution, expirations should be aligned where appropriate, and claims differentiated across the family with real technical distinctions. It should be considered whether to enter a terminal disclaimer before allowance to avoid later uncertainty. Remember that a terminal disclaimer typically conditions enforceability on common ownership, which can affect licensing and corporate transactions. [102, 103]. In litigation and PTAB practice, challengers should map the family tree, including expiration dates that account for Patent Term Adjustment, identify the closest earlier-expiring family member, and perform a side-by-side comparison to demonstrate a lack of patentable distinctness. Patent owners should test whether the posture resembles *Allergan*, emphasize genuine claim-scope differences with technical support, and be prepared with terminal-disclaimer options where they preserve value [98].

The through-line is straightforward. Obviousness-type double patenting is about patent term and fairness to the public. After *Cellect*, the adjusted expiration date governs the analysis for patents with Patent Term Adjustment, and earlier-expiring family members can invalidate later-expiring relatives absent a terminal disclaimer or real claim-scope distinctions. After *Allergan*, timing and posture matter so that obviousness-type double patenting does not lop off the full, lawfully adjusted term of a first-filed, first-issued patent.

For medical devices and SaMD, the obviousness-type double patenting and terminal-disclaimer calculus is not academic. Portfolios often include parallel system, method, and computer-readable-medium claims spread across continuations that expire at different times, which makes PTA-driven misalignment common and heightens exposure under *In re Cellect* because the adjusted expiration is the yardstick [97, pp. 1224–1231]. On the other side of the ledger, certain device patents may qualify for Patent Term Extension after regulatory review under 35 U.S.C. § 156; those extensions attach to the extended patent itself and stand on a different footing for obvious type double patenting (ODP) purposes, which is why *Novartis v. Ezra* [100, pp. 1371–1376] and *Merck v. Hi-Tech* [101, pp. 1321–1326] treat PTE differently from PTA. In practice, that means planning which patent is the PTE candidate, aligning the rest of the family with terminal disclaimers where claims are not meaningfully distinct, and recognizing that a terminal disclaimer will condition enforceability on common ownership, which can affect co-development deals, field-of-use licenses, and later corporate transactions. When a first-filed, first-issued device patent sits at the center of the family, *Allergan* reminds courts not to shorten its lawful term simply because later relatives expire earlier, so challengers must still respect the timing and posture [98, pp. 1364–1370]. The practical takeaway for med-device and SaMD teams is to design continuation strategy, claim differentiation, and licensing structures with both PTA-based ODP risk and § 156 opportunities in view, and to use terminal disclaimers proactively when they preserve real value without undermining regulatory-driven term on the keystone asset.

6.9 Patent Term Restoration

6.9.1 Getting Patent Term Adjustment

Patent term adjustment (PTA) is Congress's mechanism for restoring life to a patent when the USPTO takes longer than the statute allows to examine and issue it. For medical device and SaMD portfolios, PTA can add months or even years to a late-issuing continuation that captures software features, cloud workflows, or post-market improvements. It is also a source of portfolio risk because the Federal Circuit measures obviousness-type double patenting against the adjusted expiration, not the unadjusted 20-year date. Understanding how PTA accrues, how applicant actions can reduce it, and how to preserve or challenge it is, therefore, a core skill for health-tech practitioners.

PTA is codified at 35 U.S.C. § 154(b). It adds three categories of USPTO delay, then subtracts overlaps and any delays caused by the applicant.

- **A-delay (the 14-4-4-4 guarantees).** The Office must issue a first Office action within 14 months of the actual U.S. filing date and act within 4 months on most responses, PTAB decisions, and notices of allowance. It must also issue the patent within 4 months after the issue fee is paid. Missed statutory targets generate a day-for-day A-delay—named after the code section § 154(b)(1)(A).
- **B-delay (the 3-year guarantee).** If a patent issues more than three years after the actual U.S. filing date or national-stage entry, the time beyond three years can count as B-delay, subject to carve-outs such as time consumed by continued examination under 35 U.S.C. § 132(b). The Federal Circuit clarified two key points: A and B delays are not fully overlapping and must be calculated separately before overlap is subtracted [104, pp. 1370–1372], and no B-delay accrues while an application is in continued examination, although time after allowance does count even if an RCE occurred earlier [105, pp. 602–605].
- **C-delay.** Time lost to interferences, secrecy orders, and successful appeals adds C-delay, named after code section § 154(b)(1)(C). In practice, many device and

SaMD cases see C-delay when an appeal to the PTAB results in a reversal or in prosecution being reopened.

The Office totals A, B, and C, then subtracts overlapping days and any "applicant delay." The statute reduces PTA for periods when the applicant failed to engage in "reasonable efforts" to conclude prosecution. The implementing rule, 37 C.F.R. § 1.704, lists common examples. These include extensions of time, incomplete or non-compliant submissions that require correction, late after-final practice that does not advance prosecution, and certain supplemental papers that follow an initial response. For global device companies, a common pitfall is information disclosure: submitting an IDS long after a foreign office action or PCT search can result in applicant delay. There is, however, a safe harbor for IDS submissions made within 30 days of first learning of the prior-art citation in a counterpart proceeding if the submission meets § 1.704(d).

Applicant delay cannot include periods when the applicant was unable to act. The Federal Circuit reversed a 646-day charge attributed to the patentee by the USPTO, as there was nothing the applicant could have done to advance the case during that time [106, pp. 1358–1361]. The message for practitioners is simple: calendar response obligations carefully, use the safe harbors when applicable, and challenge any charge that counts time when no applicant action was possible.

The USPTO provides an interim PTA calculation near allowance and a final calculation on the patent's face at grant. If you disagree with the determination, you must first request reconsideration at the USPTO under 37 C.F.R. § 1.705. That request is due within two months of the grant date and may be extended by up to five months with the usual extension fees. If still aggrieved, you may file a civil action in the Eastern District of Virginia within 180 days of the patent's grant [107]. These timelines are unforgiving, so docket them early and verify the Office's day-counts against the file history.

A few calculation details matter in daily practice. For B-delay, the three-year clock runs from the application's actual U.S. filing date or the date of national-stage entry under § 371, not from the earliest priority date. Holidays, mail dates, and entry dates can move the numbers, so check the transaction history carefully. For A-delay, only statutory "actions" count; for example, a timely restriction requirement qualifies as a first action on the merits for the 14-month guarantee, which can erase what looks like delay on the surface.

PTA changes the expiration date of an issued patent, and that adjusted date controls in an obviousness-type double patenting analysis. The Federal Circuit held that earlier-expiring family members can be used as obvious type double patenting references against a later-expiring relative whose term was lengthened by PTA unless the claims are patentably distinct or a terminal disclaimer ties term and enforceability to the earlier patent [97, pp. 1224–1231]. A later case cautions that this logic does not automatically lop off the full, lawfully adjusted term of a first-filed, first-issued patent, which depends on family posture [98, pp. 1364–1370]. For device and SaMD portfolios, that means celebrating PTA on late-issuing continuations, while managing ODP exposure across the family through real claim differentiation or well-timed terminal disclaimers.

Device and digital-health applications often face multiple Office actions, complex § 101 or § 112 issues, and art units with heavy dockets, all of which can significantly delay the issuance of a patent beyond three years. PTA can become a significant component of the commercial runway for features that evolve late, such as cloud-based analytics, alert triage pipelines, or firmware that implements closed-loop control. At the same time, large families of patents with method, system, and computer-readable medium claims tend to issue at different times. PTA on one sibling can extend its term past earlier-filed family members and thereby invite ODP unless the claims are meaningfully distinct or aligned with a terminal disclaimer.

6.9.2 *Getting Patent Term Extension for Medical Devices*

The Hatch–Waxman Act allows for Patent Term Extension (PTE) to be awarded to patent owners by the USPTO in order to extend the duration of a patent beyond the twenty (20) years typically granted for utility patents in situations where the product covered by the patent could not be commercialized due to lengthy regulatory delays. Many medical devices require FDA approval before they can be sold on the market. If a patent issues for the medical device well before the FDA grants approval for sale of the medical device, the patent owner could lose out on some of the patent's protection term. PTE is intended to restore patent owners some of the lost patent term.

In order to be eligible for PTE, a patent must be directed to:

(1) The active ingredient of a new human drug, antibiotic drug, or human biological product (as those terms are used in the Federal Food, Drug, and Cosmetic Act and the Public Health Service Act), including any salt or ester of the active ingredient, as a single entity or in combination with another active ingredient; or
(2) The active ingredient of a new animal drug or veterinary biological product (as those terms are used in the Federal Food, Drug, and Cosmetic Act and the Virus-Serum-Toxin Act) that is not primarily manufactured using recombinant DNA, recombinant RNA, hybridoma technology, or other processes including site-specific genetic manipulation techniques, including any salt or ester of the active ingredient, as a single entity or in combination with another active ingredient; or
(3) Any medical device, food additive, or color additive subject to regulation under the Federal Food, Drug, and Cosmetic Act (37 C.F.R. § 1.710).

The amount of PTE awarded by the USPTO is based on the amount of time the product covered by the patent is in clinical testing and regulatory review. The specific formula applied is based on the type of product being sold.

At a high level, PTE can be summed up as follows: based on how long the regulatory review process took, how long clinical investigations took, and how long the administrative require process took, medical device patent owners seeking PTE may qualify for an extension of their patent's term. However, the amount of PTE is limited in the following two ways:

(1) The PTE cannot extend the term of the patent beyond fourteen (14) years from the date of § 515 (under the Federal Food, Drug and Cosmetic Act) approval by the FDA (applicable to Class III devices subject to premarket approval requirements), and
(2) The PTE cannot extend the term of the patent more than 5 years from the date of the end of the patent term.

37 C.F.R. § 1.777 specifies the particular formula that is used to calculate the amount of PTE that is awarded to medical device patent owners. For medical devices, first the number of days of regulatory review length (R) must be determined. The duration of the regulatory review is determined by adding the length of the clinical investigation (C) to the length of administrative review (F), as shown in the following formula:

$$R = C + F.$$

In this formula, C represents the number of days in the period beginning on the date a clinical investigation involving humans with the device commenced and ending on the date an application was initially submitted for the device under § 515 of the FDCA. In addition, F is the number of days starting from the section § 515 submission and ending on the date such application was approved.

To determine the maximum amount of PTE that can be awarded, the following formula is applied:

$$\mathrm{PTE}_{\mathrm{max}} = \mathrm{floor}\left\{\frac{1}{2}(R - P - D)\right\},$$

where P is the number of days in C and F that occurred prior to the patent issuing. In the above equation, D is the number of days during which the applicant failed to exercise due diligence. The floor operator is being used to show that the number of days awarded will be rounded down to an integer in the event of a fractional day result.

However, the amount of PTE that is awarded cannot extend the term of the patent beyond 14 years from date of § 515 approval (F_{apr}). In addition, the amount of PTE awarded cannot exceed 5 years from the date of the end of the term (P_{term}) of the patent after terminal disclaimers are considered.[3]

[3] A terminal disclaimer is an official statement by a patent owner that the term of the patent in question is disclaimed, i.e., shortened, to less than the twenty (20)-year grant. The patent owner effectively forfeits a portion of the patent's term and allows the patent to enter the public domain.

Accordingly, the patent of a medical device will expire on the soonest of the following:

$$P_{\text{term}} + \text{PTE}_{\text{max}}; \text{ or}$$

$$P_{\text{term}} + 5; \text{ or}$$

$$F_{\text{apr}} + 14$$

A request for PTE must be filed within sixty (60) days of receiving regulatory approval of the commercial marketing application. Section 37 of the America Invents Act further clarified that in the event that the § 515 approval is transmitted after 4:30 PM Eastern Standard Time, on a business day, or is transmitted on a non-business day (such as a weekend or legal holiday), the sixty (60)-day period begins on the following business day. The request for PTE must show that the product is covered by at least one claim of a non-expired patent. The request must also state that the patent has not previously been extended based on a regulatory review. This is an important consideration, as one patent may cover several products. However, the patent will only be extended once, regardless of the number of implicated products.

The request for PTE must also identify the particulars of the regulatory review and include details about the patent and the activities undertaken to secure regulatory approval. Specifically, the PTE request must "clearly explain how each listed claim reads on the approved product [108]." Perhaps, most importantly, it is the PTE applicant's responsibility to perform the PTE calculation—the USPTO will not perform the calculations for the applicant. Once the USPTO has reviewed and approved the PTE request, an appropriate Certificate of Extension is issued.

6.9.3 Patent Term Adjustment Versus Patent Term Extension

Patent term adjustment (PTA) under 35 U.S.C. § 154(b) compensates for USPTO examination delay. It applies to any utility patent, is calculated day for day under the statute with overlap and applicant delay subtracted, and has no hard cap. PTA extends the entire patent without special use limits. Because the adjusted expiration controls for obviousness-type double patenting, PTA can create ODP exposure across a family unless claims are patentably distinct or terms are aligned by terminal disclaimer, which the Federal Circuit confirmed in *In re Cellect* [97, pp. 1224–1231].

Patent term extension (PTE) under 35 U.S.C. § 156 serves a different purpose. It restores time lost to premarket regulatory review at the FDA or USDA for an approved product, including PMA-approved Class III medical devices. At least one claim must cover the approved product, an approved use, or a qualifying manufacturing method, and only one patent can be extended per approval. PTE is based on the regulatory review period with diligence offsets, is capped at five years, and total

Table 6.1 Differences between PTA and PTE

Feature	PTA (Patent term adjustment)	PTE (patent term extension)
Statute	35 U.S.C. § 154(b)	35 U.S.C. § 156
What it fixes	USPTO examination delay	FDA (or USDA[4]) regulatory review delay
Who qualifies	Any utility patent	Only a patent that claims the FDA-approved product, an approved use, or a manufacturing method; one patent per approval
How much time	Day-for-day per statute, minus overlap and applicant delay; no hard cap	Up to 5 years, with total postapproval life capped at 14 years
Scope during added term	Extends the entire patent without special use limits	Enforceable during the extension only as limited by § 156(b) to the approved product/uses (and qualifying processes)
ODP interaction	Can create ODP exposure because the adjusted expiration controls [97]	Insulated from ODP truncation; terminal disclaimers do not cut off a valid PTE [100, 101]

postapproval life may not exceed fourteen years from the date of approval. During the extension, enforceable scope is limited by § 156(b) to the approved product and uses, and qualifying processes. Unlike PTA, a valid PTE is not cut short by ODP, and a terminal disclaimer does not disclaim the extension, as held in *Merck & Co. v. Hi-Tech Pharmacal Co.*, [101, pp. 1321–1326], and *Novartis AG v. Ezra Ventures LLC*, [100, pp. 1371–1376].

Table 6.1 highlights the differences between PTA and PTE.

For medical devices and SaMD, PTA most often lengthens the term of software-heavy continuations that endured slow prosecution, while PTE is available only when the device actually underwent a qualifying regulatory review path. PMA devices are typical PTE candidates; 510(k) clearances rarely support PTE. SaMD qualifies only if its review fits § 156. In practice, device companies plan families with both clocks in view: celebrate PTA where it accrues but manage ODP risk across related patents, and for PMA products identify a clean PTE candidate patent well before approval so the extended term can be secured and used effectively.

6.9.4 Patent Term Restoration Outside the United States: PTA and PTE Analogs in Europe, China, Japan, and Korea

Patent life can be adjusted or extended in a number of major jurisdictions, but the tools and eligibility rules differ in important ways for medical devices and SaMD. In

[4] For veterinary biologics.

Europe, the European Patent Office grants a 20-year term from the filing date under Article 63 EPC and there is no EPO-level patent term adjustment for examination delay. Time lost at the Office is not restored. The European analogue to a U.S. patent term extension is the Supplementary Protection Certificate (SPC), which is granted nationally and attaches to authorized medicinal products for human or veterinary use and to plant protection products under the EU SPC Regulations. An SPC can add up to five years beyond the basic patent term, with a further six-month pediatric add-on in some cases, but its enforceable scope during the extension is limited to the authorized product and uses. Standalone medical devices and SaMD do not qualify because SPCs protect medicinal or plant protection products, not devices regulated under the Medical Devices Regulation. Drug-device combinations can sometimes support an SPC where the medicinal substance provides the principal intended action and the marketing authorization is for a medicinal product, although the protection centers on the medicinal component. The arrival of the Unitary Patent has not altered SPC practice, which remains national, despite the EU's efforts toward a possible Unitary SPC framework.

China now provides both a patent term adjustment and a patent term extension following the 2021 amendments to the Patent Law. Patent term adjustment is available for invention patents that experience unreasonable examination delays, with the typical trigger being a grant more than four years from filing and more than three years from the request for substantive examination, excluding periods caused by the applicant. Applicants must request the adjustment from CNIPA. Patent term extension in China is confined to new drug patents to compensate for the time consumed by NMPA review. The extension can be up to five years, and the total postapproval life may not exceed fourteen years from the marketing approval date. At present, China does not extend the term for medical devices or SaMD, although devices can still benefit from PTA when Office delay is present.

Japan provides both a mechanism to adjust term for Office delay and a robust extension system tied to regulatory review. The JPO grants an adjustment for unreasonable examination delay once statutory thresholds are exceeded, excluding periods attributable to the applicant, and the adjustment is effectively day-for-day for delay caused by the Office. Japan's patent term extension covers pharmaceuticals, agrochemicals, and medical devices that require approval under the Pharmaceuticals and Medical Devices Act. Up to five years can be added, measured by the time during which the patented invention could not be practiced because regulatory approval was pending. Enforcement during the extended period is limited to the approved product, approved indications, and approved methods. Because software can be authorized as a medical device in Japan, SaMD can qualify for an extension where the regulatory pathway fits and at least one claim covers the approved device configuration or method of use.

Korea likewise recognizes both adjustments for office delays and extensions for regulatory delays. KIPO adjusts term when prosecution exceeds prescribed benchmarks and excludes applicant delay. Korean patent term extension is available for pharmaceuticals, agrochemicals, and medical devices that required marketing authorization from the Ministry of Food and Drug Safety, with a maximum of five years of added time. As in Japan, the enforceable scope during the extended period is confined to the approved product and approved uses, and case law addresses how closely the accused product must track the approved configuration. SaMD can be eligible where it is regulated and approved as a device and the patent claims read on the approved article or method.

Table 6.2 summarizes the PTA and PTE Analogs in Europe, China, Japan, and Korea.

Taken together, these regimes create a predictable pattern for device and digital-health portfolios. Europe offers no examination-delay adjustment and reserves SPCs for medicinal products, so standalone devices and SaMD generally receive no post-term relief. China offers both tools, but its extension is drug-only, leaving devices and SaMD to rely on PTA when grant is slow. Japan and Korea offer both adjustment and extension and expressly include medical devices in their extension statutes, which means SaMD may benefit when it is approved as a device and the patent is drafted to cover the approved configuration or use. Across all four jurisdictions, careful docketing to preserve adjustment, early identification of the best extension candidate patent, and claim drafting that maps cleanly to the ultimately approved product or method are the practical levers for maximizing postgrant life outside the United States.

Table 6.2 PTA and PTE analogs in Europe, China, Japan, and Korea

Jurisdiction	PTA for office delay	PTE-type remedy	Devices eligible for PTE	SaMD eligible for PTE
Europe (EPO/EU)	No PTA at EPO	National SPCs for medicinal and plant protection products; up to 5 years (+ 6-month paediatric add-on)	Generally no, unless part of a drug–device combination authorized as a medicinal product	No
China (CNIPA)	Yes, request-based PTA for unreasonable delay	Yes, for **new drugs** only; up to 5 years; total postapproval $\leq$ 14 years	No	No
Japan (JPO)	Yes	Yes, for pharmaceuticals, agrochemicals, and **medical devices**; up to 5 years	Yes	Yes, when approved as a device
Korea (KIPO)	Yes	Yes, for pharmaceuticals, agrochemicals, and **medical devices**; up to 5 years	Yes	Possible, if regulated and approved as a device

References

1. United States Constitution, Article I, Section 8.
2. 35 U.S.C. § 101.
3. *Diamond v. Chakrabarty*, 447 U.S. 303 (1980).
4. *Mayo Collaborative Services v. Prometheus Laboratories, Inc.*, 566 U.S. 66, 72–80 (2012).
5. *Alice Corp. v. CLS Bank Int'l*, 573 U.S. 208, 217–22 (2014).
6. *CardioNet, LLC v. InfoBionic, Inc.*, 955 F.3d 1358, 1368–71 (Fed. Cir. 2020).
7. *Contour IP Holding LLC v. GoPro, Inc.*, 113 F.4th 1373, 1379–87 (Fed. Cir. 2024).
8. *Recentive Analytics, Inc. v. Fox Corp.*, 134 F.4th 1205, 1209–17 (Fed. Cir. 2025).
9. *United States v. Telectronics, Inc.*, 857 F.2d 778 (Fed. Cir. 1988).
10. *In re Wands*, 858 F.2d 731 (Fed. Cir. 1988).
11. *Amgen Inc. et al. v. Sanofi et al.*, 598 U.S. 594 (2023).
12. *Chiron Corp. v. Genentech Inc.*, 363 F.3d 1247, 1254 (Fed. Cir. 2004).
13. *Manual of Patent Examining Procedure* § 2164.
14. *In re Clark*, 522 F.2d 623 (CCPA).
15. *Salazar v. AT&T Mobility, LLC*, 64 F.4th 1311, 1315–17 (Fed. Cir. 2023).
16. *Convolve, Inc. v. Compaq Comput. Corp.*, 812 F.3d 1313, 1321 (Fed. Cir. 2016).
17. *In re Varma*, 816 F.3d 1352, 1362–63 (Fed. Cir. 2016).
18. *Harari v. Lee*, 656 F.3d 1331, 1341–42 (Fed. Cir. 2011).
19. *Baldwin Graphic Sys., Inc. v. Siebert, Inc.*, 512 F.3d 1338, 1342 (Fed. Cir. 2008).
20. *KCJ Corp. v. K-Mart Corp.*, 223 F.3d 1351, 1356 (Fed. Cir. 2000).
21. *CCS Fitness, Inc. v. Brunswick Corp.*, 288 F.3d 1359, 1366 (Fed. Cir. 2002).
22. *Renishaw PLC v. Marposs Società per Azioni*, 158 F.3d 1243, 1249 (Fed. Cir. 1998).
23. *Williamson v. Citrix Online, LLC*, 792 F.3d 1339, 1348–51 (Fed. Cir. 2015) (en banc in relevant part).
24. *Zeroclick, LLC v. Apple Inc.*, 891 F.3d 1003, 1007–10 (Fed. Cir. 2018).
25. *Fintiv, Inc. v. PayPal Holdings, Inc.*, 134 F.4th 1377 (Fed. Cir. 2025).
26. *Noah Sys., Inc. v. Intuit Inc.*, 675 F.3d 1302, 1311–18 (Fed. Cir. 2012).
27. *AllVoice Computing PLC v. Nuance Commc'ns, Inc.*, 504 F.3d 1236, 1245–46 (Fed. Cir. 2007).
28. *Sisvel Int'l S.A. v. Sierra Wireless, Inc.*, 67 F.4th 1372, 1379–84 (Fed. Cir. 2023).
29. *Media Rights Techs., Inc. v. Capital One Fin. Corp.*, 800 F.3d 1366, 1372–76 (Fed. Cir. 2015).
30. *Advanced Ground Info. Sys., Inc. v. Life360, Inc.*, 830 F.3d 1341, 1347–48 (Fed. Cir. 2016).
31. *Rain Computing, Inc. v. Samsung Elecs. Am., Inc.*, 989 F.3d 1002, 1007–08 (Fed. Cir. 2021)
32. *Aristocrat Techs. Austl. Pty Ltd. v. Int'l Game Tech.*, 521 F.3d 1328, 1333–38 (Fed. Cir. 2008).
33. *Traxcell Techs., LLC v. Sprint Commc'ns Co.*, 15 F.4th 1121, 1134–35 (Fed. Cir. 2021).
34. *In re Woodruff*, 919 F.2d 1575 (Fed. Cir. 1990).
35. *In re Peterson*, 315 F.3d 1325, 1330 (Fed. Cir. 2003).
36. *Iron Grip Barbell Co., Inc. v. USA Sports, Inc.*, 392 F.3d 1317, 1322 (Fed. Cir. 2004).
37. *In re Kahn*, 441 F.3d 977, 990 (Fed. Cir. 2006).
38. *In re Cyclobenzaprine Hydrochloride Extended-Release Capsule Patent Litig. Eurand, Inc. v. Mylan Pharms., Inc.*, 676 F.3d 1063, 1082 (Fed. Cir. 2012).
39. *In Gardner v. TEC Syst., Inc.*, 725 F.2d 1338 (1984).
40. *Manual of Patent Examining Procedure* §2144.04 (IV) (A).
41. *Gentry Gallery, Inc. v. Berkline Corp.*, 134 F.3d 1473, 1480 (Fed. Cir. 1998).
42. *Laitram Corp. v. Rexnord, Inc.*, 939 F.2d 1533, 1535 (Fed. Cir. 1991).
43. *Westvaco Corp. v. Int'l Paper Co.*, 991 F.2d 735, 745 (Fed. Cir. 1993) (internal citation omitted).
44. *In re Bigio*, 381 F.3d 1320, 1325 (Fed. Cir. 2004).
45. *Sanofi-Aventis Deutschland GmbH v. Mylan Pharms. Inc.*, 66 F.4th 1373, 1377 (Fed. Cir. 2023).
46. *Airbus S.A.S. v. Firepass Corp.*, 941 F.3d 1374, 1382 (Fed. Cir. 2019).
47. *Scientific Plastic Prods., Inc. v. Biotage AB*, 766 F.3d 1355, 1359–60 (Fed. Cir. 2014).
48. *Donner Tech., LLC v. Pro Stage Gear, LLC*, 979 F.3d 1353, 1360 (Fed. Cir. 2020).

49. *Elekta Ltd. v. Zap Surgical Systems, Inc.*, 81 F.4th 1368 (Fed. Cir. 2023)).
50. *KSR International Co. v. Teleflex Inc.*, 550 U.S. 398, 421 (2007).
51. *Fleming v. Cirrus Design Corp.*, 28 F.4th 1214, 1221–22 (Fed. Cir. 2022).
52. *Ariad Pharmaceuticals, Inc. v. Eli Lilly & Co.*, 598 F.3d 1336, 1351 (Fed. Cir. 2010) (en banc).
53. *Dyfan, LLC v. Target Corp.*, 28 F.4th 1360, 1366–71 (Fed. Cir. 2022).
54. *In re GPAC Inc.*, 57 F.3d 1573, 1579 (Fed. Cir. 1995).
55. *Daiichi Sankyo Co. v. Apotex, Inc.*, 501 F.3d 1254, 1256–57 (Fed. Cir. 2007).
56. *IPXL Holdings, L.L.C. v. Amazon.com, Inc.*, 430 F.3d 1377, 1384 (Fed. Cir. 2005).
57. *Mayo Collaborative Servs. v. Prometheus Labs., Inc.*, 566 U.S. 66, 79 (2012).
58. *Ariosa Diagnostics, Inc. v. Sequenom, Inc.*, 788 F.3d 1371 (Fed. Cir. 2015).
59. *Genetic Techs., Ltd. v. Merial LLC*, 818 F.3d 1369 (Fed. Cir. 2016).
60. *Exergen Corp. v. Kaz USA, Inc.*, Nos. 2016–2315, 2016–2341, 2018 U.S. App. LEXIS 6004 (Fed. Cir. Mar. 8, 2018).
61. *Rapid Litig. Mgmt. v. CellzDirect, Inc.*, 827 F.3d 1042 (Fed. Cir. 2016).
62. *Vanda Pharm. Inc. v. W.-Ward Pharm. Int'l Ltd.*, Nos. 2016–2707, 2016–2708, 2018 U.S. App. LEXIS 9360 (Fed. Cir. Apr. 13, 2018).
63. FDA's guidance for Industry and FDA Staff: *Software as a Medical Device (SAMD): Clinical Evaluation Guidance for Industry and Food and Drug Administration Staff* – FINAL GUIDANCE (December 2017). Available online at https://www.fda.gov/regulatory-information/search-fda-guidance-documents/software-medical-device-samd-clinical-evaluation. Accessed 6 July 2025.
64. *Enfish, LLC v. Microsoft Corp.*, 822 F.3d 1327 (Fed. Cir. 2016).
65. *Intellectual Ventures I LLC v. Capital One Fin. Corp.*, 850 F.3d 1332 (Fed. Cir. 2017).
66. *Elec. Power Grp., LLC v. Alstom S.A.*, 830 F.3d 1350 (Fed. Cir. 2016).
67. See *In re DiStefano*, 808 F.3d 845, 848–51 (Fed. Cir. 2015).
68. *C R Bard Inc. v. AngioDynamics, Inc.*, 979 F.3d 1372, 1381 (Fed. Cir. 2020).
69. *IOENGINE, LLC v. Ingenico Inc.*, 100 F.4th 1395, 1404–12 (Fed. Cir. 2024).
70. *In re Bernhart*, 417 F.2d 1395, 1399–1400 (C.C.P.A. 1969).
71. *In re Lowry*, 32 F.3d 1579, 1583–84 (Fed. Cir. 1994).
72. *Praxair Distrib., Inc. v. Mallinckrodt Hosp. Prods. IP Ltd.*, 890 F.3d 1024, 1032–35 (Fed. Cir. 2018).
73. *In re Ngai*, 367 F.3d 1336, 1338–39 (Fed. Cir. 2004).
74. *King Pharmaceuticals, Inc. v. Eon Labs, Inc.*, 616 F.3d 1267, 1278–80 (Fed. Cir. 2010).
75. C.F.R. § 801.20.
76. C.F.R. § 801.40.
77. C.F.R. § 801.45.
78. FDA Webpage: *Unique Device Identification System (UDI System).* Available online at https://www.fda.gov/medical-devices/device-advice-comprehensive-regulatory-assistance/unique-device-identification-system-udi-system. Updated 22 July 2022. Accessed 30 Aug. 2025.
79. *In Re Gulack*, 703 F.2d 1381, 1386–87 (Fed. Cir. 1983).
80. *Genentech, Inc. v. Hospira*, 946 F.3d 1333, 1341 (Fed. Cir. 2020).
81. *In re Applied Materials, Inc.*, 692 F.3d 1289, 1297 (Fed. Cir. 2012).
82. *E.I. du Pont de Nemours & Co. v. Synvina C.V.*, 904 F.3d 996, 1009 (Fed. Cir. 2018).
83. *In re Antonie*, 559 F.2d 618, 618 (C.C.P.A. 1977).
84. *In re Yates*, 663 F.2d 1054, 1056 (C.C.P.A. 1981).
85. *Genetics Institute, LLC v. Novartis Vaccines & Diagnostics, Inc.*, 655 F.3d 1291, 1306 (Fed. Cir. 2011).
86. *MagSil Corp. v. Hitachi Glob. Storage Techs., Inc.*, 687 F.3d 1377, 1384–85 (Fed. Cir. 2012).
87. *Wyeth & Cordis Corp. v. Abbott Labs*, 720 F.3d 1380, 1384–86 (Fed. Cir. 2013).
88. *Genetics Inst., LLC v. Novartis Vaccines & Diagnostics, Inc.*, 655 F.3d 1291, 1306 (Fed. Cir. 2011).
89. *LizardTech, Inc. v. Earth Res. Mapping, Inc.*, 424 F.3d 1336, 1344–46 (Fed. Cir. 2005).
90. *Fujikawa v. Wattanasin*, 93 F.3d 1559, 1570 (Fed. Cir. 1996).

91. *Interval Licensing LLC v. AOL, Inc.,* 896 F.3d 1335, 1343–45 (Fed. Cir. 2018).
92. *American Axle & Mfg., Inc. v. Neapco Holdings LLC,* 967 F.3d 1285, 1292–99 (Fed. Cir. 2020).
93. *Free Stream Media Corp. v. Alphonso Inc.,* 96 F.3d 1355, 1363–69 (Fed. Cir. 2021).
94. *In re Killian,* 45 F.4th 1373, 1380–87 (Fed. Cir. 2022).
95. *ChargePoint, Inc. v. SemaConnect, Inc.*, 920 F.3d 759, 769–75 (Fed. Cir. 2019).
96. *Accenture Glob. Servs., GmbH v. Guidewire Software, Inc.*, 728 F.3d 1336, 1345–47 (Fed. Cir. 2013).
97. *In re Cellect, LLC,* 81 F.4th 1216, 1224–31 (Fed. Cir. 2023).
98. *Allergan USA, Inc. v. MSN Laboratories Private Ltd.,* 111 F.4th 1358, 1364–70 (Fed. Cir. 2024).
99. *Gilead Scis., Inc. v. Natco Pharma Ltd.,*753 F.3d 1208 (Fed. Cir. 2014).
100. *Novartis AG v. Ezra Ventures LLC*, 909 F.3d 1367, 1371–76 (Fed. Cir. 2018).
101. *Merck & Co. v. Hi-Tech Pharmacal Co.*, 482 F.3d 1317, 1321–26 (Fed. Cir. 2007).
102. C.F.R. § 1.321.
103. *Manual of Patent Examining Procedure* § 804.
104. *Wyeth v. Kappos*, 591 F.3d 1364, 1370–72 (Fed. Cir. 2010).
105. *Novartis AG v. Lee*, 740 F.3d 593, 602–05 (Fed. Cir. 2014)).
106. *Supernus Pharms., Inc. v. Iancu*, 913 F.3d 1351, 1358–61 (Fed. Cir. 2019).
107. U.S.C. § 154(b)(4)(A).
108. C.F.R. § 1.740.

Chapter 7
Trademarks

7.1 Trademarks in Medical Device Technology

Just like any other consumable product or service, medical devices and surgical procedures can be protected with trademarks. Trademarks play a significant role in medical device branding. For instance, a medical device company can build its brand around providing a suite of medical device implants that become well-known for their durability, ease of implantation, and low instance of degradation in vitro. Associating a trademark with this brand will indicate to hospitals, healthcare professionals, and patients the source of the medical devices, and over time will help increase recognizably of the product among consumers and will help consumers to differentiate the product in the mind of consumers.

Another added benefit of developing a strong, trustworthy brand around medical devices is that the knowledge that the brand is reliable and of high quality can influence prescribing decisions. Trademarks associated with medical devices help foster customer loyalty even after patent protections expire.

7.2 What is a Trademark?

The use of markings such as symbols or words to identify the source of goods has been common practice since people first started trading and selling goods such as pottery, weapons, and clothing thousands of years ago. The purpose of these markings, to indicate a product's source and authenticity, has not changed to this day. What has changed, especially in the last one hundred years, is the protection afforded to trademarks. Currently, trademark protection is provided in the United States under common law, state law, and federal law.

The Lanham Act [1] defines federal trademark protection and trademark registration rules. Under the Lanham Act, a trademark is any word, name, symbol, device,

G. B. Halt Jr. et al., *FDA and Intellectual Property Strategies for Medical Device Technologies*, https://doi.org/10.1007/978-3-032-12649-8_7

or any combination thereof that is used to identify and distinguish goods or services of one source from those of another source. In short, a trademark indicates the source of the goods or services. The law also provides protection for other types of marks that are directed to different types of uses. Many of these different types of marks are common in startups, and the below table notes some of the key features of these different types of marks. It is common to informally refer to all of the above terms as trademarks or "brands" (Table 7.1).

A trademark normally consists of a word, logo or some combination of the two. A word mark can include known terms, abbreviations, something coined by the owner, or some combination of letters and numbers. A logo can be a design, stylized lettering, or a drawing of an object. However, there are other types of trademarks, including the following:

- Symbol (e.g., Philips shield emblem [4], or Apple's iconic Apple logo [9]);
- Shape (e.g., the shape of an iPod [10]);
- Slogan (e.g., "A promise for life" [11], or "Just Do It" [12]);
- Sound (e.g., "Hisamitsu" sung over the sound of four musical notes for medicated transdermal patches [13], or the NBC chime [14]);
- Color (e.g., the color red for medical connectors [15] or 3 M yellow Post-it notes [16]).
- Scents (e.g., minty scent for pain-relief patches [17] or fruit-scented industrial lubricants [18]);
- Touch (e.g., "velvet textured" feel on bottle surface for wines [19]);
- Distinctive Packaging (e.g., FlexShield thermoplastic polyurethane film pouch for medical device packaging [20] or black and white cow print pattern on Gateway computer box [21] and uniquely shaped or designed packaging for medications); and
- Building Design (e.g., University of Texas's UT Tower—a clock tower with an observation deck overlooking the UT campus and the city of Austin, Texas [22]).

7.3 Non-protectable Subject Matter

Trademarks are only protectable if they are capable of distinguishing the goods or services of one owner from those of another. Therefore, generic marks cannot be protected and descriptive marks are only protectable with a specific showing that they have acquired distinctiveness. However, there are also additional types of marks that are not protectable.

- **Generic Marks**. A mark that is merely a generic name for the general class of product or service that is synonymous with the general class of product or service is a generic mark.
- **Descriptive Marks**. Marks that merely describe a characteristic of a good or service are consider to be merely descriptive marks. Descriptive marks are often an adjective for the good or service.

Table 7.1 Types of marks

Type of mark	Key features	Example
Service mark	Used to identify and distinguish the source of services	*Stericycle* [2]
House mark	A "house mark" generally refers to a trademark that is used in all facets of a company's business, including business cards, letterhead, packaging, and advertising. Typically, a house mark is also used with a secondary mark or can be used as a primary trademark	The *PHILIPS* word mark [3] is used in conjunction with the Philips shield emblem [4], to create its iconic house mark
Trade dress	Trade dress refers to the overall impression created by a product that can be comprised of any combination of shape, color, design, and wording. If trade dress is functional it cannot be registered or protected. Product design trade dress is not registerable until there is secondary meaning	*Texas Instruments* [5] use of the outline of the state of Texas on its medical device pro-ducts and product packaging
Collective mark	Service mark used by the members of a cooperative, an association, or other collective group or organization, which indicates membership in a union, an association, or other organization	*The Institute of Electrical and Electronics Engineers* (*IEEE*) [6] indicating membership in a professional organization that is scientific and educational. The group works toward advancing the theory and practice of Electrical, Electronics, Communications and Computer Engineering, as well as Computer Science
Certification mark	Mark used to certify regional or other geographic origin, material, mode of manufacture, quality, accuracy, or other characteristics of someone's goods or services, or that the work or labor on the goods or services was performed by members of a union or other organization	*ENERGY STAR* [7] for an international standard for energy-efficient consumer products
Trade names	Used to identify a business or vocation. Trade names that merely identify a business are not registerable under the Lanham Act for federal registration. A trade name can also be a trademark if *used* as a trademark to indicate source. For example, TYLENOL [8] can be both a trademark and a trade name	Trade names are often referred to as brand names in the realm of pharmaceuticals

- **Functional Marks**. Non-traditional marks such as colors and product designs are not protectable if they are, on the whole, functional. For example, the color yellow has been found to be functional when used in conjunction with safety products. Similarly, the ISO color standards for needle gauges are functional colors.
- **Surname**. Marks that are primarily merely surnames are not protectable and absent, a showing of acquired distinctiveness. MCDONALDS and FORD are examples of surnames that have acquired distinctiveness. The VICKS brand trademark is also derived from a surname.
- **Immoral or Scandalous Marks**. The Lanham Act specifically bars immoral or scandalous matter from protection.
- **Likelihood of Confusion**. Proposed use or registration of a trademark may be blocked by a prior trademark holder under common law or federal registration rights if the proposed use is identical or is likely to cause confusion with the existing trademark rights.
- **Geographic Descriptiveness**. Marks that are geographically descriptive cannot be registered until they acquire secondary meaning. An example of such a mark that includes a geographical location, but has acquired secondary meaning is TEXAS INSTRUMENTS.
- **Names and Portraits**. Trademark protection is not available for marks that consist of the name, portrait or signature of a living person without their consent.
- **Dilution**. A trademark that dilutes the distinctive quality of a famous mark is not registerable even if no likelihood of confusion is present.

7.4 Selecting a Trademark

In most cases, the process for selecting a trademark is not simple, since it involves many, sometimes conflicting, priorities. Whether a medical device company is a newly formed startup looking for a new house mark or a multinational organization looking for a mark for a new niche consumer technology product, a process for the selection of a trademark must be in place. Moreover, it is critically important to coordinate marketing efforts with the trademark selection process. That is, there should be no resources spent on marketing for any particular potential trademarks until those trademarks are put through a selection process, and properly protected. The basic tenants of a typical trademark selection process are discussed below.

Example Let's take a look at the trademark selection process that ultimately produced the CardioColander trademark.

7.4.1 Brainstorming Phase

During this phase, creative, marketing, and technical people develop a list of potential marks. If a company does not have the time or personnel to engage in a brainstorming session, it may hire an outside company that specializes in developing potential marks. Although such companies may be a good option, they can be very expensive.

Example Potential trademark names for the LDL cholesterol filter device that were proposed during the brainstorming phase of the selection process included:

- LDL Blood Filter
- Cho-LESS-terol
- LDL Strainer
- The Percolator
- CardioColander

7.4.2 Narrowing Phase

During this phase, either a person or committee narrows the list of marks. Those on the committee should consider at least the following:

- The commercial appeal of the proposed mark.
- The legal strength of the proposed marks. As noted, arbitrary or suggestive marks are typically more easily registered and defended.
- Whether to add a design feature or stylization.
- Whether competitors have similar marks.

Although the marks should be ranked in order of preference, each of the marks on the final list should be acceptable to the startup company as a trademark.

Example Two potential trademark names for the LDL cholesterol filter device were eliminated from the list of potential trademark name options for the following reasons:

- ~~LDL Blood Filter~~—Removed from list because the trademark "LDL Blood Filter" is merely descriptive of the medical device product, thus it is not a very strong trademark. Since the name of the product would not have secondary meaning to consumers, the mark would likely not be protectable.
- Cho-LESS-terol
- LDL Strainer

- ~~The Percolator~~—The trademark name "The Percolator" was removed from the list because the marketing department thought consumers would be confused by the name, since the word "percolation" is usually associated

7.4.3 Knockout Phase

Before going through the expense of a trademark search (which will be described at the next step), it is advisable to perform a "knockout" search to identify any obvious conflicts. A knockout search is a determination as to whether the mark will be difficult to register (because, for example, the mark is scandalous, likely to be confused with some other mark, or the mark is already registered). A knockout search may be performed in-house or by an outside attorney on the available databases of registered marks to eliminate marks that may be difficult to register or that conflict with an existing third party's marks.

Example Two more potential trademark names for the LDL cholesterol filter device were eliminated during the knockout phase:

- ~~Cho-LESS-terol~~—It was discovered during the knockout phase that the trademark "Cho-LESS-terol" was already in use by a small nutraceutical company that uses the trademark on a line of cholesterol reducing food products.
- ~~LDL Strainer~~—Upon further consideration by an IP attorney, it was determined that the trademark "LDL Strainer" is a suggestive trademark name for the LDL cholesterol filter device.
- CardioColander—Everyone involved in the trademark selection process agreed that "CardioColander" would be a strong, suggestive trademark for the LDL cholesterol filter device.

7.4.4 Clearance Search

If a mark passes the knockout phase, a clearance search should be conducted in the countries or jurisdictions of interest. A clearance search assesses the availability of the mark for use or registration. These searches are typically conducted by either outside law firms or in-house personnel with trademark expertise. A search should include federal registrations, state registrations, common law marks, internet domains, and websites.

When considering whether to conduct a full comprehensive search, consider the following points:

- How will the mark be used?
- How widely will the mark be used?
- Will the mark be used on tooling for the product or just on advertisements?
- Will the mark be used on television, radio, social media, etc.?
- How important is the mark for the startup company?
- Will the mark be used in different countries or in different languages?

In setting a budget for these clearance searches, costs are directly proportional to the number of countries for which protection is sought. In most cases, the difficulty of clearing a mark increases as the pool of potential problem marks increases. Therefore, a decision should be made early on as to the intended countries in which the mark will be used. Although using an in-house search may be cost-effective, it is generally recommended to have the clearance search performed by a law firm.

7.4.5 Obtaining a Legal Opinion

The final stage in the selection process is deciding whether to obtain a legal opinion as to the availability of a mark for use and registration in conjunction with a desired mark. Although obtaining an opinion based on the results of a search is not required, there are compelling reasons to obtain such an opinion.

First, it can be difficult to assess search results. This is especially true in an industry where there are many competing products in other industries or where the mark may be considered descriptive.

Second, although the clearance process can be expensive, the failure to clear a mark can be even more costly. The ramifications of selecting a potentially confusing mark are serious. An infringer faces both the prospect of monetary damages and attorney's fees if found to infringe. The resulting embarrassment and loss of marketing momentum, and the finding of infringement would likely result in an injunction prohibiting use of the mark.

Third, obtaining a favorable opinion is strong evidence that there was no bad faith in using the mark and also shows that the brand manager is exercising due diligence. Both of these factors are important to a court when assessing willfulness and to others who may decide to second-guess a trademark's selection.

7.5 Protecting a Trademark

7.5.1 Common Law of Trademark

In the United States, unlike most countries, unregistered trademarks and names enjoy common law protection. This means that the party who adopts and uses a mark in a particular geographic territory is entitled to protection against a subsequent user who adopts the same or similar mark in that same territory. The concept of "territory" is a relatively nebulous concept that depends on the nature and extent of the use of the mark. For example, a medical device business located in New Jersey that has limited distribution or recognition in New England may only acquire trademark rights within a limited radius of its location under common law of trademarks. By contrast, a large medical device company that advertises nationally and has sales throughout the United States may conceivably claim trademark rights throughout the entire United States.

Although reliance on common law rights may offer an initial cost savings, common law rights have the several limitations set forth below:

- Limited to the particular territory where the mark was used.
- An innocent user who obtains a federal trademark registration may take over the rest of the country.
- Establishing common law rights is extremely fact sensitive. Accordingly, such rights can be difficult and extremely expensive to prove in court.

Example Medicus Societas announced in July 2018 its entry into the cardio stent industry. Medicus Societas now offers patients the patented Cardio-Colander LDL cholesterol blood filter device. The product was first launched in Philadelphia, Pennsylvania. Medicus Societas typically used the Cardio-Colander to describe its LDL cholesterol blood filter device.

Medicus Societas did a trademark search of the word "CardioColander" but decides to delay filing a federal trademark application for the product name. Instead, Medicus Societas immediately starts selling the product to hospitals and treatment centers in Philadelphia and New England. After six months of better than expected sales and positive patient and physician reviews, Medicus Societas decides to launch the product nationally and seek federal trademark protection. Unbeknownst to Medicus Societas, however, a small biotechnology startup company based in Seattle innocently launches LDL cholesterol blood testing product that they call the CardioColander, because their LDL cholesterol blood testing product filters through the components of blood to detect LDL levels.

As part of its marketing efforts, the small startup biotechnology company advertises in all the major medical publications in Washington State. Due to demand, all major hospitals and laboratories in Washington carry the Seattle biotechnology company's blood testing product.

Under this scenario, even though Medicus Societas is the senior party with first use, it is likely that the Seattle company will be able to continue to use of the "CardioColander" mark in the limited market of Washington and perhaps in some areas of the surrounding states.

7.5.2 Federal Trademark Protection

Although a trademark owner can simply acquire common law geographic trademark rights through the use of a mark, there are a number of advantages in filing for a federal trademark registration:

- Right to use the ® symbol with all federally registered marks. This symbol can have potent deterrent effects.
- Provides constructive notice to the public of the claim to ownership of the mark. Federal registration makes it much more difficult for a party to plead innocent infringement. Also, federal registration provides a basis for the United States Patent and Trademark Office (USPTO) to reject confusingly similar marks. Common law marks cannot be cited by the USPTO to deny registration.
- Confers nationwide priority of rights effective from the U.S. application filing date. This may be the most important advantage. With this right, unlike with common law, a trademark owner does not have to prove use in a particular state or states in order to claim trademark rights.
- The legal presumption of the registrant's ownership of the mark, its validity, and the registrant's exclusive right to use the mark nationwide. With this right, the trademark owner once again does not have to prove rights in the mark, as they are presumed.
- Ability to bring an action concerning the mark in federal court and the possible recovery of treble (i.e., triple) damages and attorney's fees.
- U.S. registration may serve as a basis to obtain registration in foreign countries without first using the mark in that foreign country.
- Ability to file the U.S. registration with the U.S. Customs Service to prevent importation of infringing foreign goods.
- Availability of incontestability status after five (5) years of continuous use and registration.

Although a trademark owner may gain some rights through use of a mark, it is recommended that medical device companies file a trademark application as soon as possible and early in the development of their product.

7.5.2.1 The Federal Trademark Application Requirements

The requirements for filing a trademark application are relatively straightforward. In order for an application to receive a filing date, it must include:

(1) the required filing fee for at least one class of goods or services;
(2) the name of the applicant;
(3) the name and address for the applicant or attorney for communication;
(4) a clear drawing of mark to be registered; and
(5) the identification of the goods and/or services that the mark will be used with.

The most complex part of filing an application is preparing the identification of goods and/or services for which trademark protection is being sought. The goal is to draft an identification of goods and services as broadly as possible because the identification cannot later be expanded. The USPTO requires that the identification be specific and definite. Moreover, if use is claimed, it is important that the mark be used on all of the goods or services, and that there be a *bona fide* intent to use the mark if the application is filed on an intent-to-use basis. If these requirements are not met, any subsequent registration could be subject to cancellation for fraud on the USPTO.

There are four bases on which to register a mark:

(1) Actual use in commerce
(2) *Bona fide* intent to use the mark in commerce
(3) Foreign registration—this is only available to companies domiciled outside the United States, and
(4) Under the Madrid Protocol—this is only available to companies domiciled outside the United States.

In order for an applicant to claim use in commerce as the basis for an application, the mark must have been in use as of the application date. The date of first use is the date on which the goods were first sold or transported in interstate commerce or the services first rendered anywhere in the world in an arm's length transaction. The date of first use in commerce is the date that the goods are either sold or transported in commerce, such that they could be regulated by applicable laws. For services, the date of first use in commerce is the date the mark is first used or displayed in sales or advertising of services, and the services are rendered in interstate commerce.

Once the trademark owner decides to adopt and use a specific trademark, an intent-to-use application should be filed as soon as possible in order to gain the advantage of a constructive use date. Constructive use refers to the presumption that is afforded by the act of registration that the applicant is treated as though the mark was in use across the country in connection with the registered goods or services identified in the application at the time that the application was made.

7.5.2.2 Examination

Following the filing of a trademark application, the USPTO assigns it to a trademark examining attorney who examines the mark to determine whether it is entitled to registration. The trademark examining attorney conducts a trademark search to determine if the mark is likely to cause confusion with any other mark on the Principal Register and reviews the application for compliance with the Trademark Act and USPTO Rules.

If for any reason the trademark examining attorney determines that the mark is not registerable for any reason, the trademark examining attorney will issue an Office Action that advises the applicant of all grounds of refusal and all matters that require further action. The applicant has six (6) months to respond to the Office Action. This six (6)-month period runs from the mailing date of the Office Action. Failure to fully respond to the Office Action within the statutory period results in the application becoming abandoned.

Once the applicant has had the opportunity to respond to all issues raised in the Office Action, the examining attorney issues a final communication that either allows or finally rejects the application. After a negative final office action, the applicant may file a request for reconsideration and submit additional evidence and argument in order to persuade the examining attorney to withdraw the final office action. If the examiner fails to withdraw the final office action within six (6) months, the applicant must either meet every requirement of the office action or appeal to the Trademark Trial and Appeal Board in order to avoid abandonment of the application.

Following examination, if it appears that a mark is entitled to registration on the Principal Register and there are no outstanding requirements or refusals, the examining attorney will approve the mark for publication in the USPTO *Official Gazette*. If a third party does not file an opposition within thirty (30) days of publication or request a time extension to file an opposition, registration of the application will proceed.

For applications based on intent to use and where no Amendment to Allege Use has been filed before publication, the USPTO will issue a Notice of Allowance. The application will proceed to registration upon the filing of a Statement of Use. A Statement of Use or request for extension of time to file a statement of use must be filed within six (6) months of the mailing date of the Notice of Allowance. The applicant may request an extension for a six (6)-month period without showing good cause. Thereafter, the applicant may receive an additional six (6)-month extension upon request and by a showing of good cause.

7.5.3 State Registration

State laws also provide for trademark registrations. However, given the many advantages to Federal Registration, there is little point to obtaining a State Registration unless the trademark owner cannot establish use in interstate commerce (i.e., all of the applicants' use of the mark is within a single state), or has a specific legal need to take advantage of that state's trademark or anti-dilution remedies.

7.5.4 International Protection

Trademark protection is also available internationally. Unlike the U.S., most countries award trademark rights solely on a first-to-file basis. Therefore, it is crucial to consider filing a trademark application in other countries. A U.S. registration can form the basis of an application in a foreign country without the necessity of first having used the mark in that country. In most countries, a U.S. company can file a trademark application that claims the same U.S. application date if the trademark application is filed within six (6) months of the U.S. filing date. However, foreign trademark protection is often very expensive and the costs multiply depending on the jurisdictions in which protection is sought. Accordingly, a company should develop a list of countries for which registration will be sought.

When considering which countries to pursue protection, the following factors should be evaluated.

- Serious consideration should be given to registering in those countries where the mark is used, or will be used in the near term.
- Consideration should also be given to countries where the startup company is planning on expanding in the next 3–5 years. This is especially true for large markets such as China, which uses a first-to-file system.
- Finally, if counterfeiting is a problem, consideration should be given to defensive filings in some of the key counterfeiting source nations such as Taiwan, China, and Viet Nam.

Once the decision is made, a trademark owner has several options for filing overseas. In many cases, the filing is done in each individual country, which requires hiring a trademark attorney in each country and paying filing fees in each country. However, there are some international treaties that allow a trademark owner to avoid some of these fees and costs.

7.5.4.1 Madrid Protocol

Using the Madrid Protocol can reduce foreign trademark filing costs. The Madrid Protocol is an international trademark treaty that permits the owner of a "home country" registration to file an international application with its national trademark office that designates other member countries. The Madrid Protocol offers cost savings and increased efficiency for U.S. trademark holders. The International Trademark Association [23] has aptly summarized the benefits of the Madrid Protocol as offering:

- one application
- in one place
- with one set of documents
- in one language
- with one fee
- resulting in one registration
- with one number
- and one renewal date
- covering more than one country.[1]

The cost savings of registration through the Madrid Protocol are significant. Another advantage of the Madrid Protocol is the simplicity in filing application amendments. Without the Madrid Protocol, applications would have to be filed and prosecuted individually in every country in which the mark is registered. However, the Madrid Protocol simplifies this process at a reduced cost.

Other advantages of an International Registration under the Madrid Protocol include having priority of protection in all designated countries from the date of international registration, as opposed to the date of registration in the individual countries. Also, the Madrid Protocol limits the time a national office has to act once it receives a request for extension of a Madrid registration. If the office does not act to oppose protection during the allotted time, the registration is automatically granted.

By offering simultaneous registrations in the U.S. and foreign countries, the Madrid Protocol also reduces trademark piracy. Without the Madrid Protocol, individuals in foreign countries are often free to register a U.S. company's trademark, and attempt to later sell the mark to the U.S. company at highly inflated prices. Registration under the Madrid Protocol reduces such piracy since all designated countries are given the same priority date.

Notwithstanding the cost savings and increased efficiency associated with the Madrid Protocol, there are some drawbacks for U.S. applicants. First, the Madrid Protocol requires that the scope of goods or services covered by the registration be limited to the home country's registration rules. U.S. applicants who seek registration through the Madrid Protocol will be prejudiced in this respect since the USPTO

[1] The member countries to the Madrid Protocol include: Afghanistan, African Intellectual Property Organization, Albania, Algeria, Antigua and Barbuda, Armenia, Australia, Austria, Azerbaijan, Bahrain, Belarus, Belgium, Bhutan, Bosnia and Herzegovina, Botswana, Brunei Darussalam, Bulgaria, Cabo Verde, Cambodia, Canada, Chile, China, Columbia, Croatia, Cuba, Curacao, Cyprus, Czech Republic, Democratic People's Republic of Korea, Denmark, Egypt, Estonia, Eswatini, European Union, Finland, France, Gambia, Georgia, Germany, Ghana, Greece, Hungary, Iceland, India, Indonesia, Iran, Ireland, Israel, Italy, Jamaica, Japan, Kazakhstan, Kenya, Kyrgyzstan, Lao People's Democratic Republic, Latvia, Lesotho, Liberia, Liechtenstein, Lithuania, Luxembourg, Madagascar, Malawi, Malaysia, Mauritius, Mexico, Monaco, Mongolia, Montenegro, Morocco, Mozambique, Namibia, Netherlands, New Zealand, North Macedonia, Norway, Oman, Pakistan, Philippines, Poland, Portugal, Qatar, Republic of Korea, Republic of Moldova, Romania, Russian Federation, Rwanda, Samoa, San Marino, San Tome and Principe, Serbia, Sierra Leone, Singapore, Sint Maarten, Slovakia, Slovenia, Spain, Sudan, Sweden, Switzerland, Syrian Arab Republic, Tajikistan, Thailand, Trinidad and Tobago, Tunisia, Turkey, Turkmenistan, Ukraine, United Arab Emirates, United Kingdom, United States of America, Uzbekistan, Viet Nam, Zambia, and Zimbabwe.

requires more detailed identification of goods and services than most other countries. Unlike some other countries, the USPTO will not accept registration of marks for broad classes of goods and services. Other countries, such as Egypt, allow for registration of a mark for a whole class of goods or services (i.e., clothing generally), without specifying a good or service within the class (i.e., t-shirts specifically). Therefore, U.S. companies may limit the scope of protection that could otherwise be obtained in other countries by filing an International Registration as opposed to filing individual national applications.

Another potential consequence of the Madrid Protocol for U.S. trademark owners is the limitation that the USPTO imposes on applicants to provide a statement of use or *bona fide* intent to use the mark in commerce before obtaining a filing date, and proof of use in commerce before a registration will issue. Most other countries do not require a similar statement of use or intent to use the mark in commerce, or proof of such use. Therefore, U.S. trademark applicants may be disadvantaged by their inability to "reserve" a trademark under USPTO procedure.

Trademark owners filing under the Madrid Protocol are also subject to "central attack." If a home country application or registration is cancelled or abandoned during the first five (5) years of registration, whether completely or partially, the home country must notify World Intellectual Property Organization (WIPO). The International Registration then lapses with respect to all designated countries. This is particularly disadvantageous to U.S. trademark owners because there are usually more grounds for challenging registrations under U.S. law than in other countries.

The Madrid Protocol provides for a partial safeguard against "central attack" by providing a three (3)-month grace period for the owner of a cancelled registration to file national applications in designated countries that enjoy the same priority as the International Registration. This process, however, can be costly and time-consuming.

Unlike national application systems, under the Madrid Protocol an assignment may only be recorded if the assignee is itself qualified to file a Madrid Protocol application. Although this only affects the recordation of the assignment, and national laws will govern the legal effect of the assignment, member countries such as the U.S. have passed laws that make assignments to non-member citizens invalid. This assignment provision may be problematic in cases where a U.S. citizen or corporation wishes to assign a registration to a non-member citizen or corporation for tax or other purposes.

The final drawback of implementation of the Madrid Protocol for U.S. trademark owners is that the system is outside of the U.S. and the procedures can seem unfair for those used to U.S. filings. For example, the time period for responding to office actions under the Madrid Protocol may be quite short due to the fact that an office action is sent from the national office to WIPO and WIPO sends it to the trademark owner. Moreover, many Madrid Protocol countries do not send a Registration Certificate or other notice once the registration issues. Therefore, the trademark owner is left in the uncertain position of not knowing for sure if the registration has actually issued.

7.5.4.2 European Union Trademark

The European Union offers a trademark system that allows for registration of a trademark in all of the twenty-seven (27) member countries for one application filing fee.[2] If the trademark owner intends to use the mark in more than two (2) European Union member countries, it is typically more cost-effective to file for a European Union Trademark registration. Although a European Union trademark can be canceled for five (5) years of non-use, use in one country is enough to satisfy the use requirement.

7.6 Maintaining Trademark Rights

Trademark rights can be one of a medical device company's most valuable assets, especially early on in the development of the company. Like any asset, these rights need to be maintained. Trademark rights are lost when the mark no longer acts as an identifier of the goods or services. This can occur through abandonment from non-use or through a course of conduct, including acts of commission or omission, which allow a mark to become the generic name of goods or services or lose significance as a mark.

7.6.1 Maintaining Federal Registration

Section 8 of the Lanham Act requires that an affidavit or declaration verifying continued use in commerce or excusable non-use due to special circumstances be filed with the USPTO between the fifth and sixth anniversary and on or between the ninth and tenth anniversary date of the registration. This requirement applies to all registrations. Failure to meet this requirement will result in the cancellation of a registration.

The duration of a trademark registration has varied over the years. Since November 16, 1989, however, all registrations that have been issued or renewed after that date have only a ten (10)-year term. Therefore, in order to maintain a federal trademark registration, a renewal application must be filed on or between the ninth and tenth anniversary date of the registration.

---

[2] Member countries include: Austria, Belgium, Bulgaria, Croatia, Cyprus, Czech Republic, Denmark, Estonia, Finland, France, Germany, Greece, Hungary, Ireland, Italy, Latvia, Lithuania, Luxembourg, Malta, Netherlands, Poland, Portugal, Romania, Slovakia, Slovenia, Spain, and Sweden.

7.6.2 Licensing Trademarks

A trademark owner can license its trademark to a third party. However, a trademark owner must reserve the power to exercise quality control over the nature and quantity of the goods and services in a license. If the trademark owner fails to exercise such quality control, the license is considered a "naked license" and the mark may be abandoned.

7.6.3 Assignment of Trademarks

A trademark may only be assigned to another party with the goodwill of the business in which the mark is used. Failure to meet this requirement can result in trademark becoming void. Similarly, intent to use applications cannot be assigned prior to the filing of a Statement of Use or Amendment to Allege use unless the assignment is to the successor to the ongoing and existing business of the applicant or to the portion to which the mark pertains.

7.6.4 Policing Trademark Use

The following tips can help a medical device company protect the value of a trademark.

- Do not trademark the generic name of your product. If you cannot provide a generic name of your product without referring to your brand, then the trademark may become generic. For example, aspirin used to be a trademark name for a non-steroidal anti-inflammatory drug product, but over time people began generically referring to the compound of salicylate as "aspirin," and the mark became generic. The product name aspirin no longer carries trademark protection.
- When using a trademark in a sentence, always use the trademark as an adjective. Examples of this include Stryker's IDEAL EYES [24] Laparoscopes, EPIPEN [25] Auto-injectors, and VENTILAIR® [26] II Compressor. If this rule is not followed, the public may come to see the trademark as the generic name of the product or service, which is what happened to former trademarks such as escalator, cellophane, and kerosene. Use of the word "brand" can also help emphasize that a term is a mark and not a generic descriptor, for example: ARROW® [27] brand medical and surgical procedure products. There is one caveat to this rule—many companies use their trade names as trademarks. In those instances where the trade name is being used, it is a proper noun, not an adjective. For example, one of Takeda's marketing slogans incorporates its trade name OTSUKA [28] into the slogan as a proper noun: "Otsuka-people creating new products for better health worldwide."

- In a sentence, the trademark should be set apart from the text in some fashion. This can be accomplished by using ALL CAPITAL LETTERS; **bold face type**; Initial Capital Letter; *italics*; or through the use of a unique font.
- Monitor Use of Your Mark. If you notice your mark appearing in all lower-case lettering in publications, it is a danger sign. Steps should be taken to send notifications to advertisers using the mark in that way.
- Police Misuse. If your trademark is found in a dictionary, whether intentional or unintentional, it is strong evidence that the trademark is generic. Corrective action in the form of a letter to the publisher should be taken immediately. Likewise, if you see your mark appear in lower case letters, this is a danger sign.
- Use the brand in a consistent manner. Not only is recognition of the mark enhanced through consistent use, but inconsistent use may also confuse consumers, dilute the distinctiveness of the mark, and lead to abandonment of the mark.
- Use the appropriate trademark designation. In the United States, the symbol TM (^{TM}) can be used to identify an unregistered trademark, SM (^{SM}) can be used to identify an unregistered service mark, and ® can be used to identify a registered trademark or service mark. Many foreign countries use similar terms. Local laws should be consulted because many countries require proper use of the symbol in their country in order to collect damages for infringement. Such symbols may not be the same as those accepted in the United States.
- Develop a trademark usage manual. All medical device companies should have a manual containing trademark usage guidelines to advise employees and others on the proper use of the company's trademarks.
- Non-use. A trademark owner must use the mark in order to maintain protection. Three (3) years of non-use results in a presumption of abandonment.
- Infringement. The key function of a trademark is as a source identifier. If the same or similar trademark is used by more than one (1) company on the same or related goods, the mark may cease to be a source identifier. Therefore, it is important for a company to police third-party usage of its marks and take appropriate action ranging from cease and desist letters to legal action.

7.6.4.1 Genericide

There are many examples where once valuable trademarks have become generic (i.e., the trademarks have ceased to function as a source indicator and have become the name of a particular type of a product). The following terms were at one time a company's trademark:

- ALE HOUSE (for restaurant and bar services) [29]
- ASPIRIN (for acetyl salicylic acid pain reliever) [30]
- CELLOPHANE (for transparent cellulose sheets and films) [31]
- COLA (for soft drink) [32]
- CRAB HOUSE (for seafood restaurant) [33]
- CUBE STEAK (for steaks) [34]
- DRY ICE (for the solid form of carbon dioxide) [35]

- ESCALATOR (for moving stairs) [36]
- FONTINA (for cheese) [37]
- HEROIN (for injectable diacetylmorphine opioid)
- HONEY BROWN (for a brown ale made with honey) [38]
- JUJUBES (for gum candy) [39]
- LIGHT BEER (for beer with fewer calories) [40]
- MONTESSORI (for educational services) [41]
- MURPHY BED (for folding bed) [42]
- SOFTCHEWS (for chewable medical pills) [43]
- SUPER GLUE (for glue) [44]
- SURGICENTER (for surgical center) [45]
- TOUCH TONE (for phones) [46]
- TRAMPOLINE (for jumping and gymnastic equipment) [47]
- ZIP CODE (for mailing designations) [48].

These terms are now generic and available for use by all. If a trademark owner wishes to avoid that result, it is essential that all those associated with a trademark understand the requirements for maintaining trademark protection and distinctiveness.

References

1. U.S.C. §§ 1051–1127.
2. U.S. Registration No. 4,498,292. Owner: Stericycle, Inc.
3. U.S. Registration No. 3,864,315. Owner: Koninklijke Philips N.V. Limited Liability Company.
4. U.S. Serial No. 86,895,722. Owner: Koninklijke Philips N.V. Limited Liability Company.
5. U.S. Registration No. 3,717,043. Owner: Texas Instruments, Inc.
6. U.S. Registration No. 1,770,511. Owner: Institute of Electrical and Electronics Engineers, Inc.
7. U.S. Registration No. 1,999,485. Owner: Environmental Protection Agency.
8. U.S. Registration No. 890,360. Owner: McNeil Laboratories, Inc.
9. U.S. Registration No. 3,078,580. Owner: Apple Computers, Inc.
10. U.S. Registration No. 3,341,214. Owner: Apple, Inc.
11. U.S. Registration No. 4,064,074. Owner: Abbot Laboratories.
12. U.S. Registration Nos. 1,875,307; 1,931,937 and 1,817,919. Owner: Nike, Inc.
13. U.S. Registration No. 12,814,082. Owner: Hisamitsu Pharmaceutical Co., Inc.
14. U.S. Registration No. 916,522. Owner: NBC Universal Media, LLC.
15. U.S. Supplemental Registration No. 3,976,092. Owner: Masimo Corporation.
16. U.S. Registration Nos. 2,402,722; 2,371,084. Owner: 3M Company.
17. U.S. Registration No. 3,589,348. Owner: Hisamitsu Pharmaceutical Co.
18. U.S. Registration No. 2,463,044. Owner: Mike Mantel.
19. U.S. Registration No. 3,155,702. Owner: American Wholesale Wine & Spirits, Inc.
20. U.S. Registration No. 4,456,543. Owner: UFP Technologies, Inc.
21. U.S. Registration No. 1,725,231. Owner: Acer Inc.
22. U.S. Registration Nos. 1,230,438 and 3,148,092. Owner: The Board of Regents of The University of Texas System.
23. International Trademark Association, *The Madrid Protocol: Impact of U.S. Adherence on Trademark Law and Practice*, at p. 1 (Revised April 2003).
24. U.S. Registration No. 3,764,320. Owner: Stryker Corp.

25. U.S. Registration No. 1,124,454. Owner: Mylan Inc.
26. U.S. Registration No. 3,910,329. Owner: Hamilton Medical AG.
27. U.S. Registration No. 2,176,791. Owner: Arrow International Investment Corp.
28. U.S. Registration No. 3,578,198. Owner: Otsuka Holdings Co., LTD.
29. *Ale House Management, Inc. v. Raleigh Ale House, Inc.*, 205 F.3d 137 (4th Cir. 2000).
30. *Bayer Co. v. United Drug Co.*, 272 F. 505 (S.D.N.Y. 1921).
31. *DuPont Cellophane Co., Inc. v. Waxed Products Co.*, Inc., 85 F.2d 75 (2d Cir. 1936).
32. *Dixi-Cola Laboratories v. Coca-Cola Co.*, 117 F.2d 352 (4th Cir. 1941).
33. *Hunt Masters, Inc. v. Landry's Seafood Rest., Inc.*, 240 F.3d 251 (4th Cir. 2001).
34. *Spang v. Marzall*, 104 F. Supp. 126 (D.D.C. 1952).
35. *Dryice Corp. of Am. v. Louisiana Dry Ice Corp.*, 54 F.2d 882 (5th Cir. 1932).
36. *Haughton Elevator Co. v. Seeberger*, 85 U.S.P.Q. 80 (Comm'r. Pat. 1950).
37. *In re* Cooperativa Produttori Latte E Fontina Valle D'Acosta, 230 USPQ 131 (TTAB 1986).
38. *Genesee Brewing v. Stroh Brewing*, 124 F.3d 137 (2d Cir. 1997).
39. *Henry Heide, Inc. v. George Ziegler Co.*, 354 F.2d 574 (7th Cir. 1965).
40. *Miller Brewing Co. v. G. Heileman Brewing Co.*, Inc., 561 F. 2d 75, 79–81 (7th Cir. 1977).
41. *American Montessori Society, Inc. v. Association Montessori Internationale*, 155 U.S.P.Q. 591 (1967).
42. *Murphy Door Bed Co., Inc. v. Interior Sleep Systems, Inc.*, 874 F.2d 95, 101 (2d Cir. 1989).
43. *Novartis Consumer Health, Inc. v. McNeil-PPC, Inc.*, 1999 U.S. Dist. LEXIS 20981 (D.N.J. 1999).
44. *Loctite Corp. v. National Starch & Chem. Co.*, 516 F.Supp. 190 (S.D.N.Y. 1981).
45. *Surgicenters of America, Inc. v. Medical Dental Surgeries, Co.*, 601 F.2d 1011 (9th Cir. 1979).
46. U.S. Registration No. 0,737,312. Owner: AT&T Co. Canceled: March 13, 1984.
47. *Nissen Trampoline Co. v. American Trampoline Co.*, 193 F. Supp. 745 (S.D. Iowa 1961).
48. U.S. Registration No. 1,042,499. Owner: United States Postal Service. Expired: April 7, 1997.

Chapter 8
Trade Secrets

8.1 Secrecy Creates Value: How Medical Device Innovators Utilize Trade Secret Protection

A trade secret can have immense value to a medical device company. Trade secret protection is one of the most affordable forms of intellectual property (IP) protection. When used appropriately, trade secret information can increase the value of a medical device business.

Trade secrets can generate value for medical device manufacturers in a number of ways. The value could be derived from having information that has economic value on its own so long as few people know about it (for example, having a secret formulation for a coating on a medical implant). But sometimes value can be derived from what others do not know does not work. Years of hard work determining what does not work can have its own value (for example, years of trial and error refined a manufacturing process down to an economic and ecological success). Trade secrets can create cost savings, can provide a medical device company with a strategic advantage in the marketplace over competitors, and can even be used as a marketing tool.

For many medical device makers—especially for those that are just starting out, trade secret protection is an attractive intellectual property protection strategy since it is affordable compared to patent protection. Trade secret protection could be useful at the outset of a company's development, but certain IP, such as inventions, may need to be protected by a patent later down the road. In this way, trade secret protection of an invention may serve as a stopgap measure until the company has finalized or perfected the invention that needs to be patented, or until the company can secure financing to file for a patent application. Trade secret protection is also a good option when other forms of IP protection are unavailable for the specific information being protected.

G. B. Halt Jr. et al., *FDA and Intellectual Property Strategies for Medical Device Technologies*, https://doi.org/10.1007/978-3-032-12649-8_8

8.2 What Are Trade Secrets?

Unlike patent law, which has its roots firmly grounded in federal constitutional and statutory law, trade secret law is a state law doctrine that developed out of the common law doctrine of unfair competition and unfair business practices. Until passage of the Uniform Trade Secrets Act (UTSA) in 1985, trade secret law varied significantly from state to state. The UTSA is a model law that provides a uniform definition of trade secrets and misappropriation, and 48 states, the U.S. Virgin Islands, and the District of Columbia, have adopted it.

The UTSA defines a trade secret as "information, including a formula, pattern, compilation, program, device, method, technique, or process, that: (i) derives independent economic value, actual or potential, from not being generally known to, and not being readily ascertainable by proper means by, other persons who can obtain economic value from its disclosure or use, and (ii) is the subject of efforts that are reasonable under the circumstances to maintain its secrecy." This broad definition maintains the common law that nearly any type of business information can qualify as a trade secret. Thus, information that is not otherwise patentable can be trade secret. Examples of information that can be protected by trade secret include:

- Computer programs
- Client identities
- Product pricing
- Manufacturing processes
- Technical information
- Drawings
- Strategic plans
- Schematics
- Financial statements
- Employee salaries
- Customer lists
- Vendors
- Market analysis
- Formulas
- Product testing results
- Prototypes
- Company manuals
- Product ingredients
- Employee records
- Market strategies

Because information of nearly any type of subject matter can qualify as a trade secret, the UTSA definition of a trade secret focuses on:

(1) The economic value of the trade secret
(2) Whether the trade secret is generally known or readily ascertainable, and
(3) The efforts taken to maintain secrecy.

The "economic value" requirement under the UTSA refers to whether a competitor would obtain an economic benefit if the trade secret information became readily accessible. "Economic value" can be shown by the time and effort utilized in creating the trade secret, or by showing that a third party would have to spend time and effort in creating the same trade secret.

The second requirement for a trade secret under the UTSA is that the information cannot be "generally known or readily ascertainable." This means that the information cannot be already known by the public or by competitors. Whether a trade secret is "generally known or readily ascertainable" is a factual inquiry that depends on the amount of time, effort, and money required to independently produce the trade secret, or to reverse engineer the trade secret. Information cannot be protected by a trade secret if it can be discovered by examining a commercially available product that incorporates the information. If the trade secret is hidden in a commercially available product, then the trade secret can be maintained. A trade secret that consists of the amounts and ratios of individual ingredients in a product or code embedded in a software program is not lost just because the product becomes publically available.

Published information, such as that disclosed in a book, magazine, trade publication, website, or other media, cannot be maintained as a trade secret because it is "generally known" and readily ascertainable. This can be particularly important when deciding whether to keep information as a trade secret or to pursue patent protection for that information. Anything disclosed in a patent or published patent application is generally known and readily ascertainable and cannot be protected as a trade secret. Thus, trade secret protection and patent protection are usually mutually exclusive IP strategies for the same invention/information.

The final and often most important criterion for a trade secret under the UTSA is that reasonable efforts must be taken to maintain the secrecy of the information. Maintaining the secrecy of a trade secret is viewed under a reasonable standard that does not require absolute secrecy. A court considers several factual inquiries when considering reasonable secrecy:

- Whether employees and/or contractors have executed confidentiality or non-disclosure agreements,
- Whether the company's confidentiality policy is memorialized in writing,
- Whether access to the trade secret has been limited to essential employees/contractors,
- Whether employees who are privy to the trade secret are aware that it is to be maintained as a trade secret,
- Whether the information is kept in a restricted area such as a locked file, within security-encrypted software, protected by a password, in a restricted location within a physical office space or laboratory, etc.,
- Whether documents containing information that are a trade secret are properly labeled, and
- Whether the company actively screens employee publications, presentations, etc. for disclosure of trade secret information.

In addition to these factors, it is important that the owner of the trade secret take reasonable steps to enforce the secrecy of the information. Mere intent to keep information secret, without affirmative acts, is typically insufficient to maintain a trade secret.

Example A startup medical device company, Medicus Societas, has a great new health mobile app that uses a new algorithm to enhance data compression.

The algorithm is what makes the app work, and the startup has decided to protect the algorithm by keeping it confidential as a trade secret.

Merely deciding to keep the algorithm a secret is not enough. Medicus Societas must take reasonable steps to keep the algorithm secret. This could include keeping the algorithm from being seen by employees, or limiting knowledge about the algorithm to employees who have a need to know the algorithm. The startup could also take steps to encrypt the algorithm so that the algorithm is not easily discovered by others or easily uncovered through reverse engineering.

8.3 Independent Discovery and Reverse Engineering of Trade Secrets

Under patent law, a subsequent inventor can be liable for patent infringement even though the invention was developed completely independently and without knowledge of the patented invention. Under trade secret law, independent discovery and use of the trade secret is not a violation. Further, competitors often try to uncover and trade-off on one another's trade secrets by "reverse engineering" the trade secret, which is a legally acceptable practice. The comments to the UTSA state that "reverse engineering" is a proper means of discovering a trade secret and identify reverse engineering as "starting with the known product and working backward to find the method by which it was developed. The acquisition of the known product must, of course, also be by a fair and honest means, such as purchase of the item on the open market for reverse engineering to be lawful..." Thus, the discovery of another's trade secret requires proper acquisition of the information and ethical business practices.

Example Code Rebel is a remote-access software solutions company that allegedly downloaded a trial version of Aqua Connect's "ACTS" software and then reverse engineered the program code, which Code Rebel then used to create its own iRAPP Terminal Server solution. According to Aqua Connect, Code Rebel misappropriated Aqua Connects trade secret information.

Despite the fact that the trial version of the ACTS software required acceptance of a clickwrap End User Licensing Agreement to access the program, with one of the terms being that the end-user will not reverse engineer the program, the U.S. District Court for the Central District of California held that reverse engineering software in violation of a form end-user license agreement, without more, does not rise to the level misappropriation of a trade secret [1].

8.4 Theft of Trade Secrets: Misappropriation

A trade secret owner has the right to prevent others from misappropriating the trade secret. The UTSA defines misappropriation of a trade secret as:

(i) acquisition of a trade secret of another by a person who knows or has reason to know that the trade secret was acquired by improper means; or
(ii) disclosure or use of a trade secret of another without express or implied consent by a person who
  (A) used improper means to acquire knowledge of the trade secret; or
  (B) at the time of disclosure or use knew or had reason to know that his knowledge of the trade secret was
    (I) derived from or through a person who has utilized improper means to acquire it;
    (II) acquired under circumstances giving rise to a duty to maintain its secrecy or limit its use; or
    (III) derived from or through a person who owed a duty to the person seeking relief to maintain its secrecy or limit its use; or
  (C) before a material change of his position, knew or had reason to know that it was a trade secret ad that knowledge of it had been acquired by accident or mistake.

In summary, misappropriation is the improper acquisition, disclosure, or use of a trade secret. A trade secret can be misappropriated even if the misappropriating party is not identically duplicating the trade secret.

Trade secrets can be lost or stolen from medical device companies in a variety of ways: theft, bribery, misrepresentation, or breach of a duty to maintain secrecy, are all common acts that trigger the loss of a trade secret. Violating a confidentiality or non-disclosure agreement, or obtaining a trade secret from a third party that is bound by a duty of confidentiality, can give rise to an action for misappropriation. For example, a common means by which trade secrets can be lost or stolen is typically through unhappy or former employees who use or disclose the trade secret information.

When a company discloses its trade secret information to others, such as employees, manufacturers, suppliers, or consultants, etc., those disclosures should be made under a written duty of confidentiality. This is typically done by requiring the party to execute a confidentiality or non-disclosure agreement, by way of employment contract, or third-party consulting, or supplier agreement. If a party under a duty of confidentiality with the trade secret owner breaches that duty, the trade secret owner's enforcement effort will benefit from a written agreement that clearly recognizes the trade secret status of the information.

The UTSA identifies a number of remedies for misappropriation of trade secrets including injunctions, damages, and attorney's fees. The UTSA even permits recovery of both the actual loss created by the misappropriation and any unjust enrichment resulting from the misappropriation that is not included in the "actual loss" portion of the damages. If actual loss for the misappropriation is difficult to prove, the trade secret owner may seek a "reasonable royalty" as compensation for the misappropriation. If the acts resulting in the trade secret misappropriation are willful or malicious, the UTSA grants the court discretion to award attorney's fees to the trade secret owner.

8.4.1 Defense of Trade Secrets Act

In 2016, federal legislation was passed for the defense of trade secrets by way of the Defend Trade Secrets Act (DTSA) of 2016. Under the DTSA, trade secret owners can bring a civil suit for misappropriation of trade secret information as a federal cause of action [2]. Even though the DTSA is federal law, it does not preempt state trade secret laws. Rather, trade secret holders have the option to bring suit under federal law, which is accomplished via the DTSA, or state law, which is accomplished either under the UTSA or state trade secret protection laws. The DTSA is largely modeled on the UTSA, and the two share similar definitions of trade secrets, similar elements for establishing a misappropriation of trade secrets claim, and similar remedies—namely equitable remedies, actual damages, punitive damages, and reasonable attorneys' fees. While trade secret misappropriation claims brought under state law have varying statute of limitations based on the state in which the claim is brought, misappropriation claims brought under the DTSA have a three-year statute of limitations.

The DTSA also offers a new and unique remedy that is not available under the UTSA. Namely, injured parties can seek an ex parte seizure order, which allow for the seizure of trade secret intellectual property when seizure is necessary to prevent the dissemination or propagation of the trade secret information [3]. This remedy is only available in limited circumstances, where:

- An alternative form of equitable relief is unavailable or inadequate because the offending party would simply evade, avoid, or otherwise not comply,
- The seizure order is necessary to prevent immediate and irreparable injury,
- The harm to the trade secret holder outweighs the interests of the party to be enjoined by the seizure order and substantially outweighs potential harm to other third parties,
- The trade secret owner is likely to succeed on the merits,
- The offending party to be enjoined by the seizure order has actual possession of the trade secret in question,
- The application for the seizure order describes the subject matter and/or content to be seized with reasonable particularity,
- The party to be enjoined by the seizure order would destroy, move, hide, or otherwise make the trade secret information in question inaccessible to the court, and
- The trade secret holder has not publicized the requested seizure.

Because the seizure order is in fact a seizure of property, whether the property is in fact misappropriated or not, the application for the seizure order must also satisfy the basic requirements for a lawful seizure of property, such as the application for the seizure order must contain a detailed description of the trade secret property to be seized such that the seizure of property is as narrow as possible.

8.4.2 Corporate Espionage

There is a difference between gathering competitive intelligence on a competitor and corporate espionage. Competitive intelligence is the gathering of information on a competitor legally and ethically while corporate espionage involves illegal activities, such as tapping a competitor's telephone lines or stealing their garbage to learn secret information about the competitor. Abusing a position of power or special credentials at one's old company to provide information to a new employer is also a form of corporate espionage. Theft of trade secrets is often a serious problem when employees leave one company and move to a competitor.

Example The CEO of Medicus Societas was packing up for the night to leave the office. As she walked past a conference room, she noticed a stranger at the table using two open laptops and a tablet computer. The CEO confronted the unfamiliar man who appeared to be trying to gain unauthorized access to the company's network and computer systems.

The stranger was later identified as an employee of a competitor company, ConglomoMed, who claimed to be at the Medicus Societas' facility to conduct business, but the CEO of Medicus Societas had no reason to believe that the stranger was in her facility after hours for legitimate purposes. The police were called and the strange man was arrested for trespassing. Later, the stranger was charged with attempting to steal trade secrets.

8.4.3 Employee Poaching

"Employee poaching" is a controversial business practice where companies are accused of recruiting employees from competitors so that they can gain access to the competitor's trade secrets or technical know-how. The competing company dangles a tantalizing carrot for the employee it seeks to poach; better compensation and/or great benefits, for example; in an effort to get the employee to defect. To prevent employee poaching, companies often impose restrictions on their employees to protect confidential company information. It is common for employment agreements to include non-compete clauses or a covenant not to compete. These clauses are designed to prevent employees from taking up employment with competitors.

Larger medical device companies are often more at risk for employee poaching. Many smaller companies offer workers significantly more freedom than larger corporations, which is why many workers make the leap to startups. However, they may inadvertently or deliberately, take confidential information, such as trade secret information, with them to the startup.

However, in recent years as the excitement surrounding startups has heated up, especially in the medical tech industry, large companies have been accused of poaching top talent from small startups. Employees often go to the big companies because they can offer more stability than startup life.

No company—startup or conglomerate—wants to be labeled as "anti-employee mobility." It is very bad for business, and for attracting new, talented employees. As such, a company needs to carefully evaluate how to go about preventing and discouraging employee poaching. Litigation for employee poaching can be an expense that a small medical device startup cannot endure financially, so resorting to litigation over a poached employee should be reserved as a last resort to protect stolen confidential information. Other measures could be implemented to reduce the chance of employee poaching and protect confidential company information, such as:

- **Effective use of employment agreements**. As previously mentioned, incorporating non-disclosure agreements and reasonable non-compete agreements into employment contracts can be an effect deterrent to employee poaching. By imposing a legal obligation on employees to remain loyal and keep company information confidential, a smaller medical device startup can demonstrate how significantly employee loyalty is valued at the startup.
- **If it is feasible, compensation packages can be persuasive in retaining employee loyalty**. When employees feel adequately compensated for their work, they are more likely to stay with their current employer. Startups are generally tight on cash, but an investment in retaining employees, such as with stock options or the like, is almost always a good investment.
- **Bonus compensation for staying with the company**. Perhaps it's not feasible to pay employees a premium salary initially, but offering bonus compensation could help retain employee loyalty as the company achieves growth milestones.
- **Vesting equity options**. Alternatively, small medical device companies could grant shares of equity in the company with a vesting period in an effort to retain employees.

Example Employee poaching is a problem in the area of medical device sales. For instance, in early 2017, Howmedica Osteonics Corp., a subsidiary of Stryker, filed a lawsuit that accused Zimmer Biomet of employee poaching Texas medical device sales representatives [4]. Allegedly Zimmer Biomet incentivized the Stryker sales workers to defect, and in turn the poached Stryker sales employees allegedly breached their employment agreements, solicited Stryker customers, and misused Stryker's confidential and trade secret information, all to the benefit of Zimmer Biomet. By the end of 2017, the parties had settled their dispute [5].

8.4.4 Cyber Theft

Cybercrime has been on the rise for the past few years, and companies have increasingly become targets for cyber theft. Flawed networks, weaknesses in cybersecurity measures, and careless employees can all contribute to system vulnerabilities that make a company an easy target for a cyberattack. Cyber criminals often target company information that is valuable so long as it remains secret, such as customer information, client lists, and trade secret information.

One example where cybercrime is becoming increasingly more common is among entities that use cloud computing. Cloud computing is a very popular tool used by many businesses and startups to manage their information systems. Cloud computing is a strategic technology designed to deliver computing and storage capacity as a service that is provided by a third-party vendor. A business and its employees are the end users of cloud computing services. All of the storage space and software used by end users is stored remotely on servers that are owned and maintained by the third party. The end users typically access the services and storage space of the cloud in exchange for a fee. Cloud computing is an attractive and cost-effective alternative to managing and maintaining information systems in-house. While moving business operations to the cloud may be more convenient and efficient than traditional means, there are some potential risks to be considered before making the business decision to move to the cloud.

Trade secrets are only protected by their secrecy, and once the information is disclosed, protection is lost forever. Efforts to protect trade secrets should be vigilant and cautious. It is never a good assumption that information uploaded to the cloud will be safe without first learning about the safety mechanisms put in place by the cloud vendor.

- Does the cloud vendor employ data encryption on its servers?
- What sort of anti-hacking precautions are taken by the cloud vendor?
- Will the cloud provider submit to security audits?

Businesses should be aware that there is no guarantee that a hack or inadvertent leak of its confidential information will not occur. Prospective end-user companies that are interested in moving their business to the cloud should first investigate the technical details of how the cloud provider protects client information.

Whether a startup company's trade secret information, which is stored in a cloud, is likely to lose its trade secret status as a result of being stored in the cloud (by a third party) is a legal question that is yet to be determined by the courts. Because the trade secret owner must take reasonable efforts to maintain the secrecy of the trade secret, the cloud vendor, acting as the agent for the company, must also take reasonable efforts to maintain its end-user clients' trade secret information. In order to best protect the interests of the company, the company should investigate in detail the reasonable efforts claimed to be taken by the cloud vendor for protecting the information of its end-user clients.

References

1. *Aqua Connect, Inc. v. Code Rebel, LLC et al.*, Case No. CV 11–5764-RSWL (C.D. Cal. 2012) (Lew, J.).
2. 18 U.S.C. § 1836.
3. 18 U.S.C. § 1836(b)(2)(A)(i).
4. *Howmedica Osteonics Corp. v. Biomet South Texas, Inc. et al.*, 4:17-cv-00742 (S.D. Tex. 2017) (Complaint).
5. *Howmedica Osteonics Corp. v. Biomet South Texas, Inc. et al.*, 4:17-cv-00742 (S.D. Tex. 2017) (Conditional Order of Dismissal).

Chapter 9
Copyrights in Medical Device Technology

9.1 Copyrightable Subject Matter in MedTech

The federal Copyright Act protects authors' creative works. "Copyright protection subsists… in original works of authorship… fixed in a tangible medium of expression…." [1]. Copyright automatically arises upon the creation and fixation of an original work. For the medical device industry, fixation of an original work may include advertising and marketing materials or packaging, engineering schematics and software that is used with medical devices.

There are limitations on the scope of copyright protection. First, while registration is not required to secure copyright protection, it is a prerequisite to litigating a copyright claim and is desirable to preserve the remedies for attorney's fees and statutory damages. Second, "[i]n no case does copyright protection for an original work of authorship extend to any idea, procedure, process, system, method of operation, concept, principle, or discovery…" [2]. Third, where a "useful article" is concerned, copyright protection is only afforded to the extent the original creative expression is physically or conceptually separable from the utility of the useful article. Finally, only certain subject matter is subject to copyright.

9.2 Scope of Protection

9.2.1 Software and Apps as Copyrightable Matter

As technology in the medical device industry has advanced, software has become increasingly more integral, and important, in medical device technology. Software is significantly improving both user and patient outcomes with medical devices.

G. B. Halt Jr. et al., *FDA and Intellectual Property Strategies for Medical Device Technologies*, https://doi.org/10.1007/978-3-032-12649-8_9

Implants and other medical devices communicate with computers wirelessly, software is currently being used to make prescribing, diagnostic and treatment recommendations, and artificial intelligence is poised to revolutionize patient care. In fact, software has become so enmeshed in medical device technology that the FDA released regulatory guidance at the close of 2017 for software as a medical device [3].

Software, computer programs, and apps are considered literary works for the purposes of copyright protection. Copyright protection only protects expressive elements of the program code and cannot be used to protect the functional aspects of the program. For example, methods of operation within the software, such as menu commands, for the most part are not copyrightable unless they are novel or consist of new and original artistic elements. Graphical User Interfaces (GUIs) elements are generally copyrightable.

Copyright can also be afforded to both the source code and the object code that make up a computer program. The author may choose whether he or she would like to seek copyright protection for the entirety of the program code, or only a portion of the code. Sometimes, graphic displays or wireframes are used to illustrate the underlying operation of a program's function. Like schematics, these graphic representations, such as a flowchart or diagram, may be protected under a separate copyright in addition to the software code.

A copyright is created in a work once it is fixed in a tangible medium of expression. For example, a programmer creates copyright protection in the code he or she is developing the moment he or she hits the "save" button, i.e., fixing it in a tangible medium. The copyright also attaches if the author physically prints a copy of the code. Obtaining a copyright is essentially "free" since registration of the copyright is not required, but is highly recommended.

As part of the registration process of the code, the author must provide the Copyright Office with a copy of the identifiable portions of the source code, that is, the first 25 pages of code and the last 25 pages of code. If the code is less than 50 pages, then the author must submit the entire code. If portions of the code are to be protected as a trade secret, the code may be submitted with a letter stating that it is a trade secret. For trade secret-protected code, the author must submit the first 25 pages of code and the last 25 pages of code (or the entirety of the code if it is less than 50 pages in length), but the trade secret portions may be blocked out. Both the source code and the object code are copyrightable.

9.2.2 Advertising, Marketing Materials, and Product Packaging

Original authorship is often embodied in advertising, marketing materials, and packaging. The primary issues of concern in these areas are copyright ownership, avoiding infringement, and preserving remedies through registration.

Through registration, a copyright owner preserves its ability to recover attorney's fees and statutory damages, remedies that are not available if infringement commences before registration. The registration process also compels an examination of authorship/ownership issues, as well as necessitating a review of whether any preexisting material is embodied in a work, which may give rise to a need for obtaining permission for use of such material. Pursuing registration of the copyright in a commercial should raise the issue of identifying all of the "authors" of the commercial and the use of preexisting material, if any. This will normally bring to light the need for appropriate assignments and clearances.

To the extent that packaging embodies creative expression, packaging is protectable under copyright. This can extend to the shape of packaging (so long as the shape is not functional) or designs that are printed on the packaging. Containers are, however, generally useful articles that would be more appropriately protected by patents.

9.2.3 Training Manuals, User Manuals, and the like

Training materials, user manuals, company brochures, client alerts, instruction guides and company handbooks are all copyrightable if they demonstrate the requisite minimal degree of creativity for copyright protection. These types of documents qualify for copyright protection as "literary works." Copyright in such materials is limited to the description of the processes described in the materials, and does not protect the process itself [4]. For example, if a training manual provides a step-by-step description for how an employee is to procure a reimbursement from the employer company, the written description, along with any flowcharts or diagrams, is copyrightable. However, the process itself cannot be protected. Other companies could use the same process to reimburse employees.

Employer companies should be mindful of the need for assignment of IP rights from those employees who produce these kinds of documents are part of their employment. If an employee creates a training module or presentation on performing a surgical procedure with a new medical device product, or a user manual as a byproduct of designing a new medical device, the employer company should ensure that the IP rights to the manual are assigned to the company.

9.2.4 Secret and Other Materials, Including Software

In many instances, a business may develop software that is incorporated into high-tech medical device products. To the extent such works contain original expression, they are protected by copyright law. However, as with all works protected by copyright, the underlying methods and procedures are not protected by copyright.

Software owners often want to ensure secrecy, and copyright allows such secrecy. When an owner needs secrecy of works, such as software source code, it can register the work by providing "sufficient identifying material," instead of the normally required copy of the work. The identifying material may have confidential portions redacted to preserve secrecy.

Example An example of trade secret or confidential programming code blocked out.

```
    \\use 50% one-shot finish, updated to 1.5 cool down after patch on 10/28
DC
    do {
        int targetA, actionA, actionB;
        if (targetA <834782634876328762837482= 50)
            {
                        actionA=1;
            }
        else
            {
            if (cd >1.5)
            {
            at=1;
            cd=T-1.5;
            }
            else
                        {
            at=2;
            cd=T=0.5;
            }
            }
            else
            { ...
```

9.2.5 *Data and Databases*

Many medical devices and technologies collect vast amounts of data and compile that data into databases. Data is a commodity that increasingly has commercial value in the right situations. Furthermore, it takes a lot of time and effort to compile a well-organized database. Valuable assets, like data compiled into databases, need to be protected.

Copyright protection does not extend to data itself, as data is often considered to consist of facts. Another way to think of this is that facts cannot be protected by copyright because facts lack creativity and do not have an author. However, copyright protection does extend to how data is compiled in a database, because compilation of data in a database can be a creative expression of the facts or data in the database. Database copyright protections are afforded to the selection of data compiled in the database, the coordination of the data in the database, and the arrangement of the data in the database.

Databases are protectable under copyright law as compilations, i.e., a work that is formed by the collection and assembling of preexisting data or materials [5], so long as the organization of the data in the database involves an element of creative originality [6]. It does not take much creativity in the formation of a database to rise to the requisite level of creative originality for copyright protection purposes.

Example *Feist Publications, Inc. v. Rural Tel. Serv. Co.* is a Supreme Court decision that sheds some light on the level of creative originality that is required to make a database copyright eligible. In that case, Rural Telephone Service Co. produced a telephone directory alphabetically listing names and telephone numbers for residents in its service areas. Feist, looking to make a competing telephone directory, copied Rural's listings without permission. Rural sued for copyright infringement of its compiled alphabetical list of names and telephone numbers.

The Supreme Court held that the alphabetically listing of names and telephone numbers was entirely typical and devoid of any trace of creativity. Names and their corresponding telephone numbers contained in the directory merely constituted a listing of facts. The court noted that copyright in a factual compilation is thin, and that someone who comes along later is free to use the same facts in their own compilation, so long as the subsequent work does not feature the same data selection arranged in the same way [6]. Rural's alphabetized telephone directory was held as uncopyrightable.

To enhance the value of a database, many companies creatively select data to compile in the database, arrange data in an unusually way (i.e., in some fashion that is not alphabetical), or incorporate additional copyrightable elements or text into the database. Examples of companies that have copyright-protected databases include West Publishing, Lexis-Nexis, Dun & Bradstreet, and Elsevier Science.

9.3 Ownership and Authorship

Three simple rules control the initial ownership of copyright. First, in general, ownership of copyright "vests initially in the author or authors of the work [7]." Thus, a person who creates an original work automatically becomes its copyright owner. Second, and a major exception to the first rule, where a work is "made for hire," the employer or commissioning party is deemed to be the author. Third, where there is more than one author, "the authors of a joint work are co-owner of copyright in the work [7]."

Although these rules are relatively simple, determining authorship can be difficult. When authors cannot agree on authorship, the Copyright Office will register conflicting claims to copyright in the same work. To avoid such disputes, it is best to resolve authorship/ownership issues at an early stage or to prevent any issues with agreements before they arise. For an employer, employment contracts should be used to define the scope of employment, and any work being done by independent contractors should be covered by an agreement that assigns any copyrights.

9.3.1 Works for Hire

The Copyright Act in 17 U.S.C. § 101 defines:

A "work made for hire" is—

(1) a work prepared by an employee within the scope of his or her employment; or
(2) a work specially ordered or commissioned for use as... [specific list of types of items]... if the parties expressly agree in a written instrument signed by them that the work shall be considered a work made for hire.

Employers need not have an agreement with their employees to create a "work made for hire" obligation. Employment contracts, however, are often useful in settling questions involving whether or not a particular work was prepared within the "scope of employment." Clear company policies often avoid problems in this respect. On the other hand, the law requires specially ordered and commissioned works to be identified in writing as "work for hire." Additionally, even if there is a written agreement as to "work for hire" status, only the specifically enumerated types of works are eligible to be considered "work for hire."

9.3.2 Jointly Authored Works

17 U.S.C. § 101 of the Copyright Act defines a "joint work" as "a work prepared by two or more authors with the intention that their contributions be merged into inseparable or interdependent parts of a unitary whole." The authors of a joint work

may be natural persons, persons, or other entities by virtue of "works for hire" or a combination of both where some contributions are "work for hire" and other contributions are not "works for hire." In the case of multiple employees creating a work for the same employer, there is only one author, the employer, so that such a work is not a joint work.

For works that are jointly owned, unless there is an agreement to the contrary, any joint owner is free to exploit the copyright in the entire work. The other joint owners, however, have a right of contribution to the profits made from the exploitation of a work.

9.4 Licensing and Assignment of Copyrights

After initial ownership of copyright is established through authorship, copyrights may be transferred. The law requires a writing, however, to assign a copyright. 17 U.S.C. § 204(a) provides:

> A transfer of copyright ownership, other than by operation of law, is not valid unless an instrument of conveyance, or a note or memorandum of the transfer, is in writing and signed by the owner of the rights conveyed or such owner's duly authorized agent.

Proper recordation of copyright assignments with the Copyright Office should be made to perfect the title to copyrights. The requirement for a writing under 17 U.S.C. § 204 does not extend to copyright licenses. Accordingly, there can be oral and implied copyright licenses. Recordation of license rights is not required. In the case of works not made for hire, both assignments and licenses can be terminated after 35 years [8].

Example Medicus Societas is a small startup medical device company that has developed an innovative new blood glucose monitoring wearable device called the On-Board. The company created marketing materials and user manuals for the On-Board and obtained copyrights for these works. After successful sales of the On-Board product, a large company called ConglomoMed, acquires Medicus Societas. As part of the acquisition, Medicus Societas is required to assign all of the IP rights associated with the On-Board product line to ConglomoMed. ConglomoMed also plans to partner with a third party to make the On-Board line in the future. ConglomoMed will license the copyrights and other IP associated with the On-Board technology to the third party.

9.5 Derivative Works

One of the informational requirements in registering copyright in a work is to identify preexisting copyrighted material that is present in the copyright application. This is important for registration purposes because copyright in a derivative work is limited to the newly added material; the preexisting material is protected by its own prior copyright. To the extent that a work "unlawfully" uses preexisting material, the law may invalidate copyright protection.

Example Once ConglomoMed acquires Medicus Societas, ConglomoMed plans to expand the On-Board product line by introducing newer models with additional health monitoring features. ConglomoMed plans to use the marketing materials and user manual from Medicus Societas's original On-Board product as base material, and to add new content according to the new features in the expanded products. ConglomoMed will be making derivative works in the new marketing materials and user manual.

9.6 Fair Use

The Copyright Act permits certain copying under the doctrine of fair use. The Copyright Act defines a "fair use" balancing test in 17 U.S.C. § 109 that provides:

> ...the fair use of a copyrighted work, including such use by reproduction in copies or phonorecords or by any other means specified by that section, for purposes such as criticism, comment, news reporting, teaching (including multiple copies for classroom use), scholarship, or research, is not an infringement of copyright. In determining whether the use made of a work in any particular case is a fair use the factors to be considered shall include—
>
> (1) the purpose and character of the use, including whether such use is of a commercial nature or is for nonprofit educational purposes;
> (2) the nature of the copyrighted work;
> (3) the amount and substantiality of the portion used in relation to the copyrighted work as a whole; and
> (4) the effect of the use upon the potential market for or value of the copyrighted work.
>
> The fact that a work is unpublished shall not itself bar a finding of fair use if such finding is made upon consideration of all the above factors.

What does and does not constitute "fair use" is often subject to debate, and, therefore, a lot of litigation. A good rule of thumb is that if you would object to someone copying your work, you may not want to copy without consulting an attorney. A few examples of what may constitute fair use include:

- Small excerpts in a review or criticism for purposes of illustration or comment
- A parody which incorporates some elements of the work being parodied
- Quotations from a speech in a news report, and
- Limited copying for use by a student for educational purposes.

Where there is an intention to use a significant portion of another person's work, it is generally advisable to obtain permission or a license. For music, the Copyright Act provides for compulsory licensing. For major projects, clearance activity may entail contacting multiple parties from collective rights groups such as the Copyright Clearance Center, the American Society of Composers, Authors and Publishers (ASCAP), and Broadcast Music Inc. (BMI) to track down individual authors.

9.7 Registration Issues

Copyright registrations provide an invaluable source of information about copyrighted works and aid in the preservations of works and enhancement of the Library of Congress's collection. Moreover, for copyright owners, the registration process provides a valuable procedure to compel an examination of the issues of copyright ownership and the use of third party materials. Copyright registration is a deceptively simple procedure that leads to questions of authorship and ownership. Registration of copyright preserves important remedies, including statutory (mandatory) damages and attorney fees; in fact, a plaintiff cannot even bring a copyright action without a copyright registration.

9.8 International Copyright Protection

Medical device companies with international reach need international IP protection. Just as international patent protection does not exist, international copyright protection does not exist. Rather, copyright protection, much like patent protection, is protection that is granted by each country individually. However, many countries have elected to enter into international agreements or conventions that afford legal protections to creative works that are fixed in a tangible medium of expression in multiple foreign jurisdictions (similar to the Patent Cooperation Treaty for patents discussed in Chap. 6). The two most important international agreements related to copyright protection are the Berne Convention for the Protection of Literary and Artistic Works, and the Universal Copyright Convention. There are other bilateral agreements between various countries as well.

9.8.1 The Berne Convention for the Protection of Literary and Artistic Works

To obtain copyright protection under the Berne Convention for the Protection of Literary and Artistic Works, authors either must be domiciled in a country that is a member state[1] of the Berne Convention, or must publish the work that is to be protected in a Berne Convention member state within thirty (30) days of the first publication. Member states that participate in the Berne Convention cannot require authors to comply with formalities in order to obtain copyright protection; formalities can only be imposed in the country of origin of the protected work.

9.8.2 Universal Copyright Convention

Copyright protection in Universal Copyright Convention (UCC) member states[2] is afforded by using the proper UCC copyright notice on the protected work. A UCC

---

[1] Member states of the Berne Convention include: Afghanistan, Albania, Algeria, Andorra, Antigua and Barbuda, Argentina, Armenia, Australia, Austria, Azerbaijan, Bahamas, Bahrain, Bangladesh, Barbados, Belarus, Belgium, Belize, Benin, Bhutan, Bolivia, Bosnia and Herzegovina, Botswana, Brazil, Brunei Darussalam, Bulgaria, Burkina Faso, Burundi, Cameroon, Cambodia, Canada, Cape Verde, Central African Republic, Chad, Chile, China, Colombia, Comoros, Democratic Republic of the Congo, Republic of Congo, Cook Islands, Costa Rica, Cote d'Ivoire, Croatia, Cuba, Cyprus, Czech Republic, Denmark, Djibouti, Dominica, Dominican Republic, Ecuador, Egypt, El Salvador, Equatorial Guinea, Estonia, Eswatini, Fiji, Finland, France, Gabon, Gambia, Georgia, Germany, Ghana, Greece, Grenada, Guatemala, Guinea, Guinea-Bissau, Guyana, Haiti, Holy See, Honduras, Hungary, Iceland, India, Indonesia, Ireland, Israel, Italy, Jamaica, Japan, Jordan, Kazakhstan, Kenya, Kiribati, Kuwait, Kyrgyzstan, Lao, Latvia, Lebanon, Lesotho, Liberia, Libya, Liechtenstein, Lithuania, Luxembourg,, Madagascar, Malawi, Malaysia, Mali, Malta, Mauritania, Mauritius, Mexico, Micronesia, Moldova, Monaco, Mongolia, Montenegro, Morocco, Mozambique, Namibia, Nauru, Nepal, Netherlands, New Zealand, Nicaragua, Niger, Nigeria, Niue, North Korea, North Macedonia, Norway, Oman, Pakistan, Panama, Paraguay, Peru, Philippines, Poland, Portugal, Qatar, Romania, Russian Federation, Rwanda, Saint Kitts and Nevis, Saint Lucia, Saint Vincent and the Grenadines, Samoa, Sao Tome and Principe, Saudi Arabia, Senegal, Serbia, Singapore, Slovakia, Slovenia, Solomon Islands, South Africa, South Korea, Spain, Sri Lanka, Sudan, Suriname, Sweden, Switzerland, Syrian Arab Republic, Tajikistan, Tanzania, Thailand, Togo, Tonga, Trinidad and Tobago, Tunisia, Turkey, Turkmenistan, Tuvalu, Uganda, Ukraine, United Arab Emirates, United Kingdom, United States of America, Uruguay, Uzbekistan, Vanuatu, Venezuela, Viet Nam, Yemen, Zambia, and Zimbabwe.

[2] UCC member states include: Albania, Algeria, Andorra, Argentina, Australia, Austria, Azerbaijan, Bahamas, Bangladesh, Barbados, Belarus, Belgium, Belize, Bolivia, Bosnia and Herzegovina, Brazil, Bulgaria, Cambodia, Cameroon, Canada, Chile, China, Colombia, Costa Rica, Croatia, Cuba, Cyprus, Czech Republic, Denmark, Dominican Republic, Ecuador, El Salvador, Fiji, Finland, France, Germany, Ghana, Greece, Guatemala, Guinea, Haiti, Holy See, Hungary, Iceland, India, Ireland, Israel, Italy, Japan, Kazakhstan, Kenya, Lao People's Democratic Republic, Lebanon, Liberia, Liechtenstein, Luxembourg, Malawi, Malta, Mauritius, Mexico, Monaco, Montenegro, Morocco, Netherlands, New Zealand, Nicaragua, Niger, Nigeria, North Macedonia, Norway, Pakistan, Panama, Paraguay, Peru, Poland, Portugal, Republic of Korea, Republic of Moldova, Russian Federation, Rwanda, Saint Vincent and the Grenadines, Saudi Arabia, Senegal, Serbia,

copyright notice takes the form of a © symbol, the year of first publication of the protected work, followed by the mane of the copyright owner. An example would be: © 2026 ConglomoMed LLC.

References

1. 17 U.S.C. § 102(a).
2. 17 U.S.C. § 102(b).
3. FDA's guidance for Industry and FDA Staff: *Software as a Medical Device (SAMD): Clinical Evaluation Guidance for Industry and Food and Drug Administration Staff* – FINAL GUIDANCE (December 2017). Available online at https://www.fda.gov/regulatory-information/search-fda-guidance-documents/software-medical-device-samd-clinical-evaluation. Accessed 6 July 2025.
4. *Situation Management Systems v. ASP Consulting LLC*, 90 USPQ 2d 1095 (1st Cir. 2009).
5. 17 U.S.C. § 101.
6. *Feist Publications, Inc. v. Rural Tel. Serv. Co.*, 499 U.S. 340 (1991).
7. 17 U.S.C. § 201(a).
8. 17 U.S.C. § 203.

Slovakia, Slovenia, Spain, Sri Lanka, Sweden, Switzerland, Tajikistan, Togo, Trinidad and Tobago, Tunisia, Ukraine, United Kingdom, United States of America, Uruguay, Venezuela, and Zambia.

Chapter 10
Design Protection for Medical Devices

10.1 Design Patents for Medical Devices

In the United States, a design patent can be granted for the ornamental design of an article of manufacture. 35 U.S.C. § 171 is the statute focused on design patents, which states that "[w]hoever invents any new, original and ornamental design for an article of manufacture may obtain a patent."

Notice how similar the above statute is to 35 U.S.C. § 101 for utility patents, previously discussed in Chap. 6, which states "[w]however invents or discovers any new and useful process, machine, manufacture, or composition of matter, or any new and useful improvement thereof, may obtain a patent therefore, subject to the conditions and requirements of this title."

Both statutes require that the invention be new (i.e., novel) and original (i.e., non-obvious), and each specifically identify what the invention can be. 35 U.S.C. § 101 requires that the invention must be directed to a process, machine, manufacture, or composition of matter; while 35 U.S.C. § 171 requires that the ornamental design must be incorporated into an article of manufacture. Since a medical device invention can have utility and can have an ornamental design, in many cases both a design patent and a utility patent can be obtained for different aspects of the same invention.

The United States Patent and Trademark Office (USPTO) recognizes thirty-five (35) classes of articles of manufacture under the Nice Agreement concerning the International Classification of Goods and Services. Designs related to medical devices are often classed in D24: Medical and Laboratory Equipment. Design patents are also organized using the international Locarno Classification administered by WIPO. Medical and laboratory equipment fall within Locarno Class 24. Some legacy U.S. materials and searches make reference to the U.S. "D-class" system; if used, that label should be identified as a historical classification rather than the current scheme.

G. B. Halt Jr. et al., *FDA and Intellectual Property Strategies for Medical Device Technologies*, https://doi.org/10.1007/978-3-032-12649-8_10

10.1.1 Requirements for Design Patents

Although different in scope and duration (which is described in detail below), design patents share many of the same filing requirements as utility patent applications. For instance, design patent applications, just like utility patent applications, must be directed to an inventive design that is novel under 35 U.S.C. §102 (see Chap. 6) and non-obviousness under 35 U.S.C. §103 (see Chap. 6), and must include a written description as required by 35 U.S.C. §112 (see Chap. 6). The written description, however, is different for a design patent application because the specification and claims are almost exclusively directed to the drawings included in the application, which depict the claimed design.

In the case of design patents, it is difficult to capture the nuances of an ornamental design with words [1] and therefore the drawings are critically important. In fact, in 1886 the Supreme Court concluded that a design is better represented by a photographic illustration than it could ever be by description using words, and even if a description were provided, it would probably not make much sense without a corresponding illustration [2]. Consequently, the drawings of the design patent are relied upon to provide the description of the invention [3]. The drawings included in design patent applications are critically important.

Design patent protection can be secured for individual elements of a medical device invention, or for the overall design of a medical device that is produced from a combination of individual design elements. For instance, in the Federal Circuit case of *Ethicon Endo-Surgery, Inc. v. Covidien, Inc.*, several U.S. design patents related to ornamental features of a surgical instrument (ultrasonic surgical shears) used to cut tissue and blood vessels were involved in an infringement dispute [4]. The ornamental design features of the surgical instrument included several different aspects:

An inverted "U"-shaped trigger for the surgical instruments, as shown in Fig. 10.1.

The inverted "U"-shaped trigger for the surgical instruments in conjunction with a rounded and fluted torque knob positioned above and forward from the trigger, as shown in Fig. 10.2.

The inverted "U"-shaped trigger for the surgical instruments in conjunction with a rounded activation button positioned directly above the trigger, as shown in Fig. 10.3.

The inverted "U"-shaped trigger for the surgical instruments in conjunction with the fluted torque knob, and the rounded activation button, with the torque knob and the button positioned relative to the trigger, as shown in Fig. 10.4.

To be patentable, a design must be "primarily ornamental." Thus, to determine the validity of the design patents, the *Ethicon* court considered whether the claimed design was "primarily ornamental" or "primarily functional," but cautioned that "the function of the article itself must not to be confused with 'functionality' of the design of the article" [4, p. 1328].. In their opinion, the *Ethicon* court instructed that "the utility of each of the various elements that comprise the design is not the relevant inquiry with respect to a design patent because whether a design is primarily functional or primarily ornamental requires viewing the claimed design <u>in</u>

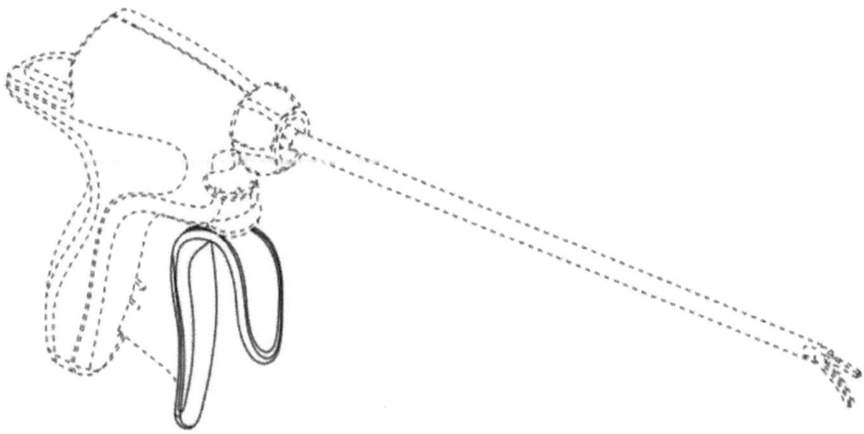

Fig. 10.1 U.S. Design Patent No. D661,801

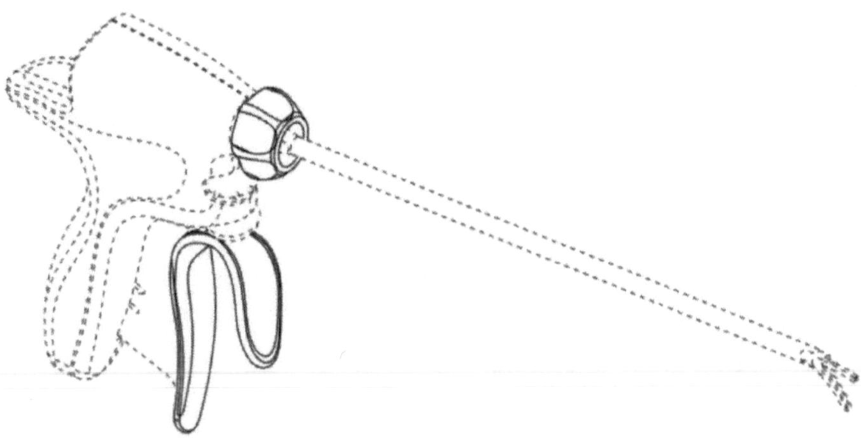

Fig. 10.2 U.S. Design Patent No. D661,802

its entirety" [4, p. 1329; quoting *L.A. Gear, Inc. v. Thom McAn Shoe Co.*,]. In their discussion, the *Ethicon* court noted "[w]e have not mandated applying any particular test for determining whether a claimed design is dictated by its function and therefore impermissibly functional. We have often focused, however, on the availability of alternative designs as an important—if not dispositive—factor in evaluating the legal functionality of a claimed design" [4, pp. 1329–1330]. Accordingly, the *Ethicon* court held that "the underlying ultrasonic shears could still function in the same manner with a differently-shaped open trigger, activation button, and torque knob, and different relative locations of the trigger, button, and torque knob" and thus the design, when viewed as a whole, constituted protectable ornamental design elements [4, p. 1331].

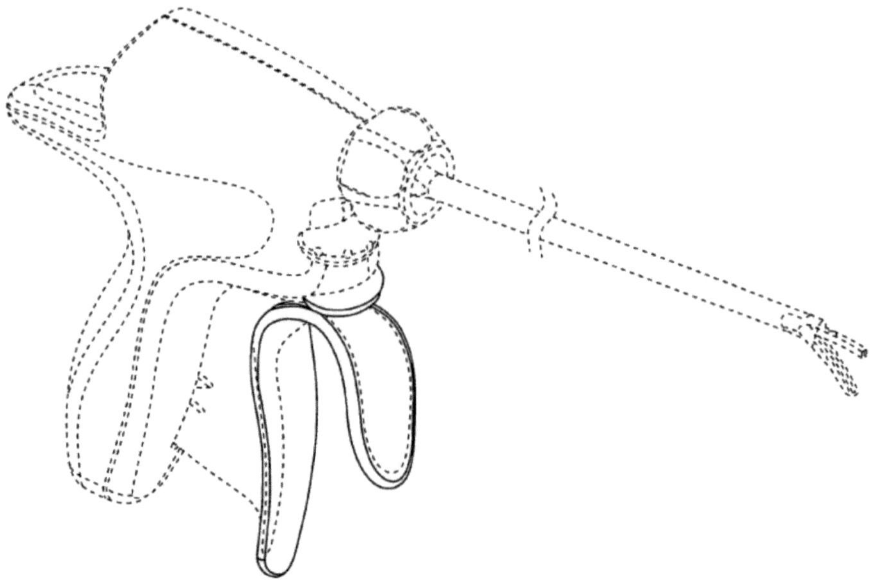

Fig. 10.3 Design Patent No. D661,803

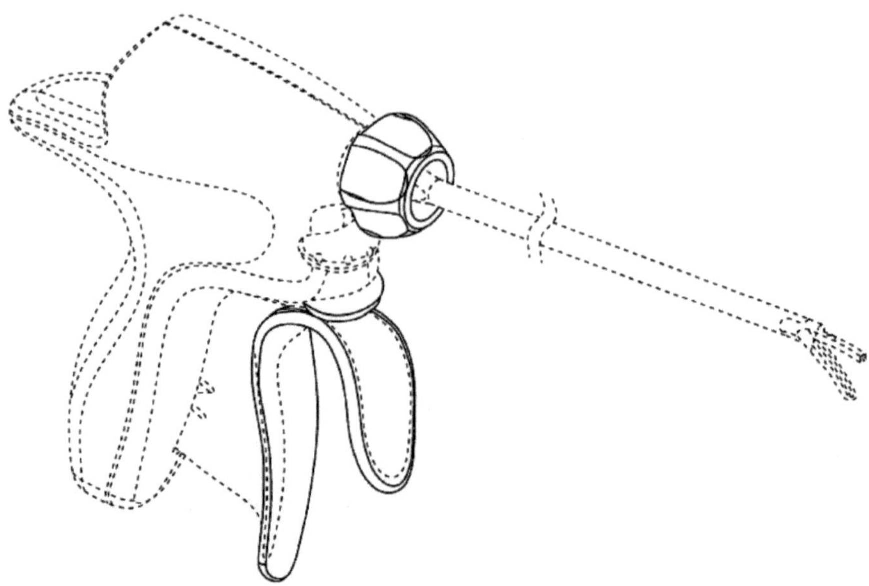

Fig. 10.4 U.S. Design Patent No. D661,804

Ethicon illustrates that a design patent can cover the ornamental design when several parts (e.g., the trigger and knob) are combined to form the ornamental design of a surgical device. However, design patents can also cover the design of the individual parts themselves.

10.1.2 Protection Afforded by Design Patents

The scope of a design patent, compared to a utility patent, is quite different. A design patent only covers how the article of manufacture looks. More precisely, "[i]f the particular design is essential to the use of the article, it cannot be the subject of a design patent" [5, p. 1123]. To determine whether a design is essential to the use of the article, alternative designs to obtain the same function are considered, as well as "if other designs could produce the same or similar functional capabilities, the design of the article in question is likely ornamental, [and] not functional" [6, p. 1378]. Accordingly, "where a design contains both functional and non-functional elements, the scope of the claim must be construed in order to identify the non-functional aspects of the design as shown in the patent" [7, p. 1405].

In addition, the duration of protection afforded to a design patent is different than a utility patent. Design patents are granted for a term of 15 years from the date of grant, for design patent applications filed on or after May 13, 2015 [8]. For design patent applications filed prior to May 13, 2015, the term of the patent is 14 years from the grant date. The change in duration was implemented in December 18, 2012 by the Patent Law Treaties Implementation Act of 2012 to harmonize the U.S. patent laws with the Hague Agreement concerning the International Registration of Industrial Designs (the Hague Agreement is discussed in more detail below in Sect. 10.2.3).

10.1.2.1 Priority Claims for Design Patents

A design application can be filed as a continuing application of a utility application, provided that the utility application shows the ornamental designs in sufficient detail. Consequently, a design patent application may claim the benefit of a utility patent application's filing date, yet the term of the design patent is not limited by the priority date. This potentially allows the design patent to be enforceable for several years after the term of the utility patent has expired, since the term of the design application is calculated from the grant date and not the effective filing date.

10.1.3 Design Patent Disputes

Infringement of a U.S. design patent is judged by the ordinary observer, evaluating the overall visual impression in light of the prior art [9]. Claim scope tracks what the drawings show and the article named; titles and figure context matter [10]. Anticipation likewise requires the prior art to be directed to the same article of manufacture [11]. Logos and other visual context can influence the ordinary-observer analysis [12].

10.1.3.1 Apple v. Samsung

Design patents are often an overlooked form of patent protection, but can hold immense value, especially when a patented design feature is copied by another. One of the biggest design patent disputes in U.S. history is *Samsung Electronics Co., Ltd. v. Apple Inc., which is a design patent case that made it all the way to the U.S. Supreme Court* [13]. *In the Samsung case, Apple alleged that Samsung copied patented design features from Apple's iPhone products and incorporated those protected designs into a number of Samsung products. Apple's iconic design, which is a minimalist phone having a rectangle-shape with rounded corners paired with a distinct user interface, is protected by a number of U.S. design patents, including U.S. Design Patent Nos.* D593,087; D618,677 and D604,305 [9, p. 433] (see Figs. 10.5, 10.6 and 10.7 for representative Figures from each of these patents).

Fig. 10.5 U.S. Design Patent No. D593,087

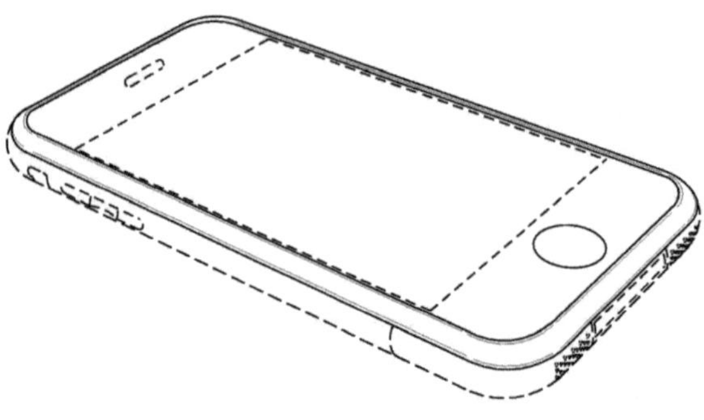

Fig. 10.6 U.S. Design Patent No. D618,677

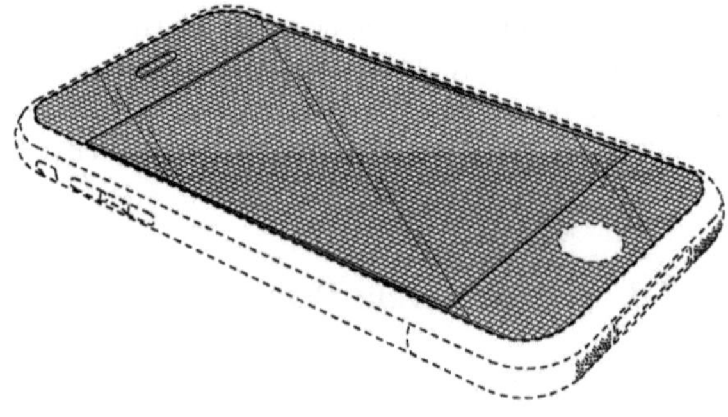

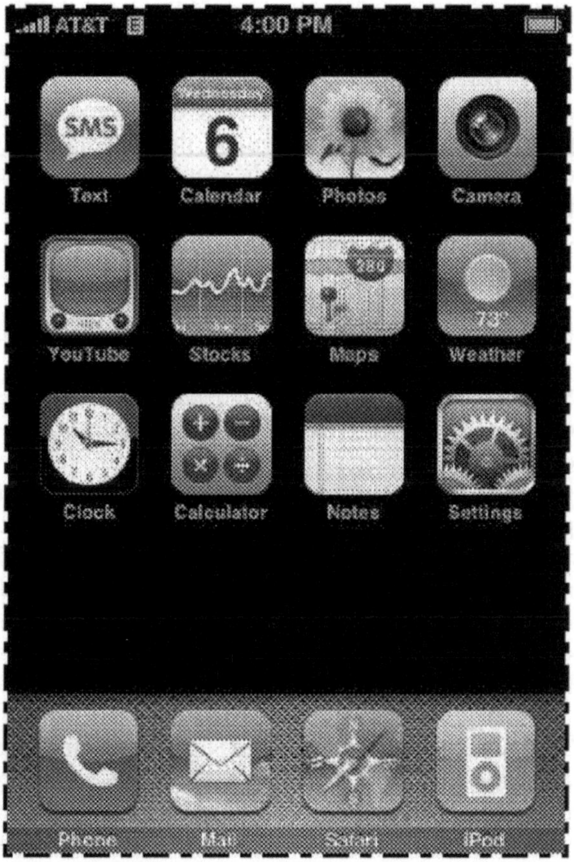

Fig. 10.7 U.S. Design Patent No. D604,305

At the lower court, a jury had awarded Apple $399 million dollars in damages after concluding that Samsung infringed on Apple-owned design patents; however, those damages were calculated based on the entire profit Samsung acquired from the sales of each phone that bore the Apple-infringed designs. The case ultimately ended up at the Supreme Court after the Federal Circuit rejected Samsung's argument regarding 35 U.S.C. § 289, which covers liability and damages for design patents, that the profits awarded should be limited to the infringing article of manufacture, not the entire device. To say this another way, if Apple owned design patents on discrete components of a phone design (such as a rounded corner of the phone housing, or a particular design of the graphical user interface), why should Apple be awarded damages for Samsung's sales of each phone Samsung sold as a whole?

The Supreme Court reversed the Federal Circuit's decision and remanded the case, holding that the term "article of manufacture" is broad enough to cover both a product sold to a consumer and a component of the product, regardless of whether or not the component is or even can be sold separately [13, p. 436]. To say this plainly, under the 35 U.S.C. § 289, a design patent can cover both the ornamental overall design of the complete product (e.g., a design patent for the overall look of a phone) as well as the ornamental design of individual components of the product (e.g., a design patent protection only the rounded corner design feature of a phone housing), and liability and damages are to be considered and determined based on which of these is being asserted. In *Samsung*, the Supreme Court also declined to set forth any type of rule or test regarding the damages inquiry under 35 U.S.C. § 289, which has now left this determination and potential creation of test(s) up to the various lower courts that may encounter similar issues.

LKQ Corp. et al. v. GM Global Technology Operations LLC

Similar to utility patents, design patents must be novel (35 U.S.C. § 102) and non-obvious (35 U.S.C. § 103). However, for decades, the test for non-obviousness for design patents was applied differently than the test used for utility patents. As noted previously, U.S. courts applied the Rosen-Durling Test, which required that the primary reference being used was an earlier design that had "basically the same" visual impression as the patented design, to which features could be added from additional prior art designs that are "so related" to it that the appearance of features in one design suggested using those features in the other design. In 2024, in *LKQ Corp. et al. v. GM Global Technology Operations LLC*, the Federal Circuit overturned the decades-long applied *Rosen-Durling* Test [14].

In *LKQ*, the Federal Circuit questioned the ongoing validity of the *Rosen-Durling* Test, particularly in light of the Supreme Court's guidance on obviousness in *KSR International Co. v. Teleflex Inc.*, [15]. The court suggested that the rigid requirement for a "primary reference" may not align with modern obviousness principles, signaling a loosening of the standards for proving invalidity of design patents. In place of *Rosen-Durling*, the Federal Circuit held that obviousness would now be determined under the same guidelines used for utility patents as set forth in the Supreme Court's precedent in *Graham v. John Deere* [16], and as further discussed in *KSR International Co. v. Teleflex Inc.*, [15]. The Federal Circuit determined that this change would allow the use of "analogous art" that is not required to be from the same field of endeavor, stating, "we do not delineate the full and precise contours of the analogous art test for design patents. ... Whether a prior art design is analogous to the claimed design for an article of manufacture is a fact question to be addressed on a case-by-case basis."

The decision in *LKQ* marked a considerable shift in the evaluation of design patent invalidity under 35 U.S.C. § 103. Given the recent change in the test, it is still uncertain exactly what impact the *LKQ* decision will have on design patents. However, this more flexible standard will likely have profound implications for U.S. design patent law by making it easier to invalidate designs that do not exhibit a sufficient leap of creativity over prior art. This may lead to an increase in invalidity challenges, particularly in industries like automotive and consumer electronics, where designs often evolve incrementally.

10.2 International Design Protection

10.2.1 Design Protection in Europe

In the European Union (EU), which is not be confused with the thirty-eight (38) contracting countries and three (3) extension countries of European Patent Convention (EPC), "the appearance of the whole or a part of a product resulting from the features of, in particular, the lines, contours, colors, shape, texture, and/or materials of the product itself and/or its ornamentation" may be protected by a Community Design Right [17, Art. 3(a)].

However, a design is only protectable "to the extent that it is new and has individual character [17, Art. 4.1]." A design is considered to be "new" if "no identical design has been made available to the public [17, Art. 5.1]." In addition, a design is to have "individual character" when "the overall impression it produces on the informed user differs from overall impression produced on such a user by any design which has been made available to the public [17, Art. 6.1]." If a design meets these two requirements, then the design may qualify for two different types of design rights that are available in the EU: Unregistered Community Designs (UCD) protection and Registered Community Designs (RCD) protection.

A UCD, like a copyright under U.S. law (see Chap. 9), does not need to be registered, and infringement requires actual copying. The UCD right arises as soon as the design is first disclosed to the public within the territory of the EU. The right then lasts for a non-extendable period of three years from the date on which the design was first made available to the public within the territory of the EU. The date of disclosure can be demonstrated in several ways, such as by a printed review bearing a date or mass publicity of the design. The date of disclosure could also be evidenced by publication of the design in a national IP Office Bulletin, display of the design at an international exhibition or by a dated mailing containing the design to all trade associations belonging to given industrial sector. Once a UCD has been established, there is a one-year grace period to obtain an RCD.

A RCD offers similar protections to those offered by a U.S. Design patent. The RCD provides protection against similar designs even when the similar design was not the result of physical, direct copying. The term of the RCD is five (5) years from the date of filing of the application and can be renewed every five (5) years up to a maximum of twenty-five (25) years.

While the use of UCD and RCD is useful in protecting many designs in the EU, their use to protect replacement parts is potentially problematic in Europe. Like in the US, European Design regulations prohibit the protection of designs that "subsist in features of appearance of a product which are solely dictated by its technical function [17, Art. 8.1]." However, the European design law also provides a "Must-Fit" exception. The "Must-Fit" exception prohibits designs from being registered for the design of a product "which must necessarily be reproduced in their exact form and dimensions in order to permit the product in which the design is incorporated or to which it is applied to be mechanically connected to or placed in, around or against another product so that either product may perform its function [17, Art. 8.2]." Accordingly, a registered design for something like a surgical cassette may not be as readily available in the EU as it would be in the U.S., since the cassette must fit into its mating receptacle in order to function.

In addition, European Design protections offered for replacement parts are further limited to designs that remain visible during normal use of the replacement part [17, Art. 4.2(a)]. Normal use is very narrowly construed to mean use someone who is the end user, whereas use by someone for the purposes of maintenance, servicing or repair work is not considered normal use [17, Art. 4.3]. Therefore, European UCD and RCD offer similar protections as a U.S. design patent; however, these protections are afforded to a more limited set of goods.

10.2.1.1 Comparing U.S. and EU Standards

The center of gravity for U.S. design patent invalidity has shifted. In *LKQ Corp. v. GM*, the Federal Circuit signaled that the rigid *Rosen–Durling* framework, which insisted on finding a nearly identical primary reference before considering combinations, sits uneasily with the flexible *Graham* and *KSR* approach to obviousness [14]. For medical device designs, that reorientation means obviousness will more often turn on the record as a whole, including analogous art, reasoned combinations, and overall visual impression rather than a gatekeeping search for a near twin.

Across the Atlantic, the European Union's Community Registered Design regime remains steady. Protection continues to hinge on novelty and individual character, assessed by whether the claimed design creates a different overall impression on the informed user, taking account of the designer's degree of freedom in the field. There is no primary-reference requirement; tribunals may consider the full landscape of prior designs when assessing both novelty and individual character. For medical device forms that are constrained by standards or function, small but perceptible departures can carry weight if they alter the informed user's overall impression.

For practitioners, the practical implications are immediate. In the United States, post-*LKQ* challenges are more likely to stitch together visual features from multiple references where there is a reasoned motivation, so incremental changes to housings, handpieces, cassettes, and tubing interfaces may face closer scrutiny under §103. In the European Union, the "overall impression" lodestar remains the focus, which means that in crowded subfields even modest visual shifts can matter, while in areas with broad design freedom more pronounced variation is needed to clear the bar. The result is a degree of convergence in tone but not equivalence in outcome: U.S. rights stay tightly tied to what is claimed in the drawings, while CRDs reach designs that do not produce a different overall impression. Portfolio strategy should reflect those different envelopes.

The prosecution playbook follows from those differences. Claim what you care about and disclaim what you do not by using broken lines and targeted figure sets so protection centers on the aesthetic choices that drive market differentiation, such as bezel transitions, surface textures, and handle geometry. Build a prior-art record early:

in the United States, anticipate post-*LKQ* combinations by distinguishing analogous art and documenting reasons not to combine, including incompatibility of design vocabularies or clinical-use constraints; in the European Union, curate comparators that show limited designer freedom where appropriate. In both systems, align the title and figure context with the true article of manufacture, for example an infusion-pump handpiece or a reusable endoscope handle, to reduce risk at validity and scope.

Enforcement will track the same themes. In U.S. litigation after *LKQ*, expect more section 103 challenges supported by mosaics of design references and expert analysis on analogous art and reasons to combine. Plaintiffs should be prepared to defend the coherence of the claimed visual theme and to explain why proposed combinations would not yield the claimed overall impression. In the European Union, disputes will remain impression-focused, with parties contesting the informed user's perspective, the density of the design field, and how differences read in context, including how a cartridge face or control-knob profile registers in real-world use.

10.2.2 Design Protection in Asian Countries

China grants design patents for a term of ten (10) years from the filing date of the Chinese design patent for application filed prior to June 1, 2021. For applications filed on or after June 1, 2021, the term has been extended to fifteen (15) years. In China, a design patent can be granted for "the shape, pattern, or combination thereof, or the combination of the color with shape and pattern which are rich in an aesthetic appeal and are fit for industrial application [18, Art. 2(4)]."

Japan allows for designs to be registered for "the shape, patterns or colors, or any combination thereof, of an article (including a part of an article, the same shall apply hereinafter except in Article 8), which creates an aesthetic impression through the eye [19, Art. 2(1)]." The term of a Japanese Industrial Design patent, which is comparable to a U.S. design patent, was extended to a patent term of twenty-five (25) years from the date of registration in 2020. Japan requires that annual maintenance fees on the patent are paid to maintain the patent.

South Korea (Republic of Korea) registers industrial designs for "the shape, pattern, color or combination of these in an article that produces an aesthetic impression on the sense of sight" for objects that are "movable properties which are independent and definitive [20, Art. 2(1)]." The industrial design registration in South Korea is valid for twenty (20) years from the date of filing.

The industrial design patent in India is valid for up to fifteen (15) years from the date of registration provided that a five (5) year extension, and the corresponding fee, are paid before the expiration of the initial ten (10) year term.

10.2.3 The Hague Agreement

The Hague Agreement Concerning the International Deposit of Industrial Design ("the Hague Agreement") is an agreement among eighty-two (82 nations) that allows applicants to register an industrial design by filing a single application with the World Intellectual Property Organization (WIPO) in order to protect their designs with minimum formalities in multiple countries or regions.[1] This allows the applicant to file a single International Design Application (IDA) and potentially receive design registration protection (e.g., a design patent or registered industrial design) in each of the nations or regions participating in the Hague Agreement the applicant designates in the IDA. When the IDA is registered in a particular country, the same protections are afforded to the design as if the application had been originally filed locally in the respective patent office. For example, an IDA that designates the U.S., once granted, will be treated under U.S. law as a design patent, even if the IDA was filed in the EU.

In many ways, the IDA under the Hague Agreement is similar to a PCT Application filed under the Patent Cooperation Treaty (PCT) (see Chap. 6). As with a PCT patent application, the IDA can be either filed directly with the International Bureau (IB) or with the applicant's home patent office (e.g., the USPTO, the European Patent Office, the Japan Patent Office, etc.). However, unlike the process for a PCT, regardless of where the IDA is filed, the application will be filed with the IB for examination.

Unlike in the case of a PCT application, the IB does not perform a substantive review of the IDA. Instead, the IB conducts a review of the design filing to confirm that formal requirements are satisfied. For instance, the IB ensures that the application is filed in English, French, or Spanish. The IB further confirms that the applicant has identified the Contracting Parties where the applicant seeks protection, and has paid the basic filing fee, the publication fee, and a fee for each designated Contracting Party.

Additionally, the IB confirms that a reproduction of the industrial design the applicant seeks to protect meets the WIPO publication standards and includes the required prescribed data (such as the name of the applicant and goods classification). The IB further confirms that the applicant is eligible to file an IDA. To be eligible to file an IDA, the applicant must either:

---

[1] The contracting parties to the Hague Agreement Concerning the International Deposit of Industrial Design include: the African Intellectual Property Organization (OAPI), Albania, Armenia, Azerbaijan, Belarus, Belgium, Belize, Benin, Bosnia and Herzegovina, Botswana, Brazil, Brunei Darussalam, Bulgaria, Cambodia, Canada, China, Croatia, Côte d'Ivoire, Democratic People's Republic of Korea, Denmark, Egypt, Estonia, the European Union (EU), Finland, France, Gabon, Georgia, Germany, Ghana, Greece, –**Holy See**–, Hungary, Iceland, Israel, Italy, Jamaica, Japan, Kyrgyzstan, Latvia, Liechtenstein, Lithuania, Luxembourg, Mali, Mauritius, Mexico, Monaco, Mongolia, Montenegro, Morocco, Namibia, Netherlands, Niger, North Macedonia, Norway, Oman, Poland, Portugal, Republic of Korea, Republic of Moldova, Romania, Russian Federation, Rwanda, Saint Kitts and Nevis, Samoa, San Marino, Sao Tome and Principe, Saudi Arabia, Senegal, Serbia, Singapore, Slovenia, Spain, Suriname, Switzerland, Syrian Arab Republic, Tajikistan, Tunisia, Turkey, Turkmenistan, Ukraine, United Kingdom, United States of America, Uzbekistan, and Viet Nam.

(1) Be a national of a Contracting Party.
(2) Be a national of a member state of an intergovernmental organization that is a Contracting Party (*e.g.*, European Union).
(3) Be domiciled in a Contracting Party's territory.
(4) Have a real and effective industrial or commercial establishment in a Contracting Party's territory.

Once the IB confirms that the formal requirements of the IDA are satisfied and the formal, preliminary examination phase of the IDA is completed, the IB then publishes the IDA. Next, substantive examination of the IDA before the patent offices of the countries or regions that the applicant has designated commences.

The substantive examination of the IDA is performed in accordance with the local design laws of the particular patent office examining the IDA. However, the particular patent office has six (6) months (if the local laws permit opposition) to issue a refusal. If the local patent office issues a refusal, the patent prosecution process begins according to the local patent laws and will likely require the applicant to obtain local counsel.

An international registration under the Hague System is maintained by paying renewal fees in five-year increments, up to the maximum term allowed in each designated Contracting Party. The total available term therefore depends on local law in each designation rather than a single global cap. This allows the applicant to pay one renewal fee directly to the IB instead of having to pay individual renewal fees to each country where protection has been obtained.

10.3 Importance of Protecting Medical Device Designs

Medical devices are often designed in particular ways and decisions about the ornamental design of a medical device or its component are often chosen for a reason. Similarly, one medical device can be distinguished from another through the use of unique icons, colors, branding, and material and texture variations. The placement of buttons or interactive components might be located in a certain place on the medical device so that the user can have easy access, the shape of a component might be because it is most suitable for a particular use, or a medical device design might be tailored for user comfort during use (e.g., for ergonomics). As long as the features do not impart functionality, pursuing one or more design patents on a medical device can assist in strengthening a medical device's IP portfolio. Designs can also be important to encouraging market adoption of the medical device, ensuring that users use the device properly, or conveying the brand of the medical device product to the user.

Since design is so important in the medical device field, it is not surprising that medical device ornamental designs serve as one of the leading technology categories for design patent filings and grants. Design protection is sought for various aspects of medical devices. The overall design of the medical device could be patentable, as well as the design of individual components of a medical device or unique icons used on the medical device or display screen.

10.3.1 Graphical User Interfaces and Icon Designs

There are many computer-based medical device technologies available in the health-care field. It is typical for computerized medical devices allow for user interaction with the device through a graphical user interface (GUI). Many features of a GUI, such as unique icons and animations of how icons transition from one state to another, can be protected by a design patent.

Computer-generated that are incorporated into an article of manufacture, such as a display screen, can be protected by a design patent. An example is shown in Fig. 10.8, assigned to Uber Technologies Inc. In the design patent application, the icon must be shown as incorporated into a display, or a portion of a display. Figure 10.8 shows an interactive icon (shown in dark solid lines) at the bottom of a display screen (shown in dashed, light-colored lines).

How an icon transitions from a first state to a second state on a display screen can also be protected with a design patent. For example, Fig. 10.9 shows the transition of an Apple window icon on a display screen in U.S. Design Patent No. D496,941 owned by Apple, Inc.

10.3.2 Replacement/Consumable Parts

Medical devices often require a large number of consumable replacement parts in order to maintain the sterility of the device. Similarly, many medical device components need to be replaced once they wear out. Design patents may be utilized to protect these consumable replacement parts.

For example, U.S. Design Patent No. D501,605 was granted on February 8, 2005 to The Gillette Company for a "Toothbrush head." Gillette is the parent company of the popular toothbrushes sold under the Oral-B brand. U.S. Design Patent No. D501,605 patent claims a toothbrush head as shown in Fig. 10.10.

Similarly, U.S. Design Patent No. D733,867, granted to Abbott Medical Optics Inc. on July 7, 2015, is for a "Surgical cassette for a surgical system." The surgical cassette is covered by U.S. Design Patent No. D733,867, shown in Fig. 10.11, captures liquids used during phacoemulsification eye surgery, and a new cassette is required for each patient.

By employing design patents, both Gillette and Abbott Medical Optics are able to protect the replaceable consumables used in their respective medical device systems.

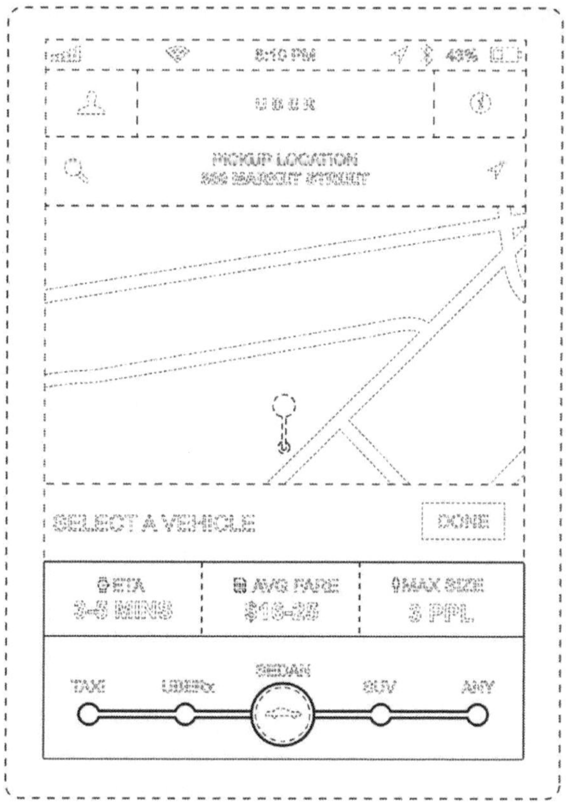

Fig. 10.8 U.S. Design Patent No. D763,294

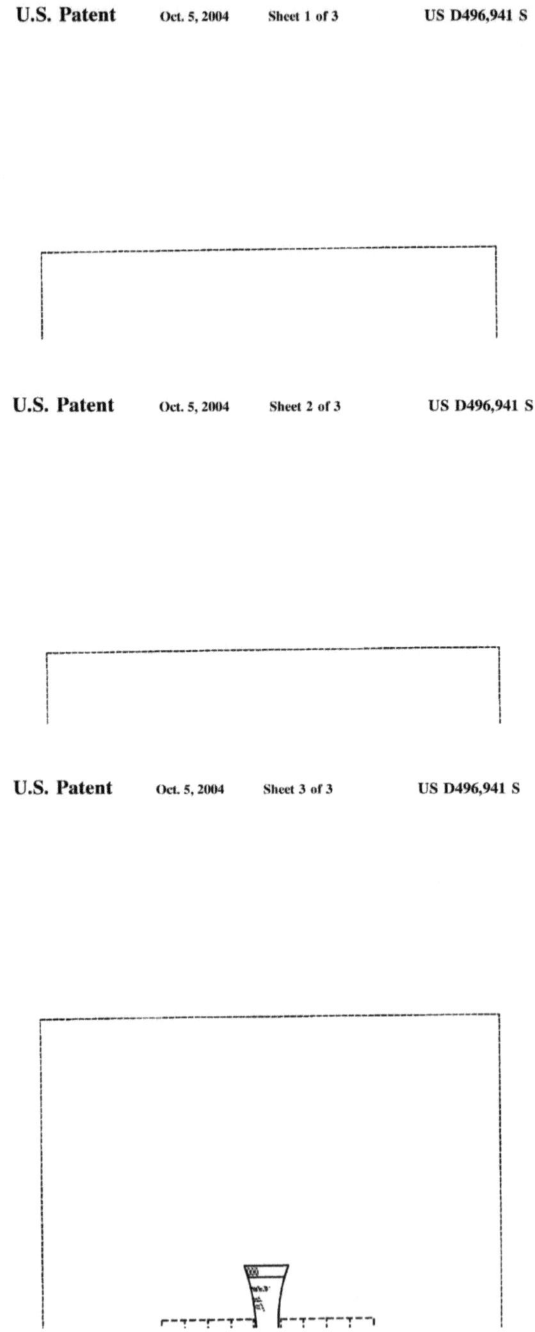

Fig. 10.9 Design Patent No. D496,941

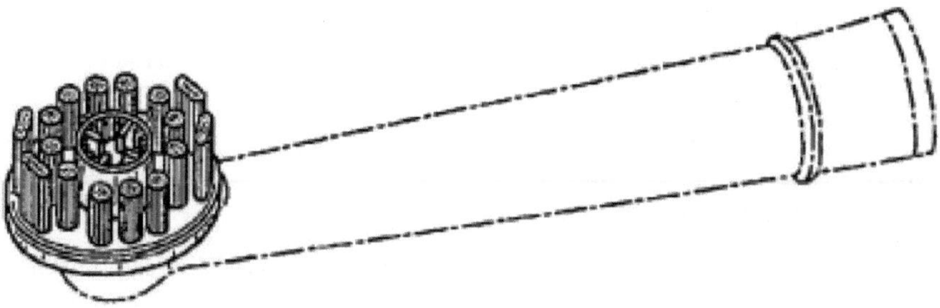

Fig. 10.10 U.S. Design Patent No. D501,605

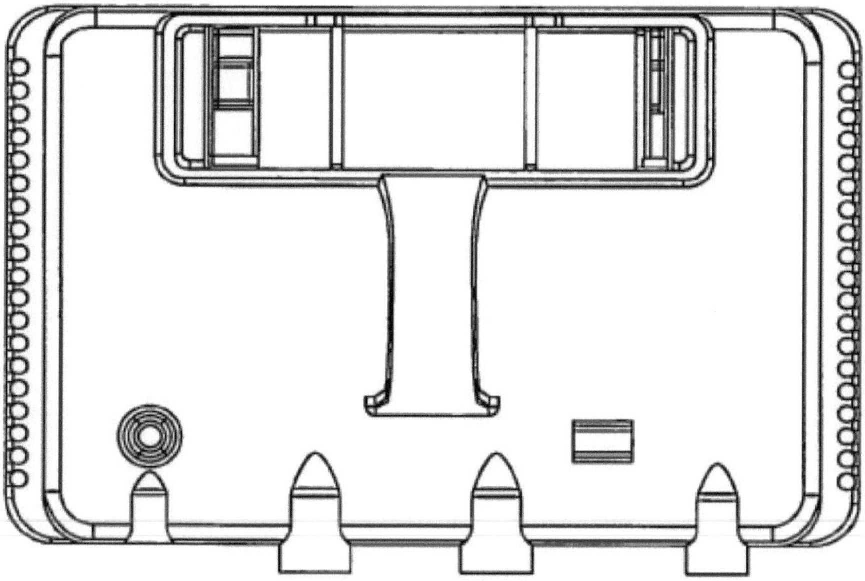

Fig. 10.11 U.S. Design Patent No. D733,867

References

1. *Sport Dimension, Inc. v. Coleman Co.*, 820 F.3d 1316, 1320 (Fed. Cir. 2016).
2. *Dobson v. Dornan*, 118 U.S. 10, 6 S. Ct. 946, 30 L. Ed. 63, 1886 Dec. Comm'r Pat. 202 (1886).
3. *Pac. Coast Marine Windshields Ltd. v. Malibu Boats, LLC*, 739 F.3d 694 (Fed. Cir. 2014).
4. *Ethicon Endo-Surgery, Inc. v. Covidien, Inc.*, 796 F.3d 1312 (Fed. Cir. 2015).
5. *L.A. Gear, Inc. v. Thom McAn Shoe Co.*, 988 F.2d 1117 (Fed. Cir. 1993).
6. *Rosco, Inc. v. Mirror Lite Co.*, 304 F.3d 1373 (Fed. Cir. 2002).
7. *OddzOn Prods. v. Just Toys*, 122 F.3d 1396 (Fed. Cir. 1997).
8. U.S. Code § 173.
9. *Egyptian Goddess, Inc. v. Swisa, Inc.*, 543 F.3d 665 (Fed. Cir. 2008) (en banc).
10. *Curver Luxembourg, SARL v. Home Expressions Inc.*, 938 F.3d 1334 (Fed. Cir. 2019) (claim limited to the recited article of manufacture).

11. *In re Surgisil, L.L.P.,* 14 F.4th 1380 (Fed. Cir. 2021).
12. *Columbia Sportswear N. Am., Inc. v. Seirus Innovative Accessories, Inc.*, 80 F.4th 1363 (Fed. Cir. 2023).
13. *Samsung Elecs. Co. Ltd v. Apple Inc.*, 137 S. Ct. 429 (2016).
14. *LKQ Corp. et al. v. GM Global Technology Operations LLC,* slip op. 2021–2348 (Fed. Cir. 2024).
15. *KSR International Co. v. Teleflex Inc.*, 550 U.S. 398 (2007).
16. *Graham v. John Deere*, 383 U.S. 1 (1966).
17. Council Regulation (EC) No 6/2002 of 12 December 2001 on Community designs, [2002] O.J. L3.
18. Patent Law of the People's Republic of China.
19. Japanese Patent Act.
20. Industrial Design Act of Korea.

Chapter 11
Intellectual Property Issues in Medical Device Labeling and Marketing

11.1 Government Controls Over Medical Device Advertisements and Labeling

11.1.1 Food and Drug Administration

The Food and Drug Administration (FDA) plays a significant role in regulating medical device labeling and advertising. Since labeling is defined as "all labels or other written, printed, or graphic matter, (1) upon any article or any of its containers or wrappers, or (2) accompanying such article at any time while a device is held for sale after shipment or delivery for shipment in interstate commerce" by Section of 201(m) of the Food, Drug and Cosmetics Act, advertisements and sales material that meets these criterial are generally treated the same as labels by the FDA. Any medical apparatus, instrument, implant, in vitro agent or similar component or article that is used to cure, prevent, mitigate, diagnose, or treat disease is subject to labeling requirements (FDA labeling requirements are detailed in Chap. 4).

When marketing materials qualify as labeling, then the FDA has the authority to regulate the content and exercises that authority at its discretion. Marketing claims related to medical devices must be accurate and truthful and cannot be misleading or include an omission of fact. For instance, medical device manufacturers can label a medical device as "FDA approved" only if the medical device has actually been approved by the FDA through the PMA process. Conversely, the same statement, "FDA Approved," would be inaccurate if applied to a medical device that was cleared by the FDA via a 510(k) notification.

Other forms of promotion that do not meet the definition of labeling are not generally regulated by the FDA. However, under sections 502(q) and 502(r) of the Federal Food, Drug and Cosmetic Act, the FDA is also able to impose regulations concerning the advertisement of restricted medical devices. Generally speaking,

G. B. Halt Jr. et al., *FDA and Intellectual Property Strategies for Medical Device Technologies*, https://doi.org/10.1007/978-3-032-12649-8_11

restricted devices that are subject to FDA advertising regulations are often prescriptive devices (e.g., medical devices that cannot be obtained without a prescription from a physician, dentist or veterinarian), experimental or investigative medical devices, approved or cleared devices that are being marketed for off-label uses (i.e., uses which were not cleared or approved by the FDA), and are often Class III devices. Advertising for hearing aids [1], analyte-specific reagents [2], and over-the-counter illicit drug testing kits [3] also falls under the FDA's purview.

Advertising related to restricted medical devices that fall within the FDA's jurisdiction that is false or misleading is often found by the FDA to be misbranded. Similarly, a restricted medical device is considered misbranded by the FDA if the advertisement lacks a brief statement regarding the intended use of the device, the relevant precautions, warnings, contraindications and side effects associated with the device.

Additionally, the FDA also evaluates proprietary names (i.e., trademarks) for certain regulated products, such as human drugs and biological products, to evaluate the safety of a trademark, which is relevant to medical devices that include a drug or biological product, such as a drug-eluting stent. In terms of marketing and promotional purposes, the FDA evaluates proprietary names to ensure that the names are not overly fanciful or imply that the drug does something it cannot. For instance, trademarked drug names cannot imply that the drug is more effective than it truly is, that the drug is for more indications than it is approved for, or that the drug is lower risk than is accurate (i.e., the name cannot be misleading). The trademark name also may not make unsubstantiated superiority claims either.

Furthermore, in an effort to improve patient safety by reducing medication errors, the FDA also evaluates trademark drug names to reduce the potential to confuse consumers. Similar to how a trademark search is conducted during the trademark application process, the FDA evaluates proposed proprietary names for drugs against other drug names that are already available on the market to assess the risk of consumers becoming confused by two similar trademark drug names. The goal of conducting these evaluations is to reduce the risk that a doctor prescribing the medication, a nurse administering the medication or a patient taking the medication, becomes confused due to similarities between the drug they mean to use versus a drug they do not.

When selecting a proposed proprietary name for a drug or biological product that is subject to FDA review, it is recommended that companies avoid choosing a trademark that is similar in spelling or pronunciation to an existing drug product and should refrain from using the same trademark "root" name for multiple products (e.g., a first drug product called Hiborio, and a second drug product for a different indication made with different active ingredients HiborioDulas), unless the two products share at least one common active ingredient (e.g., a first drug product called Ordio, and a second drug product having a lower dose of active ingredient called OrdioLo or OrdioLite). Similarly, Sponsors should refrain from reusing trademark names of discontinued products on new products.

11.1.2 Federal Trade Commission

Medical device advertising that does not constitute labeling falls within the jurisdiction of the Federal Trade Commission (FTC). The FTC is responsible for maintaining a competitive marketplace for both consumers and businesses and preventing unfair or deceptive trade practices. As such, the FTC administers laws and regulations ranging from the content of clothing labels to laws requiring truth in advertising and prohibiting price fixing. The FTC is authorized to regulate all medical device advertising (most class I and class II devices) except for advertising on restricted medical devices, which is reserved for regulation by the FDA.

The FTC regulates advertising under its statutory authority to prohibit deceptive acts or practices. False advertisements related to food, drugs, devices, and services are expressly prohibited by section 12 of the Federal Trade Commission Act. The FTC will find an advertisement deceptive if:

(1) It contains a representation or omission of fact.
(2) That is likely to mislead consumers acting reasonably under the circumstances.
(3) That representation or omission is material.

The first step in the analysis is to identify representations made by an advertisement. A representation may be made expressly or implicitly. An express claim directly makes a representation of fact. An implied claim is not so straightforward and requires an examination of both the representation in order to determine the overall meaning or commercial impression of an advertisement. False claims can also stem from an omission of information, which makes an affirmative representation misleading. In other words, it can be deceptive for a seller to simply remain silent if such silence constitutes an implied, but false, representation.

The second step in identifying deception in an advertisement requires the Commission to consider the representation from the perspective of a consumer acting reasonably under the circumstances.

Finally, a representation must be material, i.e., likely to affect a consumer's choice or use of a product or service. Express claims involving health, safety, price, or efficacy is presumed to be material.

Any claim by an advertiser must have a reasonable basis. Where compliance claims are made (e.g., Energy Star compliance or European Conformity markings), those claims should normally be substantiated by competent and reliable scientific evidence such as tests, analyses, research, studies, or other evidence conducted and evaluated in an objective manner by persons qualified to do so, using generally accepted scientific methods.

The FTC also monitors and reports on industry practices regarding the marketing of violent movies, music, and electronic games to children; provides guidelines regarding U.S. origin claims placed on product packaging; and provides guidelines on the use of environmental marketing claims, or the marketing of "green" products.

11.1.3 State Regulation

Individual states within the U.S. may also regulate labeling and advertisements. Many activities, but by no means all, relating to labeling are preempted by federal laws and regulations promulgated by the FTC, the FDA, and other agencies. The strength of state power lies in each state's consumer protection laws, which seek to prevent deceptive trade practices. Since the penalties for violating these laws can be severe in some cases, individual states can have a great influence on regulating the labeling and advertising of products.

States can also regulate how consumers dispose of their electronics. Many medical devices include electronic components and batteries. For example, some states have enacted laws that prohibit the disposal of consumer electronics in state landfills and have instituted state-regulated electronics recycling programs. Of note are laws such as "Producer Take Back laws," which require manufacturers of certain electronic devices to take back particular types of old electronics that they were responsible for selling in the first place. Medical device companies should take care to ensure that they do not run afoul of the law.

11.1.4 Government Controls Abroad

Regardless of one's knowledge of U.S. labeling laws and regulations, these laws and regulations are only valid within the jurisdiction of the United States. If a medical device is intended to be exported, the labeling requirements of the import country must be identified and followed. Furthermore, the U.S. Bureau of Industry and Security has policies and regulations for the export of dual-use commodities (dual-use items are predominantly for commercial uses, but also have military applications), software, and technology.

By way of example, the European Union (EU) requires the European Conformity (CE) marking to be affixed to any product subject to it before the product is to be placed on the market in the EU. Similarly, the EU, China, and Japan each require markings on products that consist of certain hazardous substances (Restriction on certain Hazardous Substances or "RoHS") to indicate compliance with each country's RoHS standards.

11.1.5 Patent Marking

A recent amendment to the patent law, entitled The America Invents Act (AIA), has changed the manner in which products may be marked with patent numbers. The purpose of placing patent numbers on a product is to put purchasers on notice as to what patents protect the product. Marking serves as "constructive notice" to potential infringers. The AIA amended 35 U.S.C. § 287(a) to allow products to be "virtually marked." Marking with "patent" or "pat." and "an address of a posting on the Internet, accessible to the public without charge for accessing the address, that associates the patented article with the number of the patent," is now a permissible form of notice.

> ***Example*** Under the AIA, a medical device company's product could be virtually marked in any of the following ways:
>
> Patents: http://www.companyname.com/patents
> Patents: http://www.companyname.com/patentnotice
> Patents: http://www.companyname.com/legal/patent

Prior to the adoption of the AIA, any individual could bring a lawsuit on behalf of the United States in what was called a qui tam action for "false patent marking," where products were marked with expired patent numbers. The AIA introduced 35 USC § 292(c), which effectively eliminates any causes of action based on marking products with expired patent numbers.

11.2 Non-government Controls

Many industries have trade associations that may have a code of conduct governing advertisements of their products. However, any such code of conduct is effectively non-binding on members. Moreover, if the code of conduct limits information provided to consumers (e.g., in the event of truthful competitive advertising), it could be considered anti-competitive and subject to challenge by the FTC. Examples of this would be an industry code of conduct that imposes a higher standard of substantiation for comparative claims than for unilateral claims.

Alternatively, false or misleading statements can be brought before the National Advertising Division ("NAD") of the Better Business Bureau. In an action before the NAD, a competitor may file a complaint to take action against false or misleading statements, while avoiding the distractions and expense of full-blown inter-party litigation or involvement with a government agency. After a complaint is initiated and accepted, the NAD will investigate the matter, and if a deceptive or misleading practice is found, it will take action, which can range from requesting corrective advertising from the publisher, publication of the NAD's findings, or referral to a regulatory authority such as the FTC for further penalties.

11.3 Comparative Advertising

Comparative advertising refers to identifying a competitor's product by its trademark and comparing it to the advertised product. Comparative advertising is widely used in medical technology, computer software, and startup industries to convey valuable information to consumers. However, using such comparative advertising runs the risk that a competitor could take action that the claims are either false or that the advertisement infringes the competitor's trademark rights.

The FTC and courts have approved the use of brand comparisons where the bases of the comparison are truthful, objective, and clearly identified. Advertisements containing truthful and non-deceptive statements that a product has certain desirable properties or qualities that a competing product does not possess are permitted. However, false or misleading comparative advertising can trigger liability for false advertising or trademark dilution. Therefore, when a competitor's trademark is used in comparative advertising, the statements used must not have a tendency or capacity to confuse or be considered to be false or deceptive.

In order to lessen the risk of litigation arising from comparative advertising, the steps outlined below should be considered before making any claims regarding a competitor's product.

- Only use the competitor's trademark to the extent necessary for comparison purposes.
- Use the competitor's trademark in the same way that the text is used throughout the advertisement. In other words, do not place emphasis on the competitor's trademark that would lead a consumer to believe that the trademark is associated with the advertiser's company.
- Do not disparage or mock the competitor or the competitor's trademark in the advertisement. Make sure any photographs of a competitor's product are accurate and do not place the product in an unfavorable light. In addition, do not use any photographs of a competitor's product that may be copyrighted by a competitor or a third party. Instead, use original photographs.
- Try to make sure your product's advertisements, packaging, and trade dress create a separate commercial impression from that of your competitor. In other words, a consumer looking at the two products or advertisements should not believe that the products are somehow related because of similarities in colors, fonts, stylization, etc.
- Include an easily visible disclaimer in any advertisement that uses the competitor's trademark. An example of such disclaimer is, "COMPETITOR'S TRADEMARK is a trademark of COMPETITOR. ADVERTISER does not make or license COMPETITOR'S TRADEMARK and is not affiliated or associated with COMPETITOR."
- Verify that any statements regarding a competitor's or advertiser's product are true both explicitly and implicitly. The following steps should be taken:

- Retain a certified, independent, and well-respected laboratory in the industry to perform the testing. Verify that the testing to be performed is the test generally accepted by technical or scientific community.
- Make sure testing is fair and impartial. For example, ensure that the appropriate comparative products from the competitor's product line are compared.
- Testing should support all statements made regarding a competitor's product.
- Make sure copies of the test methodology, test results, and samples tested are preserved in a safe place in case the tests have to be repeated at a later date.

Even if all the above suggestions are followed, there is nothing that can be done to prevent a party that feels it has been harmed from deciding to bring a court action to protect its rights. It is recommended to consult legal counsel regarding individual advertising claims before a comparative advertising campaign is launched.

References

1. C.F.R. § 801.420-801.421.
2. C.F.R. § 809.30.
3. C.F.R. § 809.40.

Chapter 12
Enforcement, Infringement, and Monetization of Intellectual Property Rights

12.1 Policing Your IP Rights

After a medical device company invests the time and resources into developing and protecting its intellectual property rights, it is important to police the market to ensure that no competitors are improperly benefiting from such intellectual property investments. Policing of intellectual property rights can be broken down into a four (4)-step process:

(1) The rights must be identified by class: trademark, trade secret, copyright, or patent. It is also important to remember that any single right may be protected by more than one type of intellectual property protection. For instance, a patent or trade secret right may have an associated trademark, or there may be copyrightable materials that accompany distribution of the patented item. Copyrighted instructions may be an important element in the commercialization of a trade secret. Consider the sale of a trade secret formula where the temperature, rate of mixing, and amount of the intermediary is critical to its performance in the end product. While the owner of such information would not want the disclosure to reveal the trade secret, the copyrighted instructions could be an important sales tool because it may be the only way to make the formula commercially useful to the customer. This identification process should be conducted carefully to consider all possibilities for protection.

(2) The protection of intellectual property rights identified through the prior analysis will require consideration of the competing marketplace of interest. If the identification step yields a product having commercial interest that is potentially patentable and marketing has identified or developed a trademark for the product, there are competing interests associated with the different forms of protection. On the one hand, the patent interest requires diligence to be sure that the invention is not disclosed or sold more than one year prior to applying for the patent. On the other hand, trademark rights are based on use of the mark and there is a desire to initiate commercial use of the mark in association with

G. B. Halt Jr. et al., *FDA and Intellectual Property Strategies for Medical Device Technologies*, https://doi.org/10.1007/978-3-032-12649-8_12

a product as soon as possible. As noted in Chap. 7 of this book, it is possible to file an "Intent to Use" trademark ("ITU") application before there is any disclosure of the invention or use of the trademark. There is a similar procedure, known as a provisional patent application, which can be filed to preserve a patent application filing date prior to any disclosure. Use of these two vehicles permits the preservation of patent and trademark rights while efforts are undertaken to gauge the market's commercial interest. This relatively simple solution to the potential problem is made possible through diligence in the identification of the concerned rights and remaining mindful of the critical first date of the underlying right.

(3) Vigilance in the marketplace is a critical component of intellectual property management. Market vigilance is the key to gathering information about competitive practices and products. This information is useful both for detecting infringement of rights and for avoiding infringement of third-party rights, i.e., the rights held by others. With respect to the issue of detecting infringement, the intellectual property owner is charged with a certain level of vigilance and long delays in detecting an infringement may result in an infringer having an equitable defense against the assertion of the infringed intellectual property right. Thus, you cannot "sleep" on your rights to the detriment of another. Conversely, one cannot count on ignorance of another's rights as a complete defense to a charge of infringement. As a general rule, a medical device company must act in a reasonably prudent manner with respect to its own rights and the rights of another. Frequently, a sales force or distribution network is an excellent source of competitive information. These individuals continuously interface with customers and interact in the marketplace, and are generally aware of competitors' products and services. While the law does not require the engagement of an investigator to search out all possible infringements, it does require an increased level of vigilance once there is a reason to believe there is an infringement.

(4) Enforcement is the end result of properly conducting the above steps and identifying an infringer. Enforcement is a process that should not be taken lightly, as it can have consequences for rights being asserted and consequences for the business entity itself. Once an infringer is discovered, there are steps that need to be taken for the purposes of evaluating the infringement and the impact on the business of the intellectual property owner. In other words, knowledge of an infringement carries with it the requirement of action or at least an informed decision not to take action. The steps needed for the purposes of evaluating the infringement and the impact on the business of the intellectual property owner should be followed in either case.

12.1.1 Medical Use Exception

Patents in the medical field can be secured on a variety of inventive medical devices, as well as for innovative medical procedures. Unique to the medical field is an exception to patent infringement known as the medical procedure exception. It is an exception from patent infringement liability for performing a patented medical procedure (e.g., a medical or surgical procedure on the body of a patient) without authorization. The exception is only available to medical practitioners. The medical practitioner must either be a licensed medical professional licensed to perform the procedure or medical activity, or someone who is performing the infringing activity under the instruction of a licensed medical professional who is licensed to perform the procedure or medical activity.

The medical procedure exception is limited only to patented medical procedures. It does not apply, for example, to infringement of a patent for a medical device. Many medical device companies file dual patent applications: one for an innovative medical device, and a second for a medical procedure using the innovative device. With protection on both the medical procedure and the medical device, infringers may be able to be excepted from liability when they satisfy the criteria for the medical procedure exception, but will still be liable for infringing the medical device patent.

12.2 What is Infringement

Every patent case involving a medical device or SaMD ultimately answers one question by following a two-step path: "Whether an accused device infringes requires a two-step analysis—the court first determines the scope and meaning of the patent claims asserted, and then the properly construed claims are compared to the allegedly infringing device" [1, p. 1295].

Step one is claim construction. The court fixes the legal meaning of terms such as "controller," "server," "catheter," or "secure transmission" in light of the claims, the specification, and the prosecution history, and, where applicable, applies 35 U.S.C. § 112(f) to tie software nouns to disclosed algorithms.

Step two is comparison. With the constructions from step one in hand, the parties map the accused pump, monitor, implant, or cloud workflow to the claim, element by element, first for literal presence and then, only if needed, under the doctrine of equivalents. In the SaMD context, that proof often turns on whether one processor or message performs all verbs tied to "said" element, whether a specific control law is actually implemented, and whether distributed microservices together meet a step the claim assigns to a single server.

The sections that follow track that sequence and add one threshold issue unique in its practical impact for health-tech litigants. The claim-construction section (Sect. 12.2.1) explains how courts read device and software terms through the eyes of a skilled team working under clinical and regulatory constraints. The infringement section (Sect. 12.2.2) then illustrates how these constructions drive the element-by-element comparison in connected-care systems, including divided-method questions and § 112(f) analyses for software claims. Finally, the venue section (Sect. 12.2.3) situates these merits battles in the correct forum under 28 U.S.C. § 1400(b), a choice that often turns on where a device company actually does business, where its SaMD architecture is supported by people on the ground, and where evidence and witnesses are located. Read together, these sections provide a practical playbook for transitioning from legal meaning to technical proof in the medical device and SaMD space.

12.2.1 Claim Construction

Claim construction is the first merits decision in most patent cases, and it sets the framework for everything that follows. The Supreme Court case *Markman v. Westview Instruments, Inc.*, provides that the court, not the jury, construes the claims as a matter of law [2]. Constructions are given the meaning that a person of ordinary skill in the art would ascribe to the claim terms at the time of the invention, in the context of the patent as a whole. The Federal Circuit's en banc decision in *Phillips v. AWH Corp.* is the lodestar: start with the claims, read them in view of the specification, and consider the prosecution history, while using extrinsic evidence cautiously and only to the extent it illuminates rather than overrides the intrinsic record [3, 4].

Courts begin with the claim language, which carries a rebuttable presumption of its ordinary and customary meaning to a skilled artisan. In med-device and SaMD disputes, context often decides the fight: terms like "electrogram," "catheter," "implantable," "biocompatible," "controller," "server," "module," and "configured to" are read in the company they keep and in view of how the specification uses them. The specification is "the single best guide to the meaning of a disputed term," and it can resolve ambiguity by showing how the inventor used a term or by characterizing the invention [3, pp. 1315–1317]. For example, if the specification describes "the present system" as a catheter-based ablation platform that performs both sensing and therapy with a single steerable catheter, that description can narrow "catheter" in a way that excludes two separate tools. Prosecution history can also narrow the scope through clear and unmistakable disclaimer, including statements made in IPRs or reexaminations [5, pp. 1323–1326; 6, pp. 1361–1362]. Dictionaries, clinical treatises, standards such as Health Level 7 (HL7) or Digital Imaging and Communications in Medicine (DICOM), and expert testimony may be consulted, but they cannot contradict the intrinsic record [3, pp. 1318–1324].

Two well-defined exceptions can move a term away from its ordinary meaning. First is lexicography: a patentee can act as its own lexicographer with "reasonable clarity, deliberateness, and precision," for example defining "secure transmission" as "TLS 1.3 with mutual authentication" in a SaMD specification [7, p. 1249; 8, pp. 1366–1367]. Second is disavowal: a patentee can clearly and unmistakably limit the claims, for example by stating that "the present invention requires closed-loop control using a proportional-integral controller," which can exclude other control laws [9, pp. 1341–1345; 5, pp. 1323–1326]. Courts are reluctant to find either lexicography or disavowal unless the record makes the intent clear [10, pp. 1365–1368].

When parties urge "plain and ordinary meaning," the court must still resolve any genuine dispute about scope. It cannot send an unresolved legal question to the jury, such as whether "server" in a remote cardiac-monitoring patent means a single physical machine or a distributed microservices cluster [11, pp. 1360–1362]. At the same time, courts will not redraft claims to correct perceived drafting errors. If the language is clear, even if awkward, it controls [12, pp. 1373–1374].

If a claim uses "means for" doing a function, or a term that is a functional nonce, § 112(f) applies. The court then identifies the claimed function and the "corresponding structure" disclosed in the specification that performs it, and limits the claim accordingly [13, pp. 1348–1351]. In medical device and SaMD claims, that often means tying "processor," "module," or "handler" terms to specific algorithms, data paths, or protocol roles. Failure to disclose sufficient corresponding structure—including an algorithm for computer-implemented functions like "detecting atrial fibrillation," "controlling infusion to maintain a therapeutic window," or "reconstructing an image"—can render the claim indefinite [14, pp. 1333–1338; 15, pp. 1007–1009].

Construction and indefiniteness are intertwined. If, after applying these tools, the court cannot reasonably assign boundaries to a term, the claim is invalid as indefinite because it fails to inform a skilled artisan, with reasonable certainty, about the scope of the invention [16, pp. 910–912]. In clinical contexts, terms like "clinically significant," "stable," or "secure" need clear and objective anchors in the specification (such as metrics, thresholds, state machines, or standards), so a skilled team knows where the line is.

The presence of a dependent claim that adds a limitation gives rise to a presumption that the corresponding independent claim lacks that limitation, although the presumption yields to contrary intrinsic evidence [17, pp. 910–912]. In device cases, if claim 1 recites a "catheter configured to sense electrograms," and claim 3 adds "and deliver radiofrequency ablation," differentiation suggests claim 1 is not limited to therapy delivery unless the intrinsic record says otherwise.

Accordingly, claim construction is not simply an early procedural step; it is the foundation for infringement and validity analysis. By requiring courts to construe claims as a matter of law, *Markman* ensures that the meaning of the patent is determined consistently and grounded in the intrinsic record. The *Phillips* framework anchors that task in the claims, the specification, and the prosecution history, while giving limited weight to extrinsic sources. That approach is particularly important in the medical device and SaMD context, where terms often bridge engineering, clinical, and regulatory domains.

12.2.2 Infringement After Claim Construction

Once the court construes the claims, infringement is a comparison exercise. The patentee must prove, by a preponderance of the evidence, that the accused product or method meets every properly construed limitation, either literally or under the doctrine of equivalents. Courts compare the accused device to the claims, not to the patentee's commercial embodiment [18, pp. 1575–1576].

Literal infringement asks whether each claim element, as construed, is present in the accused instrumentality. Open-ended "comprising" language allows the accused product to include additional features, but it does not excuse the absence of a required element [19, pp. 1371–1372; 20, p. 1535]. If even one limitation is missing, there is no literal infringement. For computer-implemented and software claims, courts apply any § 112(f) construction the court adopted: the accused software must perform the identical function using the same or an equivalent algorithmic structure disclosed in the specification [14, 21, pp. 1267–1268].

If a limitation is not literally present, a patentee may rely on the doctrine of equivalents. Under that doctrine, an element is met if the accused feature is insubstantially different from what is claimed, often tested by whether it performs substantially the same function, in substantially the same way, to achieve substantially the same result [22, pp. 608–609]. The Supreme Court reaffirmed that the all-elements rule still governs; the doctrine cannot erase or vitiate a claim limitation [23, pp. 29–40; 24, pp. 1358–1359]. Prosecution history estoppel can bar equivalents when the patentee narrowed the claim for patentability. The presumption of estoppel that follows a narrowing amendment can be overcome only in limited circumstances, such as when the rationale for the amendment is tangential to the accused equivalent or when the equivalent was unforeseeable at the time [25, pp. 736–741; 26, pp. 1366–1370; 27, pp. 1377–1379]. Because equivalence is limitation by limitation, the proof must be particularized and supported by evidence, not generalities [28, pp. 1328–1330].

Method claims raise two recurring questions. First, "who" performs all the steps. Direct infringement under 35 U.S.C. § 271(a) requires that a single entity performs every step, or that one entity directs or controls the others who perform missing steps, or that the participants form a joint enterprise. The Federal Circuit's en banc decision clarified that direction or control can exist when an alleged infringer conditions participation in an activity or receipt of a benefit upon performance of the steps and establishes the manner or timing of that performance [29, pp. 1022–1029; 30, pp. 1377–1378]. Second, for "use" of system claims, a party uses a system when it puts the invention into service, controls the system as a whole, and obtains benefit from it, even if some components are operated by others [31, pp. 1284–1285].

Indirect infringement depends on someone else's direct infringement. Induced infringement under § 271(b) requires proof that the accused inducer knew of the patent and knew that the induced acts constituted infringement, and that it specifically intended to encourage those acts [32, pp. 766–771]. A good-faith belief in invalidity is not a defense to inducement [33, pp. 642–648]. Contributory infringement under § 271(c) requires sale or offer to sell a component that is a material part of the invention, knowledge of the patent, and that the component is not a staple article of commerce suitable for substantial noninfringing use [34, pp. 484–491].

Territorial rules can affect liability. Section 271(a) covers making, using, selling, offering for sale, or importing in the United States. Supplying components from the United States for combination abroad can create liability under § 271(f) [35, pp. 441–457]. Importing a product made by a patented process infringes under § 271(g), although importing information generated by a process is not enough [36, pp. 1377–1378]. When § 271(f) applies, damages may include certain foreign lost profits [37, pp. 2136–2139]. The FDA research safe harbor exempts certain uses reasonably related to obtaining regulatory approval [38, 39, pp. 202–208].

The mapping exercise is very fact intensive. Parties typically select representative claims and accused configurations, then present element-by-element charts that track the court's constructions. For software and networked systems, contemporaneous architecture diagrams, logs, and source code help show whether a limitation is present "as construed." For medical devices and SaMD, disputes often center on whether a single processor or message performs all verbs tied to "said" element, whether a control law that maintains a physiological parameter is actually implemented on the accused device, or whether distributed services collectively meet a step that the claim assigns to one server. In divided method scenarios common to connected health, parties litigate whether a manufacturer or platform provider directs or controls clinicians or patients who perform remaining steps, or whether the provider conditions access to an app or benefit on performing the claimed steps in a prescribed manner [29, pp. 1022–1029; 40, pp. 1364–1370].

After liability comes into focus, collateral issues shape remedies but can feed back into infringement proof. Constructive marking under 35 U.S.C. § 287(a) can limit pre-suit damages unless the patentee marked or provided actual notice; the accused infringer can shift the burden by identifying unmarked patented products [41, pp. 1366–1368]. Willfulness, which can support enhanced damages, requires egregious infringement behavior beyond typical infringement [42, pp. 103–106]. Permanent injunctions depend on the four-factor test, including irreparable harm and inadequacy of legal remedies [43, pp. 391–394].

In sum, infringement analysis ultimately comes down to applying the court's constructions to the accused product or method and proving that each limitation is met literally or by an equivalent. For medical devices and SaMD, this analysis often involves translating highly technical features into the claim language. For example, one question is whether an infusion pump actually implements the claimed control law or instead approximates the clinical outcome by other means. Another question is whether a remote monitoring system uses a single processor or message that performs all of the recited steps, or whether distributed services divide the work in a way that falls short of the claim requirements. Because infringement must be proven element by element, parties in this space rely heavily on device schematics, source code, communications protocols, and regulatory records to show how the system behaves in practice.

This process is particularly consequential for medical technologies because the stakes of claim scope go beyond financial remedies. In a field where FDA clearance/gran/approval, hospital adoption, and patient trust hinge on reliability and safety, an adverse infringement ruling can reshape product design, licensing strategy, and even regulatory submissions. For connected health platforms, questions of divided infringement are now routine, especially when a manufacturer directs or controls clinicians or patients who perform certain steps. With SaMD, where functionality often sits in distributed cloud architectures, the doctrine of equivalents can become critical, capturing minor implementation differences that achieve the same clinical results in substantially the same way.

The conclusion is clear: once claims are construed, infringement is a rigorous mapping exercise that is both legally precise and technically granular. In the medical device and SaMD sectors, the analysis requires not only matching technical elements to claim limitations but also understanding the clinical and regulatory environment in which those systems operate. Courts demand evidence that a person of ordinary skill would recognize as meeting each limitation as construed, and parties who can connect that legal standard to the practical realities of device design and deployment will be best positioned to prevail.

12.2.3 Venue

Venue in patent cases is governed exclusively by 28 U.S.C. § 1400(b). A plaintiff may sue either in the judicial district where the defendant "resides" or in a district where the defendant "has committed acts of infringement and has a regular and established place of business." In 2017, the Supreme Court held that for domestic corporations "resides" means the state of incorporation, rejecting broader practice and re-centering patent venue on § 1400(b)'s narrow text [44, pp. 261–270].

For medical device and SaMD defendants incorporated in states with multiple federal districts, residence is further narrowed. A corporation does not "reside" in every district within that state. The Federal Circuit holds that such a defendant resides in only one district: the district of its principal place of business, or if none, the district of its registered office [45, pp. 982–983].

If the defendant does not "reside" in the chosen district, the plaintiff must satisfy § 1400(b)'s second prong by showing three things: a physical place in the district, that is a regular and established place of business, and that is the place "of the defendant" rather than merely a place of an employee or a third party [46, p. 1360]. In the medical context, a field engineer's home office, a hospital's server closet, or a third-party data center typically will not qualify without additional facts tying the location to the company. Equipment alone is not enough [47, pp. 1343–1347]. Likewise, a retail or service location belonging to an affiliate or independent distributor is not automatically a place of business for corporate siblings; the record must show that the specific defendant controls or operates that site [48, pp. 1287–1295]. The plaintiff bears the burden on venue [48, pp. 1287].

For years, the Eastern District of Texas (E.D. Tex.) was the nation's busiest patent venue, including for device makers and digital health platforms. After *TC Heartland* and *Cray*, its venue rulings helped define the post-2017 landscape. Two decisions in particular matter. First, *In re Google* rejected attempts to base venue on "edge" servers in third-party data centers, emphasizing the need for a physical place of the defendant and an employee or agent regularly conducting the defendant's business there [47, pp. 1343–47]. Second, *Andra Group* confirmed that a store owned by one affiliate is not necessarily a place of business of other related entities without proof of control or agency [48, pp. 1288–95]. Together these cases narrowed theories once used in E.D. Tex.

Beginning in 2018, the Waco Division of the Western District of Texas (W.D. Tex.) became a major magnet for tech cases, including suits against SaMD and connected-device companies. The Federal Circuit responded with a series of mandamus decisions enforcing 28 U.S.C. § 1404(a) transfer principles, often ordering transfer from W.D. Tex. to the Northern District of California where engineers, documents, and third-party witnesses were concentrated [49, pp. 1336–1339; 50, pp. 931–933; 51, pp. 2–4; 52, pp. 2–4; 53, pp. 2–4; 54, pp. 2–4]. These orders instruct W.D. Texto prioritize convenience factors, such as witness location and access to proof, and to resolve transfer motions before significant activity.

The bottom line for medical devices and SaMD is that patent venue is narrow and fact intensive. Post-*TC Heartland*, plaintiffs in the medical device and SaMD space must either sue where the defendant is incorporated, and in the correct district within that state, or prove past acts of infringement plus a regular, established place of business of the defendant in the forum. Texas remains important, with E.D. Tex. as the crucible for modern venue rules and W.D. Tex. as the focal point of transfer jurisprudence, but § 1400(b) and the Federal Circuit's guidance now define the outer bounds for where device and digital health cases can proceed.

12.3 Evaluating a Controversy Before Commencing Litigation

A number of inquiries should be made prior to the commencement of filing a lawsuit to enforce intellectual property rights. First, an initial assessment should be made as to whether the intellectual property right can be enforced. This involves identifying the right and checking its validity and enforceability. This evaluation needs to be made even when all steps have been taken to obtain a valid enforceable right, in order to ensure that the right has maintained its enforceability. For example, a patent may have lapsed for failure to pay maintenance fees and will no longer be enforceable under this scenario. Likewise, a trademark that has become abandoned (i.e., the owner is no longer using the mark and has no intention of resuming use) is no longer enforceable.

In cases where an intellectual property right has been lost or becomes unenforceable, in certain instances the right can be regained or made enforceable again. For example, a patent that has lapsed for failure to pay maintenance fees may be reinstated up to two (2) years after the lapse if it can be shown that the delay in payment was unintentional. With any intellectual property right, a review should be conducted to ensure that nothing has occurred to cause the right to cease being enforceable, and if it is found that such an event has occurred, a further check should be made to identify whether any steps can be taken to reinstate the right.

All elements of a legal cause of action must be present for successful enforcement. This evaluation is different for each type of intellectual property right.

- For a patent, the evaluation involves asking whether an allegedly infringing device meets each and every limitation of one or more of the patent's claims.

- For a trademark, the evaluation involves asking whether there is a likelihood of confusion between the asserted trademark and the allegedly infringing mark.
- For a copyright, the evaluation involves asking whether there was an actual copying of the work or a portion of the work and whether there is any evidence that the infringer has access to the copyrighted work.

With any cause of action, the first step is a check of whether all the elements of the potential claim can be satisfied.

It is always necessary to make a risk–benefit analysis of the economic impact of any potential litigation. Litigation can be very costly, and the potential cost of a lawsuit is often difficult to predict. The American Intellectual Property Law Association recently published a report that found that legal fees in patent litigation could be $600,000 where less than one million dollars was at stake, and $2 million or more where more than one million dollars was at stake [55]. The estimated litigation cost should be weighed against the damages sought, and the probability of actually obtaining such an amount. It should be kept in mind that in many cases, remedies other than monetary damages may be awarded. A valuation of all potential remedies should be made. For example, if an injunction to stop an infringement is the main relief sought, an analysis of the value imparted to the intellectual property owner by the cessation of the infringing behavior needs to be made before any action is taken. Litigations can be extremely disruptive to an organization. Accordingly, the analysis should include the non-monetary costs of the time and energy that litigation takes away from the normal business operations.

12.4 Litigation Outcomes

12.4.1 Injunctions

An injunction is an order by a court to the infringing party that orders them to stop an infringing action. For example, in the case of an injunction for a product that infringes a patent, the court may issue an injunction to immediately stop sales of an infringing product, or to cut off the sale of an infringing service. An injunction is a remedy that is available in most intellectual property lawsuits, especially patent and trademark lawsuits. Injunctions can be sought in addition to monetary damages. As disclosed in more detail below, an injunction may be permanent or preliminary (i.e., temporary).

A party seeking permanent injunctive relief must demonstrate by competent evidence that:

(1) It has suffered an irreparable injury.
(2) Other remedies available at law, such as monetary damages, are inadequate to compensate for the injury.

(3) The balance of hardships between the plaintiff and defendant warrant the granting of an equitable remedy.
(4) The injunction will serve the public interest.

Although permanent injunctions are more common in intellectual property lawsuits because monetary damage amounts are often difficult to ascertain, they are not considered to be automatic upon a finding of infringement. Injunctive relief in an intellectual property matter must meet the same four (4) elements noted above.

12.4.2 Payment of Royalties

One possible remedy for patent, trademark, or copyright infringement is an order that the infringer must pay the intellectual property owner a reasonable royalty. The court may determine what constitutes a reasonable royalty and order payment for past infringement, and, in certain cases involving public interest, order that royalties be paid for the future use of another's intellectual property right. An order for payment of reasonable royalties may be desirable where the actual damages are difficult to evaluate. Courts will often engage in a weighing of the harms when determining an appropriate remedy in intellectual property cases, and avoid awarding a remedy that will impose an undue burden on the infringer, particularly where the infringement was innocent. In some cases, courts will give an innocent infringer an option between an injunction and an order to pay reasonable royalties until such time as the infringer phases out the use of the owner's intellectual property without suffering excessive damage.

Example Dr. Fictum, MD had an arrangement with BigMED Company in which Dr. Fictum would share ideas with BigMED, and BigMED would evaluate the ideas for possible commercialization. One idea that Dr. Fictum memorialized in writing in an invention disclosure and pitched to BigMED was a design for an expandable urinary catheter device that was meant to be used in patients with incontinence issues. After a couple years, BigMED expressed no interest in the design. However, during the same time period, BigMED was developing a similar expandable urinary catheter device called the Intercludor.

BigMED sought patent protection on features and functions that Dr. Fictum claimed were attributable to him. Dr. Fictum sued BigMED for use of confidential information in the form of misappropriation of trade secret, along with a number of other causes of action.

A jury awarded Dr. Fictum a royalty of 7.5% of the total net sales earned by BigMED prior to trial, and the court later awarded Dr. Fictum an on-going 5.5% royalty rate on all future sales made by BigMED on two other devices made by BigMED containing Dr. Fictum's designs, limited to a period of 10 years from the date when Dr. Fictum gave the invention disclosure to BigMED.

12.4.3 Monetary Damages

Monetary damages can be difficult and expensive to prove in intellectual property litigation because it is often difficult to ascertain the actual loss attributable to, and incurred, as a result of the infringing behavior. An investigation should be made to determine what damages were incurred, and if they are recoverable. For example, in a suit for trademark infringement, the trademark owner may seek damages for loss of goodwill resulting from the infringement, but measuring those damages can be difficult and involve surveys and experts that may push the costs beyond what is recoverable.

In certain cases, such as infringement of a registered copyright, the owner of the copyright may be entitled to statutory damages without the requirement of proving actual damages if the statutory requirements are met. The copyright statute provides, upon election by the copyright owner, that the court may award anywhere from $750 to $30,000 for each work infringed upon and may further increase the award per infringement up to $150,000 for willful infringement [56].

12.4.4 Invalidity

In patent infringement actions, a common defense that is often raised by the accused infringer is patent invalidity. This defense is based on the premise that the patent that is being asserted against them is not a valid patent, i.e., the patent was issued by the United States Patent and Trademark Office in error. The court will examine the patent at issue and make a determination as to whether the patent is valid or not. If this defense is successful, the patent holder's patent will be deemed invalid and thus unenforceable, and the infringement allegations necessarily will be dismissed since there is no longer a valid patent to base an infringement claim on.

12.5 Settling Controversy Without Litigation

12.5.1 Arbitration

Arbitration is a non-judicial form of resolution where the parties select an arbitrator or panel of arbitrators who hear evidence and decide the matter. The proceeding is similar to, but not as formal or nearly as expensive, as a court trial. The parties can agree that an arbitration award can be binding and enforceable in court and appeals may be permitted.

Arbitration offers various advantages over litigation, and several particular advantages unique to resolution of intellectual property disputes. Arbitration presents the opportunity to have a dispute settled in a more expedient manner than conventional courtroom litigation. Litigation of cases involving intellectual property issues can be long, tedious, and complex. Arbitration may shorten the dispute process, and, because of the shortened time frame and limited need for discovery, reduce the costs compared to those typically associated litigation.

One particular advantage of arbitration with respect to settlement of patent disputes lies in the fact that the arbitrator is chosen by the parties, permitting selection of an arbitrator having a technical background that will facilitate understanding of the patented subject matter. In 1983, the United States Patent Act was amended to provide for the voluntary settlement of patent disputes by binding arbitration.

12.5.2 Mediation

Mediation can be an effective means of settling a dispute more quickly and at a lower cost than litigation, but it is very different from arbitration. The mediator functions differently than a judge, jury, or arbitrator by working with the parties to assist in a negotiated agreement that satisfies the interests of all without any determination of who is right or wrong.

Settlement of disputes by mediation offers several practical advantages. Litigation costs, such as discovery and motion practice, can be greatly reduced or eliminated. Another significant advantage is the abbreviated time lapse between commencement and resolution of the dispute, particularly due to the elimination of appeals, which can draw out the litigation process. This is of particular importance in the context of patent litigation, where the right could expire or the technology involved could potentially become obsolete before the dispute is resolved. In addition, parties using mediation for dispute resolution avoid the risk of a complete loss on all counts, and, hopefully, negotiate a resolution that favors the continuation of the business interests of the party.

Mediation can be particularly valuable in resolving disputes over intellectual property rights because it offers the parties an opportunity to come to a negotiated resolution that is unlikely to be obtained in court. Agreements may be reached that allow all parties to exercise an intellectual property right with minimal intrusion on the rights of the other parties. For example, in a dispute over trademark rights, the parties may come to an agreement where each agrees to keep limited use of the trademark, such as by confinement to a particular geographic boundary, or certain business verticals.

12.5.3 Licensing Agreements

Licensing agreements are often a good way to circumvent litigation, mediation, and arbitration. A licensing agreement is a contract that the parties enter into where the property right holder grants permission for the infringer to use the intellectual property going forward in exchange for something in return. The infringer becomes a licensee.

The license can place limitations on the licensee, such as requiring the licensee to pay for use of the intellectual property rights, prohibiting the licensee from modifying the intellectual property, or limiting how the licensee may use the intellectual property. Many infringers who are faced with the choice of a lawsuit or taking a license opt for the license as it is often the more affordable choice.

12.6 IP Litigation

In cases where there is no other solution, litigation may be the necessary means to resolve the dispute.

12.6.1 Selecting a Jurisdiction

Intellectual property lawsuits may be brought in either the state or federal courts. The court used is dependent on the issue and the authority for the asserted property right. Since trade secret rights most typically arise under state law, those suits are most common in state courts. Patent and copyright disputes are both the subject matter of federal statutes that invoke a federal court jurisdiction. Trademark rights may arise under state or federal law and the authority for such rights will determine whether the suit should be brought in state or federal court. Although most cases involving patent and copyright issues will be brought in federal court, there are exceptions where an intellectual property issue arises collaterally with a state law matter, such as interpretation of a license or contract.

In addition to determining whether state or federal court is the appropriate place to bring suit, the defendant must reside in or do business within the state where the suit is brought.

12.6.2 *Causes of Action*

12.6.2.1 Infringement

Infringement actions are the most common type of litigation involving patents, trademarks, and copyrights. An infringement lawsuit may be brought for either direct or contributory infringement. In a lawsuit brought for direct infringement, the party allegedly engaging in the infringing activity is named as the defendant.

For patents, infringement results from making, using, selling, offering to sell, or importing into the United States an invention within the scope of the patent claims without the patent owner's permission. For trademarks, infringement results from the unauthorized use of the same mark or a mark that has a likelihood of creating confusion between that mark and the plaintiff's mark. For copyrights, infringement results when the infringer had notice of, or access to, the copyright work and engaged in actual copying of all or part of the copyrighted work.

In some circumstances, the infringement may result from activity that is not directly infringing, but it "encourages" or "enables" infringement by a third party. In these cases, the intellectual property owner may bring a lawsuit against a party for inducing or contributory infringement (i.e., indirect infringement). Such causes of action for indirect infringement can be based upon various activities. For instance, the defendant may be making or selling product that is intended for an infringing use or is sold with instructions that provide directions for another to infringe. This situation is common in both the patent and copyright contexts. In either situation, it may be undesirable to file suit against the direct infringers as there may be a large number of defendants or the defendants may be customers, and the direct infringers are typically less likely to have recoverable funds than the indirect infringer.

Indirect infringement frequently arises where a patent contains only method claims. Someone making or selling a product that does not itself infringe the patent claims may encourage its use in a way that does infringe the claimed method. The patent owner may not be able to sue the maker or seller for direct infringement of the method, but could sue the maker or seller for indirect infringement if those activities induced others to infringe, or contributed to others infringing a patent.

12.6.2.2 Declaratory Judgments

A party having a real interest to capitalize on subject matter that is alleged to be covered by another party's intellectual property right may file suit seeking a "declaration" by the court that there is no infringement of the intellectual property right in question, the right in question is invalid, or that the right in question is not enforceable. This is known as a declaratory judgment action.

The decision to seek a declaratory judgment requires serious thought. The owner of the rights in question may not pursue the potential infringement based on its own policy reasons, or may no longer have a sufficient commercial interest in them to expend the costs to pursue litigation. Thus, any decision to file a declaratory judgment action must undergo the same analysis applied to any other litigation.

12.6.2.3 Trademark Dilution

Trademark dilution originated as a state law cause of action pertaining to common law trademark rights. In 1995, it was codified in the Federal Trademark Dilution Act and made applicable to federally registered trademarks. Dilution actions can only be brought by owners of "famous" trademarks. To make out a cause of action, the plaintiff must show that the defendant adopted the use of a mark that tends to dilute the distinctive quality of plaintiff's famous mark. If such a showing is made, the plaintiff can obtain an injunction against the defendant's use of the mark, even absent a showing of likelihood of confusion.

12.6.2.4 Misappropriation of Trade Secrets

To bring an action for misappropriation of trade secrets, a plaintiff must show the existence of a trade secret, and that the defendant misappropriated the trade secret. Under the Uniform Trade Secrets Act (UTSA), which many states follow, a trade secret is any information having independent economic value due to its nature and the fact that it is not commonly known, for which reasonable efforts are made to maintain secrecy. Misappropriation occurs where the defendant either acquired the information by improper means, or received the information with knowledge that it was derived by improper means or disclosed in breach of a duty to keep secret. The Defend Trade Secrets Act of 2016 enables trade secret owners to sue for misappropriation in federal court [57].

Trade secret law does not take the "strict liability" approach of patent and trademark law, and requires some wrongful behavior on the defendant's part in order to impose liability. It should also be kept in mind that a party who "reverse engineers" the trade secret cannot be held liable for misappropriation of trade secrets, as this is not considered an "improper means."

12.6.3 Preliminary Injunction

Seeking a preliminary injunction at the outset of the litigation may be desirable to prevent further infringement during the proceedings. Such an injunction is temporary, and is typically only awarded upon a showing that the plaintiff has a strong likelihood of success on the merits, along with the potential to suffer irreparable injury if the allegedly infringing activity is permitted to continue during the proceedings. This can be a difficult burden for a plaintiff, particularly in intellectual property lawsuits where the court is often charged with detailed factual evaluations or subjective multifactor tests. Nonetheless, a case can be presented that warrants a preliminary injunction in order to mitigate damages.

12.6.4 Discovery Stage

Discovery is a highly important phase of intellectual property litigation and its focus will vary depending on what type of intellectual property right is at issue. This is a time period where both sides of a dispute ask for, and are required to provide, information to the opposing party. The Federal Rules of Civil Procedure govern intellectual property proceedings in federal court, while state rules will govern those cases that are brought in state courts.

Depositions, interrogatories, requests for admissions, and document requests are all available as discovery tools in intellectual property litigation. The Federal Rules of Civil Procedure provide for discovery of "any non-privileged matter that is relevant to any party's claim or defense [58]."

12.6.5 Summary Judgment

Motions for summary judgment can be an efficient means for bringing litigation to an early conclusion. Federal Rule of Civil Procedure 56(c) states that a party is entitled to summary judgment where "the pleadings, the discovery and disclosure materials on file, and any affidavits show that there is no genuine issue as to any material fact and that the movant is entitled to judgment as a matter of law."

Having a motion for summary judgment granted can be challenging in intellectual property cases where a large number of complex issues are often present and the evidence may be subject to different interpretations. One example is the case in which a defendant moves for summary judgment holding that the plaintiff's intellectual property right is invalid or unenforceable. While there are some instances where the controlling law is so clear that there is no factual dispute, this is fairly rare. Most issues frequently involve the need for testimony and an evaluation of credibility by a fact finder. Although a grant of summary judgment may be difficult at times, it may be desirable as a means of framing the issues in an effort to expedite the proceedings.

12.6.6 Trial

The trial phase of litigation can be long and expensive in both time and money. The more issues and the more complex the issues are, the longer the trial is likely to take. Patent and trade secret cases can have lengthy trials due to the need to educate the fact finder (i.e., a judge or a jury) on the technology at issue. Trademark and copyright trials can also be time consuming; however, some of these cases can involve simple issues that are easily comprehended by the fact finder.

Experts are especially important in intellectual property litigation, particularly in trade secret and patent litigation where complicated technological issues are likely to be present. An expert witness in a patent case may be needed to testify as a "person having ordinary skill in the art." An expert can also help by assisting the judge and jury in gaining an understanding of technology involved in the case. Experts are also frequently used in the economic area to establish or refute damages.

12.6.7 Costs

Some statutes make attorney's fees and costs available as remedies to intellectual property owners involved in litigation. The Patent Act and Trademark Act both make attorneys' fees available in "exceptional cases." The Copyright Act permits a court, in its discretion, to award litigation costs including attorney's fees to the prevailing party. Courts usually interpret each of the provisions as applying to cases where willful infringement has been found. U.S. courts do not routinely grant fee shifting awards and litigation should not be viewed as a "loser pays" situation.

12.7 Proceedings in the U.S. Patent and Trademark Office

12.7.1 Trademark Oppositions

A trademark opposition is an *inter partes* (meaning "between the parties") proceeding that may be requested by any party who believes there is a potential to be damaged by the registration of a trademark. After a trademark application is examined and found allowable, the U.S. Patent and Trademark Office publishes the mark in its *Official Gazette* to put the public on notice of the potential registration of the mark. A potential opposer (i.e., someone who does not want the trademark to issue) has thirty (30) days from the date of publication to file an opposition, or request an extension of time for filing an opposition. Typical grounds for filing an opposition include likelihood of confusion with the opposer's mark, genericness or descriptiveness, abandonment, and fraud.

After an opposition is filed, the trademark applicant/owner responds by filing an answer asserting any defenses. Frequently, the answer to an opposition will assert a counterclaim to cancel a registered mark of the opposer, which forms the basis of the opposition. The opposition proceeding is held before the Trademark Trial and Appeal Board. This is an administrative panel, rather than a judicial panel, with expertise in trademark matters. The proceeding is conducted similar to a federal court proceeding, but is less formal. The PTO proceeding concludes with a Board decision either refusing the registration or permitting the pending registration to go forward to registration. The losing party may appeal the Board's decision to the U.S. Court of Appeals for the Federal Circuit. The denial of a registration does not mean that the mark cannot be used. The right to use the mark requires a judicial determination; however, the results of an opposition can often lead to a settlement between the parties.

12.7.2 Patent Post Grant Review Options

When a party believes that a patent is invalid and should not have been issued by the USPTO, the party can seek review of the patent in question through a post grant review proceeding. The post grant review options include Post Grant Review, *Inter Partes* Review, and Patent Reexamination.

12.7.2.1 Post Grant Review

Post Grant Review (PGR) is a trial proceeding that is conducted by the Patent Trial and Appeal Board (PTAB) that reviews the patentability of patent claims. Third parties can file a petition for Post Grant Review of an issued patent within nine (9) months of the patent being issued by the United States Patent and Trademark Office. It must be shown that it is more likely than not that the claim or claims of the patent that are being challenged are invalid or unpatentable. PGR can also be requested to resolve a novel or unsettled legal issue that is pertinent to other patents, or other patent applications.

PGR can be sought on the following bases:

(1) The issued patent does not satisfy the technical requirements for issued patents (e.g., lacks written description, lacks enablement, or is not patent eligible on other grounds, such as 35 U.S.C. §101, §102, and §103). PGR will not be initiated if the grounds for review is that the patent fails to comply with the disclosure of the best mode requirement.
(2) Unpatentability based on evidence that predates the patent filing date, such as evidence of disclosure in a publication, disclosure in a patent or patent application, a public sale or a public demonstration.

PGR proceedings are designed to be time efficient. Not only must PGR challenges be filed within nine (9) months of the issue date of the patent in question, but the PGR review conducted by the PTAB is usually completed within twelve (12) to eighteen (18) months.

There are a few drawbacks or disadvantages to using the PGR proceeding. For instance, all petitioners must be identified in PGR proceedings, i.e., the party requesting the PGR cannot remain anonymous. Another concern with PGR is that a legal estoppel that results from the PGR proceeding applies to a wide array of other forums, such as other USPTO proceedings, district court proceedings, and U.S. International Trade Commission proceedings. In PGR proceedings, estoppel attaches with the issuance of the PTAB's final written determination. Once estoppel attaches in PGR, the third party that initiated the PGR is prevented (i.e., precluded) from asserting a challenge that a claim or claims from the same patent are invalid or unpatentable on any ground that was raised or reasonably could have been raised in the PGR in a later proceeding before the USPTO, district court, or U.S. International Trade Commission.

12.7.2.2 Inter Partes Review (IPR)

Inter Partes Review is another type of trial proceeding that is conducted by the Patent Trial and Appeal Board (PTAB) that reviews the patentability of patent claims. Congress created IPR in the America Invents Act to offer a faster and more expert forum than district court for certain prior-art challenges, and the Supreme Court has upheld the regime's constitutionality. Cancelling claims in an IPR falls within the "public rights" doctrine and does not violate Article III or the Seventh Amendment [59, pp. 1372–1380].

Only certain issues are able to be raised in an IPR. A petition may challenge one or more issued claims "only on a ground that could be raised under section 102 or 103 and only on the basis of prior art consisting of patents or printed publications" [60]. That limitation matters for medical devices and SaMD because it focuses the case on documentary proof. Hardware tear-downs, on-sale bar facts, public-use demonstrations, or § 112 issues are outside the statute's scope, although those subjects can arise indirectly in claim construction or in motions to amend. In practice, the best device and SaMD petitions are built on patents, 510(k) summaries, user manuals, labeling and instructions for use, conference papers, standards documents, white papers, academic literature, archived web pages, and other materials that qualify as "printed publications" through public accessibility.

Any person other than the patent owner may file a petition, but there are timing and identification requirements that drive strategy. If the patent owner serves a district-court complaint alleging infringement, a would-be petitioner has one year from service to file an IPR, subject to a narrow joinder safety valve [60]. The petition must identify all real parties in interest and lay out the evidence with particularity. Multiple petitions are permitted, but the Board scrutinizes serial and parallel filings. When a petition is filed, the patent owner may respond with a preliminary response that can include testimonial evidence. The Board must decide whether to institute the IPR within three months of the earlier of the preliminary response's due date or its filing. The institution standard asks whether the petition and any preliminary response show a reasonable likelihood that the petitioner would prevail on at least one challenged claim. Institution decisions are final and nonappealable [61–63, p. 269–273].

If instituted, the case proceeds on a tight calendar. The Board sets a schedule with limited discovery. Routine discovery includes the cross-examination of declarants; additional discovery requires a showing that it is necessary in the interests of justice.

The petitioner bears the burden to prove unpatentability by a preponderance of the evidence, which is lower than the clear-and-convincing standard in district courts [64]. Since November 2018, the PTAB applies the same claim-construction approach as district courts, construing claims under *Phillips* in light of the intrinsic record rather than the former broadest-reasonable-interpretation standard [3, pp. 1312–1319; 65]. The statute directs the Board to issue a final written decision within one year of institution, extendable by up to six months for good cause or in cases of joinder [66].

Motions to amend are available but structured. A patent owner may file one motion to amend as of right to cancel challenged claims and, if desired, propose a reasonable number of substitute claims. After *Aqua Products, Inc. v. Matal*, the burden framework shifted. The patent owner bears the burden to show that the motion itself meets statutory and regulatory requirements, while the petitioner bears the burden to prove any proposed substitute claim unpatentable [67, pp. 1303–1315]. For medical devices and SaMD, effective amendments usually come with data that demonstrates why the added claim features distinguish over the documentary prior art and still satisfy § 112. The Board may issue preliminary guidance on a motion to amend and permit a revised motion on a defined schedule.

Discretion by the Board can often be the determinant between an IPR being instituted and not. Even if the merits threshold is met, the Board may deny institution for case-management reasons. Two recurring doctrines are the Board's consideration of parallel litigation factors and its practice under § 325(d) for art and arguments previously considered by the Office. The precedential order in *Apple Inc. v. Fintiv, Inc.* identifies factors for assessing whether a parallel district-court schedule and overlap of issues counsel against institution [68, pp. 5–16]. The precedential decision in *Advanced Bionics, LLC v. Med-El Elektromedizinische Geräte GmbH* frames the § 325(d) analysis for re-presented art [69, pp. 8–10]. Parties in device cases should address these points directly because district-court suits often proceed in parallel with IPRs and because device patents frequently revisit art cited in 510(k) files or earlier examinations.

Review and appeal are available after a final written decision. A party dissatisfied with the Board's final decision may appeal to the U.S. Court of Appeals for the Federal Circuit. See 35 U.S.C. § 319. Institution decisions remain nonappealable, and the Supreme Court has also held that the Board's application of the one-year time bar is not subject to appellate review [70, pp. 1373–1377]. After *United States v. Arthrex, Inc.*, parties may seek Director Review of certain PTAB decisions within the Office, and the Director may grant review to ensure consistency with USPTO policy [71, pp. 1985–1987].

Estoppel is the back-end check. If the Board issues a final written decision, the petitioner, its real parties in interest, and its privies are estopped from asserting in court or the U.S. International Trade Commission any invalidity ground that was raised or reasonably could have been raised in the IPR [72]. After the Supreme Court's all-claims rule, estoppel extends broadly to grounds a skilled searcher reasonably would have found and included [73, pp. 1354–1359; 74, pp. 991–999]. In medical device and SaMD disputes this means a petitioner should assemble the best patents and printed publications that describe the accused architecture or method at the outset, because late-found art may be estopped after a final written decision.

For device and SaMD portfolios, a few features recur. First, the available art is often rich and public. FDA application summaries, labeling, user manuals, operator guides, training slide decks, clinical-society guidelines, International Organization for Standard (ISO) and International Electrotechnical Commission (IEC) standards, archived product web pages, and peer-reviewed studies can all qualify as printed publications if publicly accessible. Gathering and authenticating those materials early is critical. Second, system prior art is out of scope. A petition cannot rest on a sales brochure alone if the necessary disclosures are not in a patent or printed publication. Third, claim construction can decide the case. Many device and software claims use functional language or § 112(f) terms such as "module" or "processor." Under *Phillips*, the Board will tie those to the specification's disclosed algorithms and structures, which can narrow the field of relevant art. Fourth, parallel litigation and regulatory timelines intersect. When a product is in active clinical use or under FDA review, district-court schedules can be aggressive. Addressing *Fintiv* factors and proposing realistic trial dates can be outcome-determinative at institution.

Settlements can terminate a case early, but timing matters. The parties may settle and move to terminate as to the petitioner and often as to the patent owner if the Board has not yet decided the merits. If the Board has already invested heavily or is near a decision, it may proceed to a final written decision even if the parties settle. A final written decision will trigger estoppel. For med-tech litigants, that dynamic can affect licensing leverage and the sequencing of district court and PTAB strategy.

The net effect is straightforward. IPR is a focused, fast, paper-driven validity trial before technically trained judges. Petitioners gain speed, a lower burden of proof, and expert adjudication on documentary art, but accept estoppel risk and a documentary record that excludes non-printed-publication grounds. Patent owners gain a defined timetable, targeted discovery, and opportunities to amend, but must be prepared to defend claims on the intrinsic record and the literature that surrounds their devices and software. For medical devices and SaMD, success usually turns on building or rebutting a coherent prior-art story around manuals, standards, and papers that a skilled artisan would have read, and on claim constructions that reflect how the technology actually operates. Keeping those realities front and center will align PTAB strategy with how these products are designed, cleared, and used in the real world.

12.7.2.3 Ex Parte Reexamination

An Ex Parte reexamination may be ordered at any time after a patent has been granted. Anyone, including the patent owner, may file a request for reexamination, which shows the existence of a substantial new question of patentability. The Ex Parte reexamination proceeding is between the patent owner and the USPTO. Even if the Ex Parte reexamination is requested by a third party, the proceedings will be limited to only the patent owner and the USPTO. Ex Parte reexaminations requests can be filed anonymously.

Ex Parte reexamination involves the review of a patent by a panel of three patent examiners who are part of the central reexamination unit, which is a dedicated unit of the USPTO specially trained for this purpose. The decision of the Ex Parte reexamination panel does not create legal estoppel, unlike PGR and IPR proceedings. Appeals of the decision of the Ex Parte reexamination panel can be made to the PTAB, and further appeal can be made to the U.S. Court of Appeals for the Federal Circuit.

Reexaminations are most frequently requested in two situations. First, a competitor or potential infringer seeks to have a patent invalidated and removed from the commercial arena. Second, the patent owner wants to enforce the patent, but has concerns about the patent's validity. Reexamination may resolve that issue, and it has the potential to strengthen the patent by bringing the additional prior art not applied during examination to the attention of the U.S. Patent and Trademark Office. Patent owners also frequently use Ex Parte reexamination to strengthen their patent or to address a potential issue that the patent owner has become aware of.

12.8 Monetizing IP Assets Through Licensing

Licensing IP can be a valuable tool for medical device companies of any size, and has many advantages. IP-holders may exclude others from using their protected IP. Licensing agreements are effectively grants to others made by the IP-holder for access to the protected technology, while creating a revenue stream for the IP-holder. For example, a medical device company may need to license proprietary manufacturing equipment from another company. Conversely, a medical device company may consider licensing its product along with a trade name and marketing campaign to a third party to create a larger distribution network and generate revenue from a larger market. Finally, IP rights can be licensed or, in some cases, cross-licensed to resolve litigation.

12.8.1 What is an Intellectual Property License?

An IP license is a contract between two parties allowing the licensee to use at least a portion of the licensor's intellectual property in exchange for some consideration. Some examples of what the licensor receives are: a lump sum payment, multiple payments, a royalty stream, goods, services, a cross-license to the licensee's IP, or combinations thereof. Where there are multiple payment types due during the term of the license, the payments may be tied to the sale of goods or services provided or sold (*e.g.*, as a percentage of gross or net revenue for a product that the licensee is selling). Technology transfer agreements, which are often used by universities to capitalize on research and development, often involve an up-front minimum payment as well as a stream of royalty payments.

12.8.2 Factors to Consider in an IP License

There are numerous factors that need to be considered in an IP license that are unique depending on the facts and circumstances presented. Rarely, if ever, are two licensing deals the same. Such factors may include what rights to license, term, territory considerations, exclusivity, the amount and structure of the royalty, indemnification from damage caused by the other party, etc. Special consideration is also required if a license relates to the use of a licensor's trademark. These issues are discussed briefly below.

12.8.2.1 Identification of Rights to Be Licensed

Patent rights include the right to exclude others from making, using, selling, or offering for sale a patented article or process. By licensing a patent, the patent holder is giving permission to the licensee to no longer be excluded from practicing the patented invention. Patent rights are territorial, and are limited to the country in which the patent is granted. Trademark rights are the right to use a trademark in connection with goods or services and are also territorial by country. Other rights can provide access to a proprietary or trade secret technology or know-how.

Example Medicus Societas produces a few patented medical device technologies related to its CardioColander product, which it has made available for sale either directly from its company website or through authorized distributors and resellers. While Medicus Societas holds the IP rights to a particular electronic device that is for sale, it has granted the authorized distributors and resellers the right to sell its products.

When creating a licensing agreement, it is important to identify the rights that are going into the license. This could include the rights of one or more of the patents directed at the particular device or technology, as well as the know-how related to manufacture of the product, trademarks associated with the product, whether registered or unregistered, and possibly trade secrets, such as the product formula, if there is one. As patent rights are for a limited term (the maximum life of a patent being 20 years from the earliest filing date) and any patent license would automatically terminate once the patent expires, it is beneficial to also license intellectual property rights that do not expire, such as a trademark and/or technology and know-how for manufacturing the product so that the license does not have a fixed term.

12.8.2.2 Restrictions

A license allows the licensor to restrict a licensee's activities to something less than an unlimited right to use the licensed intellectual property. For example, a patent licensor may choose to grant a licensee one or more of its rights to make, use, sell, or offer for sale a patented product.

Another restriction is whether the licensor grants an exclusive, sole, or non-exclusive license. An exclusive license is similar to an assignment of the IP rights to the licensee, and the licensor foregoes the ability to use its own IP rights in favor of the licensee.

- An exclusive license presents the most value to the licensee because it prevents all competition from using the licensed IP rights and is often granted by research institutions that have no intention of commercially exploiting an invention.
- A sole license, in contrast to an exclusive license, allows the licensor to continue to use the IP rights, but limits the licensor from granting any further licenses to third parties. Thus, in a sole license agreement, both the licensor and the licensee may use the technology under license.
- A non-exclusive license allows the licensor to grant multiple licenses to third parties, which allows for multiple companies to use the licensed technology, which increases competition.

Various types of restrictions may also be included in a licensing agreement. For example, territorial restrictions are a common restriction in license agreements. A license should explicitly define the licensed territory or geographic area where the licensee may use the licensed technology. It is very common in licensing agreements for a company to only be permitted to service a limited geographic territory. In exchange for the territorial limitation, the licensor will not make a similar grant to any third party allowing the third party to compete in this same restricted area. This is a very common practice in the operation of franchises.

Field of use restrictions are also frequently included in licenses, and may limit the use of the equipment or components contained within the product to a specific type. Field of use restrictions could limit the use of the licensed technology to certain applications such as therapeutic applications, veterinary applications, and industrial materials applications. This allows a licensor to grant multiple licenses in various fields of use, enabling the licensor to exploit its IP more effectively.

Product restrictions can also be included in a licensing agreement. This limitation restricts the licensee's use of the IP to a particular class of product. For example, a manufacturing company may license to a medical device company the right to use its patented manufacturing technique to produce a particular product.

Example In order to produce the CardioColander LDL cholesterol blood filter, Medicus Societas licensed a patented manufacturing technique from Fiberfacture. The licensing agreement between Fiberfacture and Medicus Societas includes a product restriction, stipulating that Medicus Societas is only allowed to use Fiberfacture's patented manufacturing technique to produce CardioColander filters.

12.8.2.3 Considerations

Consideration between parties to a contract regarding intellectual property rights can take many forms, but usually involves some form of payment. The most common are a lump sum payment or a stream of royalty payments (most commonly based on the number of products sold), as well as a combination of both. A lump sum payment can be used in a number of circumstances, such as settlement of past infringement, or when the technology is being transferred and the licensor is going to be working in the same field. This type of royalty shifts the entire risk of success onto the licensee since the licensor receives payment no matter what happens. As a result, an up-front lump sum payment is often discounted from the amount that could be obtained by taking a stream of royalty payments over time based on the sales volume of the product.

In contrast, a stream of royalty payments over time based on the sales volume mitigates the risk that the licensee must take on, because the licensee does not make any payments unless the product is selling. The licensor stands to receive a greater stream of royalty revenue if the product does well in commerce. A combination of some up-front lump sum payment to offset research and development costs incurred by the licensor along with a reduced stream of royalty payments is sometimes used to strike a balance for a licensor who wants some immediate payment and also wants to share in the success of the product.

Many licenses are structured so that a minimum royalty payment must be made during a given time period in order to maintain the license in force. This prevents a licensor from obtaining, for example, an exclusive license and then failing to take any action to manufacture, advertise, or sell the product. Failure to meet minimum payments can act as a trigger for automatically changing a license from exclusive to non-exclusive, or for terminating the license.

Setting royalty rates is considered to be more of an art form more than a science, and should be considered in view of the overall strategic plan for the business. While some reference materials are available that provide "typical ranges" of royalty rates for a particular field, each situation is unique and must be reviewed independently based on all of the available information, such as the predicted market size, ramp-up time, and whether production is capital intensive. Additionally, the relative size and strength of the parties is often taken into consideration as well.

In addition to royalties, other non-monetary items may form the consideration. Cross-licenses may be obtained if both parties possess intellectual property rights that the other party desires. This may replace or offset some or all of the monetary consideration.

12.8.2.4 Maintenance of IP Rights in Licensing Agreements

There are a number of terms that should be included in a license agreement that relate to the maintenance of the licensed intellectual property rights. For patent rights, this should include the requirement that the licensee include the appropriate patent marking on any goods sold that are covered by the patent. Lack of patent marking can limit claims for damages against infringers. Depending on whether the license granted is an exclusive license, the licensor may also require the licensee to pay patent maintenance fees. In some situations, a licensee may even take over prosecution of pending patent applications, which may benefit the licensor especially in the situation where the licensor is an individual or has limited resources to prosecute the patent application.

In trademark licenses, in order for the licensor to maintain its trademark rights, the agreement should require that any goods and services using the trademark identify that it is being used under license from the licensor. Additionally, to maintain the mark, the trademark license should have provisions for inspecting any products that use the trademark to ensure that the quality of the goods is consistent with the licensor's standards. If a licensee does not maintain quality control over the usage of the trademark, not only can the licensor suffer damage due to poor quality goods being associated with the mark, but the trademark right can be lost.

12.8.2.5 Other Terms in Licensing Agreements

License agreements will generally include a number of other terms, some of which are discussed below.

(1) Representations and warranties are typically included from the licensor to the licensee, and should include a statement that the licensor is, in fact, the owner of the IP rights being licensed and has the right to grant the license. This offers some protection to the licensee from fraudulent transactions. The licensor should also warrant that they believe the IP rights to be valid, and should identify any known challenges to the IP rights. While a licensee should conduct its own due diligence review of the IP rights being licensed, if the licensor fails to reveal this type of information, it can provide grounds for rescinding the license if the IP rights ultimately prove to be invalid.
(2) A license will often include terms regarding ownership and/or cross-licensing of further developments, in what is often termed a "grant-back" provision. From the licensor's perspective, this can be important if the licensor is also producing a

product under the IP rights, and wants to have the benefit of any of the licensee's improvements, which are based on the licensor's underlying IP rights. From the licensee's perspective, having rights to improvements may eliminate the need for a further license relating to the same products or services being provided under the licensed IP rights.

(3) The burden for obtaining regulatory approval, for example, from the Food and Drug Administration, should be designated in the license. For new product types, this can be a time-consuming process, and the burden for obtaining the required approvals should be specified. If the burden is placed on the licensee, this can be used as a negotiating point to obtain a reduced royalty, at least during the time it takes to obtain the approval.

(4) In the event that the parties ultimately disagree over the meaning or enforcement of any provision in the license agreement, some form of dispute resolution should be included in the agreement. This should not only include a choice of law, but a forum for any action or arbitration that will take place. In order to avoid the time and cost of a court proceeding, many agreements now call for binding arbitration of any dispute between the parties. A "loser pays" provision has also become standard in most license agreements to avoid meritless claims.

(5) A termination clause is also standard in any license agreement, and should, in addition to setting any fixed or renewable term limits for the license, include a list of circumstances or actions that will result in automatic termination of the license. Automatic termination will generally occur for non-payment of any royalties due, failure to launch or market the product within a predetermined time limit, or bankruptcy of the licensee. Termination may also occur for breach of any other terms of the license agreement, generally after a notice and cure period.

Example A sample license agreement, between Medicus Societas and ConglomoMed, which includes many of the above items, is included in Appendix B of this book.

12.8.3 Cross Licensing Intellectual Property

Cross licensing occurs between two or more parties with symmetrical interests: One company needs its competitor's patent just as badly as its competitor needs a patent belonging to the company. For example, the medical device industry utilizes cross-licensing frequently, with mutually beneficial results for the parties involved because the industry consists primarily of a limited number of players that produce similar products and hold similar IP portfolios. Major players, such as GE, Johnson &

Johnson, Siemens, Baxter International, Fresenius, and Philips, are frequently in the news because of their newest IP portfolio cross-licensing deal.

Grant-back provisions in a cross-licensing agreement require the licensee to disclose and transfer back to the licensor any and all improvements that result from the licensee's use of the licensed technology during the licensing period. If there is no grant-back provision in the license agreement, the licensee could file improvement patents of its own, rendering the licensor's technology obsolete. It would be possible for the licensee to block the licensor from commercializing its own product if the licensee attempts to incorporate the licensor's improvements into the product.

Grant backs can be exclusive, non-exclusive, or an assignment. Under an exclusive grant back, the licensor is granted the exclusive right to use or sublicense any improvements created by the licensee. Conversely, the licensee retains merely a non-exclusive right to practice the improvements that it creates. In a non-exclusive grant-back provision, the licensee retains the title and rights to his or her improvements, but the licensor is allowed to practice the improvement as well. Under an assignment grant-back provision, the licensee must assign all rights and title of any improvements to the licensor. However, the licensee still retains a non-exclusive right to practice the improvement it was responsible for creating.

12.8.4 Licensing Code and Software

12.8.4.1 When a Medical Device Startup Needs Software

There are a lot of businesses that need computer code or software in order to be successful, but they may not have the technical skills required to program their own code. This means that the startup needs to seek out a third party who can provide these services. The business might need proprietary code, customized code, or generic code, for the purposes of running their business, or for incorporation into their medical device product. In order to get the necessary code or software, the medical device startup company may need to procure it though a licensing agreement, more particularly, a software licensing agreement. A software licensing agreement is a contract that governs how the licensee can use or redistribute the software.

When the company needs software for its own use, there are two ways to get the code: either the code is available for free or open source, or the company purchases a license to use the necessary software from a software service provider.

Open source code is programming code that is openly available for use by others and in many cases users can modify the code to suit their needs.

When open source code is not an option, the company needs licensed software, also sometimes referred to as "closed source" or "proprietary" software. With these types of software, the person or entity that made the code retains control over the code, meaning that licensees cannot modify it. The license often takes the form of an end-user license agreement, which gives the purchaser/licensee the right to use the software. A good example of software that would have an end-user licensing agreement would be purchasing word processing software, docket management software, or the code necessary to develop a customer support system or an online payment system for a web-based platform can all be accessed through a licensing agreement.

Alternatively, the company may choose to hire a contractor to build software for the company in a work-for-hire arrangement. In this situation, the company would not have to worry about licensing agreements since the rights to the software would belong to the company through the use of the software contractor's employment contract.

12.8.4.2 When a Medical Device Startup Sells or Develops Software

As a startup that has built software or an app for a medical device, whether or not there is a need to consider developing licenses for the software or app will depend on how the software is made in the first place. How the software was made will determine which routes are available.

For instance, a startup that builds a program or application based on open source code, likely will not be able to control how others use their code in the future. Most open source code comes with an open source code license that often requires that the user release his or her code, which is based on the open source code, to others as a term of the license to use the open source code. To say this another way, in order for a startup to use the open source code to make their software or application, the startup must agree to a license that requires the startup to release the code that they develop as open source as well. Since the company's software is based on open source, and thus in turn will need to be released as open source, rather than making money by licensing the software, the company can instead make money by providing additional software services or support for their program.

When a startup company develops proprietary software, and the value of that software is derived from the proprietary nature of the software, the company should consider the value that licensing the software would have for the company. By retaining control to the program code, through the use of a licensing agreement, the company will have a competitive edge over competitors and customers/licensees will be prevented from depreciating or devaluing the company's software product by redistributing it or modifying it.

References

1. *CommScope Techs. LLC v. Dali Wireless Inc.*, 10 F.4th 1289, 1295 (Fed. Cir. 2021).
2. *Markman v. Westview Instruments, Inc.*, 517 U.S. 370, 372–91 (1996).
3. *Phillips v. AWH Corp.*, 415 F.3d 1303, 1312–25 (Fed. Cir. 2005) (en banc).
4. *Vitronics Corp. v. Conceptronic, Inc.*, 90 F.3d 1576, 1582–84 (Fed. Cir. 1996)).
5. *Omega Eng'g, Inc. v. Raytek Corp.*, 334 F.3d 1314, 1323–26 (Fed. Cir. 2003).
6. *Aylus Networks, Inc. v. Apple Inc.*, 856 F.3d 1353, 1361–62 (Fed. Cir. 2017).

7. *Renishaw PLC v. Marposs Società per Azioni*, 158 F.3d 1243, 1249 (Fed. Cir. 1998).
8. *CCS Fitness, Inc. v. Brunswick Corp.*, 288 F.3d 1359, 1366–67 (Fed. Cir. 2002).
9. *SciMed Life Sys., Inc. v. Advanced Cardiovascular Sys., Inc.*, 242 F.3d 1337, 1341–45 (Fed. Cir. 2001).
10. *Thorner v. Sony Comput. Entm't Am. LLC*, 669 F.3d 1362, 1365–68 (Fed. Cir. 2012).
11. *O2 Micro Int'l Ltd. v. Beyond Innovation Tech. Co.*, 521 F.3d 1351, 1360–62 (Fed. Cir. 2008).
12. *Chef Am., Inc. v. Lamb-Weston, Inc.*, 358 F.3d 1371, 1373–74 (Fed. Cir. 2004).
13. *Williamson v. Citrix Online, LLC*, 792 F.3d 1339, 1348–51 (Fed. Cir. 2015) (en banc in relevant part).
14. *Aristocrat Techs. Austl. Pty Ltd. v. Int'l Game Tech.*, 521 F.3d 1328, 1333–38 (Fed. Cir. 2008).
15. *Rain Computing, Inc. v. Samsung Elecs. Am., Inc.*, 989 F.3d 1002, 1007–09 (Fed. Cir. 2021).
16. *Nautilus, Inc. v. Biosig Instruments, Inc.*, 572 U.S. 898, 910–12 (2014).
17. *Liebel-Flarsheim Co. v. Medrad, Inc.*, 358 F.3d 898, 910–12 (Fed. Cir. 2004).
18. *Southwall Technologies, Inc. v. Cardinal IG Co.*, 54 F.3d 1570, 1575–76 (Fed. Cir. 1995).
19. *Gillette Co. v. Energizer Holdings, Inc.*, 405 F.3d 1367, 1371–72 (Fed. Cir. 2005).
20. *Laitram Corp. v. Rexnord, Inc.*, 939 F.2d 1533, 1535 (Fed. Cir. 1991).
21. *Odetics, Inc. v. Storage Tech. Corp.*, 185 F.3d 1259, 1267–68 (Fed. Cir. 1999).
22. *Graver Tank & Mfg. Co. v. Linde Air Prods. Co.*, 339 U.S. 605, 608–09 (1950).
23. *Warner–Jenkinson Co. v. Hilton Davis Chemical Co.*, 520 U.S. 17, 29–40 (1997).
24. *Freedman Seating Co. v. American Seating Co.*, 420 F.3d 1350, 1358–59 (Fed. Cir. 2005).
25. *Festo Corp. v. Shoketsu Kinzoku Kogyo Kabushiki Co.*, 535 U.S. 722, 736–41 (2002).
26. *Festo Corp. v. Shoketsu Kinzoku Kogyo Kabushiki Co.*, 344 F.3d 1359, 1366–70 (Fed. Cir. 2003) (en banc).
27. *Festo Corp. v. Shoketsu Kinzoku Kogyo Kabushiki Co.*, 493 F.3d 1368, 1377–79 (Fed. Cir. 2007).
28. *AquaTex Industries, Inc. v. Techniche Solutions*, 479 F.3d 1320, 1328–30 (Fed. Cir. 2007).
29. *Akamai Technologies, Inc. v. Limelight Networks, Inc.*, 797 F.3d 1020, 1022–29 (Fed. Cir. 2015) (en banc).
30. *Travel Sentry, Inc. v. Tropp*, 877 F.3d 1370, 1377–83 (Fed. Cir. 2017).
31. *Centillion Data Systems, LLC v. Qwest Communications Int'l, Inc.*, 631 F.3d 1279, 1284–85 (Fed. Cir. 2011).
32. *Global–Tech Appliances, Inc. v. SEB S.A.*, 563 U.S. 754, 766–71 (2011).
33. *Commil USA, LLC v. Cisco Systems, Inc.*, 575 U.S. 632, 642–48 (2015).
34. *Aro Mfg. Co. v. Convertible Top Replacement Co.*, 377 U.S. 476, 484–91 (1964).
35. *Microsoft Corp. v. AT&T Corp.*, 550 U.S. 437, 441–57 (2007).
36. *Bayer AG v. Housey Pharmaceuticals, Inc.*, 340 F.3d 1367, 1377–78 (Fed. Cir. 2003).
37. *WesternGeco LLC v. ION Geophysical Corp.*, 138 S. Ct. 2129, 2136–39 (2018).
38. U.S.C. § 271(e)(1).
39. *Merck KGaA v. Integra Lifesciences I, Ltd.*, 545 U.S. 193, 202–08 (2005).
40. *Eli Lilly & Co. v. Teva Parenteral Medicines, Inc.*, 845 F.3d 1357, 1364–70 (Fed. Cir. 2017) (inducement for method-of-treatment claims where physicians perform steps).
41. *Arctic Cat Inc. v. Bombardier Recreational Prods. Inc.*, 876 F.3d 1350, 1366–68 (Fed. Cir. 2017).
42. *Halo Electronics, Inc. v. Pulse Electronics, Inc.*, 579 U.S. 93, 103–06 (2016).
43. *eBay Inc. v. MercExchange, L.L.C.*, 547 U.S. 388, 391–94 (2006).
44. *TC Heartland LLC v. Kraft Foods Group Brands LLC*, 581 U.S. 258, 261–70 (2017) (reaffirming *Fourco Glass Co. v. Transmirra Prods. Corp.*, 353 U.S. 222 (1957)).
45. *In re BigCommerce, Inc.*, 890 F.3d 978, 982–83 (Fed. Cir. 2018).
46. *In re Cray Inc.*, 871 F.3d 1355, 1360 (Fed. Cir. 2017).
47. *In re Google LLC*, 949 F.3d 1338, 1343–47 (Fed. Cir. 2020) (Google's servers in third-party facilities in the Eastern District of Texas did not establish venue).
48. *Andra Group, LP v. Victoria's Secret Stores, LLC*, 6 F.4th 1283, 1287–95 (Fed. Cir. 2021).
49. *In re Apple Inc.*, 979 F.3d 1332, 1336–39 (Fed. Cir. 2020).
50. *In re Adobe Inc.*, 823 F. App'x 929, 931–33 (Fed. Cir. 2020) (nonprecedential).

51. *In re Google LLC*, No. 2021–170, 2021 WL 4427899, at 2–4 (Fed. Cir. Sept. 27, 2021).
52. *In re Google LLC*, No. 2021–178, 2021 WL 5292267, at 2–4 (Fed. Cir. Nov. 15, 2021).
53. *In re Samsung Elecs. Co.*, No. 21–139, 2021 WL 2672134, at 2–4 (Fed. Cir. June 30, 2021) (nonprecedential).
54. *In re Apple Inc.*, No. 2022–128, 2022 WL 1196768, at 1–3 (Fed. Cir. Apr. 22, 2022) (nonprecedential).
55. American Intellectual Property Law Association, *AIPLA Report of the Economic Survey 2017*, American Intellectual Property Law Association (2017).
56. U.S.C. § 504.
57. U.S.C. § 1836, et seq.
58. Federal Rule of Civil Procedure 26(b).
59. *Oil States Energy Services, LLC v. Greene's Energy Group, LLC*, 138 S. Ct. 1365, 1372–80 (2018).
60. U.S.C. § 311(b).
61. U.S.C. § 314(a).
62. U.S.C. § 314(d).
63. *Cuozzo Speed Technologies, LLC v. Lee*, 579 U.S. 261, 269–73 (2016).
64. U.S.C. § 316(e).
65. Fed. Reg. 51, 340; 51, 344–45 (Oct. 11, 2018).
66. U.S.C. § 316(a)(11).
67. *Aqua Products, Inc. v. Matal,* 872 F.3d 1290, 1296, 1303–15 (Fed. Cir. 2017) (en banc).
68. *Apple Inc. v. Fintiv, Inc.*, IPR2020–00019, Paper 11 at 5–16 (PTAB Mar. 20, 2020) (precedential).
69. *Advanced Bionics, LLC v. Med-El Elektromedizinische Geräte GmbH*, IPR2019–01469, Paper 6 at 8–10 (PTAB Feb. 13, 2020) (precedential).
70. *Thryv, Inc. v. Click-to-Call Technologies, LP*, 140 S. Ct. 1367, 1373–77 (2020).
71. *United States v. Arthrex, Inc.*, 141 S. Ct. 1970, 1985–87 (2021).
72. U.S.C. § 315(e).
73. *SAS Institute Inc. v. Iancu*, 138 S. Ct. 1348, 1354–59 (2018).
74. *California Institute of Technology v. Broadcom Ltd.*, 25 F.4th 976, 991–99 (Fed. Cir. 2022).

Part III
Implementation Strategies for Medical Device Innovators

Chapter 13
Successful Implementation of a Medical Device Company's IP Strategy

13.1 Talk with an Intellectual Property Lawyer

As the facts and circumstances surrounding product development and branding vary widely, consulting with an intellectual property attorney is highly recommended.

> ***Example*** Virtually all major medical device corporations who utilize and value IP rights have in-house counsel that specifically practice IP law. If a company doesn't have in-house counsel, the company certainly seeks IP counsel externally. GE Health, Abbot Laboratories, Johnson & Johnson, Medtronic Inc., Novartis AG, Baxter International Inc. and Siemens AG are just a handful of examples of medical device companies that have IP counsel on staff. Even the USPTO has general IP counsel. Whether your company is just starting up, is small but growing or is fully established and highly successful, consulting IP counsel is important for protecting your company's intellectual property rights.

13.2 Freedom to Operate Opinions

One of the first steps to successfully implementing an IP strategy is to have an IP attorney perform a Freedom to Operate (FTO) study. With an FTO study, an IP lawyer reviews prior patents within a specific technological field to ensure that you have the "freedom" to make, use, and sell your medical device in a particular technological space without risk of infringing the IP rights held by others. The IP lawyer will search for all relevant patents, and review the patents to ensure that your device does not infringe any of the identified patents.

G. B. Halt Jr. et al., *FDA and Intellectual Property Strategies for Medical Device Technologies*, https://doi.org/10.1007/978-3-032-12649-8_13

For instance, a small medical device company that invents a new lock-out feature for an opioid-antagonist administration device used for combating the effects of a drug overdose would obtain an FTO opinion from a lawyer concerning their new product. The FTO study would involve reviewing various types of IP rights held by others in the relevant technological area of drug administration devices, with particular emphasis on devices aimed at treating the effects of drug overdoses. The FTO study will evaluate the IP assets held by others to determine whether there is a place in the intellectual property landscape where the small medical device company can legally operate without infringing the rights held by others. It is highly recommended that a FTO study be performed for the key technological characteristics of a medical device or method of use invention before seeking patent protection.

An FTO opinion shares many characteristics with a prior art patentability search; however, the two are directed to different legal concepts. Both the FTO and prior art patentability search aim to locate relevant public documents that describe similar technical features to the invention. However, in the prior art patentability search, the scope of available documents is much wider and encompasses both patent applications, issued patents and any other publicly available information. In contrast, an FTO opinion only considers issued patents, and more specifically their claims, to evaluate whether the invention may infringe the existing patent rights held by others.

Even though the FTO and the prior art patentability search both consider issued patents, the two are concerned with different legal aspects of the invention. The prior art patentability search considers the entire disclosure of any identified relevant prior art (i.e., the lawyer must review all written materials that might be similar to yours and evaluate whether your invention is novel and non-obvious in view of what is disclosed in the prior art). In contrast, the FTO is exclusively concerned with comparing the features of your medical device to the claims of existing patents. When one or more of the features of your medical device are too close to the claims of an existing, valid patent, you may infringe on the patents rights held by another.

As a result, the FTO opinion considers only a subset of the information that considered in a prior art patentability search, and therefore, an FTO opinion that indicates a low infringement risk does not necessarily imply that the invention is patentable.

The FTO study is generally a bifurcated process. In the first part of the process, the FTO study identifies patents that are no longer in force (meaning that a patent has expired or has been held invalid by the court). In the second part, the FTO study seeks to identify patents that are still in force (meaning that patents are valid and enforceable) and therefore pose an infringement risk. Based on the patents identified in both parts of the FTO study process, the infringement risk can be assessed.

In the first portion of the FTO study, expired patents that cover the technical features of the medical device are identified. A patent may be considered expired if the patent can no longer be in force due to invalidation, failure to pay maintenance fees, or expiration of the term of the patent. The FTO study seeks to locate relevant expired patents because upon "the expiration of a patent the monopoly created by it ceases to exist, and the right to make the thing formerly covered by the patent becomes public property [1, p. 185]." When the patent becomes public property, "the public is free to use not only the same invention claimed in the expired patent but also obvious or patentably indistinct modifications of that invention [2, p. 1214]." Accordingly, once the patent enters the public domain, the monopoly right is exhausted, and therefore, anyone can freely perform any act that previously would have infringed the patent without fear of an infringement suit.

For instance, any manufacturer can produce a medical device for injecting a substance into a biological body that was previously covered by U.S. Patent No. 7,425,204 (the '204 Patent). The '204 patent is an example of a patent that has entered the public domain for failure to pay renewal fees. The patent was filed on April 28, 2005, and therefore would have expired on April 28, 2025. However, the patent expired on September 15, 2012, because the owner of the patent did not pay the applicable (4-year) renewal fee. Specifically, the '204 Patent claims:

1. An injector for injecting a substance into a biological body, comprising:

   a chamber adapted to hold the substance to be injected and an output port in fluid communication with the chamber.

   a drive element.

   a magnetic force able to cause movement of the drive element.

   an actuator that opposes movement of the drive element and then allows the magnetic force to cause the drive element to move and thereby cause the expulsion of the substance out of the chamber through the output port.

As a result, the patent has lapsed, and any manufacturer can make, sell, use, or import an injector that has each of the features claimed by the '204 patent. In this context, if the FTO were to uncover a patent like the '204 that had become public property that covered the medical device, the likelihood of an infringement suit from a competitor can be minimized.

In the second phase of the FTO, valid and enforceable (i.e., in force) patents are identified that potentially pose an infringement risk to the medical device. Once these in-force patents are identified, the next step is to perform a prior art patentability search in an effort to identify prior art references that could invalidate the in-force patent. Prior art that could be used to invalidate an existing in-force patent, either by proving that the in-force patented invention was not novel or was obvious, could render an existing patent invalid. The prior art required to invalidate the in-force patent can often be found among the patents that have entered public ownership, identified in the first step of the FTO.

If invalidating prior art can be located, a manufacturer can produce the medical device and rely upon the invalidating prior art as a defense to infringement accusations made by the in-force patent holder—the defense being that the alleged in-force patent is invalid in view of the invalidating prior art.

If the medical device manufacturer would like more certainty that they will not be sued for infringement, the manufacturer can proceed even further and take offensive measures to invalidate an existing patent, utilizing one of several different procedures that are available to invalidate a patent. In a first procedure, *Ex Parte* Reexamination, which is a post-grant proceeding before the USPTO (see Chap. 12), the manufacturer may provide the USPTO with the invalidating art found in the FTO. The USPTO will then essentially reexamine the in-force patent (a process in which the manufacturer is not allowed to participate) if the art raises a "substantial new question of patentability" under 35 U.S.C. §304 (see Chap. 12).

Another alternative to invalidate the in-force patent is through an *Inter Partes* Review (IPR). In an IPR, unlike in an *Ex Parte* Reexamination, the manufacturer actively participates in the invalidation proceeding (see Chap. 12). An IPR can be requested at any time and will be instituted under 35 U.S.C. §314 if there is "a reasonable likelihood" that at least one of the claims of the in-force patent is invalid based on the art found in the FTO (see Chap. 12). As a result, the invalidating prior art can be leveraged to minimize the infringement risk posed by the in-force patents found during the FTO study.

When invalidating prior art cannot be located to invalidate the in-force patents identified during the FTO study, the manufacturer can consider "designing around" the in-force patent, acquiring rights to the in-force patent through purchase or licensing, or making a business decision to produce the medical device and risk facing infringement allegations.

Even in the case where the FTO study reveals a high infringement risk, the information contained in the FTO opinion can be an extremely valuable asset. Medical device companies can make use of the patents and other prior art identified in their FTO study in other ways. For instance:

- For a medical device company that opts to design around in-force patents, various technical features from prior art that is in the public domain or from the expired patents can be incorporated into a redesigned version of the medical device, thereby designing around the in-force patents.
- In the case where the medical device company opts to acquire or license the in-force patent, the prior art from the FTO study can be leveraged to get a discounted price based on the likely cost the in-force patent holder would have to shoulder in defending against an invalidation proceeding.

However, an FTO opinion does not come without associated liabilities, particularly when the medical device is produced even after the FTO opinion has identified potential infringement. Under 35 U.S.C. §284, a court can award up to triple damages for "willful" infringement of an in-force patent, and the FTO opinion can be used as evidence to demonstrate that the medical device company's infringement was willful.

Generally, the communications between a client and an attorney are privileged, and therefore cannot be used by the opposition in litigation. However, in the case of an FTO opinion, the study is likely admissible, because the alleged infringer "waives its immunity for any document or opinion that embodies or discusses a communication to or from it concerning whether that patent is valid, enforceable, and infringed by the accused" if the medical device manufacturer were to rely upon the contents of the FTO opinion during a litigation [3, p. 1304].

The risk of not designing around or acquiring the rights to in-force patents further heightens the infringement risk. This heightened risk must be offset by the fact that 35 U.S.C. §298 provides that "[t]he failure of an infringer to obtain the advice of counsel" or "the failure of the infringer to present such advice to the court or jury, may not be used to prove that the accused infringer willfully infringed." Yet, when properly used, the benefit of having an FTO opinion exceeds the risks.

In addition, if patent protection is still sought for the medical device in question, all of the patents in the public domain, in-force patents and invalidating prior art, must be disclosed to the USPTO in an Information Disclosure Statement (IDS) during the patent application process. These documents must be disclosed under 37 C.F.R. § 1.56(a) because a patent applicant has "a duty to disclose to the Office all information known to that individual to be material to patentability." However, once disclosed and considered by the patent examiner, these documents may be unavailable to subsequently invalidate a patent for the medical device once it is granted.

Consequently, it is critically important that the FTO study be performed early in the development of the medical device, when technical features can be easily changed and the infringement risk is low. To obtain the most significant benefit by minimizing the potential infringement risk, it is best that the FTO be performed prior to the medical device being presented to the FDA for market approval.

13.3 Commercial Novelty Versus Patent Novelty

A medical device may be commercially novel by virtue of not being available in the market, yet the medical device may still not meet the novelty requirements for patentability under 35 U.S.C. §102. As discussed in Chap. 6, in order for a medical device to be novel under 35 U.S.C. §102, it cannot be previously "patented, described in a printed publication, or in public use, on sale, or otherwise available to the public [4]." As a result, there are several instances where a medical device is novel in the commercial sense, but not novel in the patent sense.

For instance, a device no longer meets the threshold for patent novelty if the technical features of the device are described in a printed publication such as on a website, in a technical journal, in a doctoral thesis found in an obscure university library, or on a poster displayed for a few days at a technical conference, and patent protection is not sought within one year of time after the public disclosure is made. This rule is "grounded on the principle that once an invention is in the public domain, it is no longer patentable by anyone [5, p. 898]."

Accordingly, printed publications are documents that are available to the public. They must be sufficiently accessible to anyone in the public who is interested in the invention. As a result, a printed publication can destroy the patent novelty of an invention if the printed publication is disseminated or made available to the public in a way that someone with an interest and reasonable diligence can find it.

In addition, a device no longer meets the threshold for patent novelty when the same technical feature is applied in a completely different context. Specifically, printed publications disclosing a technical feature, even if it is being applied in a completely different field of endeavor or is being used to solve a completely different problem, can potentially anticipate patent novelty. For example, an insulin delivery pump used for the treatment of diabetes could be rendered no longer novel by an oil pump used to circulate oil in a car, if both the insulin pump and the oil pump have the same technical features. The patentable novelty of a medical device can potentially be destroyed by a non-medical device if they operate in the same way or have the same technical feature.

On the other hand, the commercial novelty of a medical device may be the result of an existing patent on the underlying technology. To reiterate, patent law does not create any affirmative right to make, use, or sell a patented invention. Rather, patents create the right to exclude others from doing those things with the protected invention. Further, under U.S. patent law, unlike under U.S. trademark law (discussed in Chap. 7), there is no affirmative obligation for a patent owner to commercialize their patent rights. Accordingly, the commercial novelty may be the direct result of a patent owner holding a patent that covers a particular medical device waiting to extract a licensing royalty from someone who attempts to commercialize the invention. As a result, the commercial novelty may not imply patent novelty, but instead imply a potential infringement risk.

Consequently, commercial novelty is necessary for the successful monetization of a medical device, but is legally insufficient for patent novelty. Patent novelty is based on a broad scope of public knowledge beyond that which is available on market shelves and must be judged in the context of the invention's technical features, irrespective of the context in which the features are used. As a result, the only way to determine patent novelty is through a prior art patentability search, which also considers the commercial novelty of the invention.

13.4 Confidential Disclosure or Non-disclosure Agreements

Confidential Disclosure Agreements, which are also referred to as Confidentially Agreements or Non-Disclosure Agreements (NDAs), refer to a contract that protects confidential or trade secret information ("Confidential Information") from disclosure to third parties. NDAs are commonly included as a part of an employment contract or in agreements with vendors, contractors and sometimes customers.

An NDA should include the following provisions:

(a) A clear definition of the information that is to be held in confidence, as well as a provision defining how the information should be marked or identified, for example with a "Confidential" stamp or label. Employment agreements typically adopt broader definitions requiring employees to maintain any work-related information that the employee develops or has access to as confidential.

Supplemental agreements for specific employee development projects may also be used to more clearly identify the information that is subject to the agreement.

(b) A specific recitation of the limited purposes for which the Confidential Information can be used. For example, in an NDA with an outside vendor, the Confidential Information may be provided for specific testing and evaluation of a product, such as RF transmitter and receiver calibration testing for a wireless wearable health tracking device or debugging of a software application for a medical device. For an advertising or marketing agency, the Confidential Information could be limited to the specific purpose of package development, branding or other specific marketing tasks for the benefit of the disclosing party.

(c) A recitation that the receiving party cannot breach the confidential relationship, induce others to breach it or induce others to acquire the Confidential Information by improper means.

(d) A recitation of exceptions to the confidentiality requirements. This generally will exclude any information that:

   i. The receiving party can show they were already in possession of it at the time of the disclosure.
   ii. Information that is in the public domain through no fault of the receiving party.
   iii. Information that the receiving party receives without restriction from a third party.

(e) The time period which the information must be held in confidence. This can be any reasonable term agreed upon by the parties, and often falls in the range of two (2) to five (5) years, depending on the technology in question and the disclosure's purpose. For example, an advertising agency that is working on an advertising campaign that is going to be released within a year would not need an NDA term that extends beyond the advertising campaign release. However, for an outside consultant that does product testing of beta products to determine a preferred version of a product for commercialization, the NDA could justifiably require that the information be held in secrecy for five (5) years or more.

(f) A provision defining the remedy available to the non-breaching party in the event of a breach or impending breach. This should recite that the disclosing party is entitled to injunctive relief for breach of contract by the receiving party to prevent the release of the Confidential Information. It is also possible to include a liquidated damages provision as an incentive for the receiving party not to disclose the Confidential Information, although this is less typical given the circumstances surrounding most NDAs, where the disclosing party is attempting to obtain information or services from the receiving party.

(g) Other miscellaneous provisions may include:

   i. A provision for return or destruction of the Confidential Information when the task or review by the receiving party is completed.

ii. A transfer of ownership of any additional intellectual property that results directly from the Confidential Information and/or in connection with the work being done by the receiving party.
iii. A recitation of the courts or jurisdictions where any potential dispute will be resolved.

Example A sample Non-Disclosure Agreement is included in Appendix D of this book.

While NDAs have several advantages, there are a few drawbacks that they cannot address. Many large companies will not sign NDAs for any outside submissions, regardless of the purpose. In fact, some large corporations require the opposite: a signed statement saying that no information will be held in confidence and that the party submitting the information will rely exclusively on any intellectual property rights (such as patent or trademark rights) that they applied for, or may obtain, as the sole recourse against the receiving party in the event that their information is used. This typically occurs as a result of a competitor's parallel development efforts. A competitor may accuse a company receiving its confidential information of theft or misappropriation even though the receiving company was already working on a similar development. Rather than face potential lawsuits or bad publicity, the policy of such express waivers of confidentiality shields the receiving company.

Even if an NDA is signed, if the Confidential Information is purposefully or even inadvertently disclosed, it may not be possible to "put the genie back in the bottle." Although damages may be available against the discloser, an actual public disclosure cannot be undone. This could result in inadvertent loss of the ability to seek patent protection in many countries that have absolute novelty requirements. While the United States allows a one (1) year grace period from the date of first public use or disclosure of an invention to the public, if the owner of the Confidential Information is unaware of the disclosure, U.S. Patent rights could also be lost. Furthermore, any trade secret information that is publicly disclosed ceases to be a trade secret.

Example A medical technology company should require employment agreements with all employees who will have access to a new project, including its senior engineers, designers, programmers, etc. as well as anyone else that will be working on the new product, its marketing employees, and any others who will have access to the information. If confidentiality terms are not included in the company's standard employment agreements, the company can enter separate agreements, preferably at the same time as an annual review and pay raise so that there is no question regarding a potential lack of consideration for the new NDA provisions.

The confidentiality term should extend for a time period beyond the end of employment, especially if a rival competitor is known to poach employees away from the company in order to copy products or product concepts. In the event that an employee with knowledge and/or confidential information related to the project leaves the company, the NDA should provide for injunctive relief to prevent the confidential information from being improperly disclosed.

In addition, any outside vendors used for the new project should sign a project-specific NDA with the company. The vendor agreements should also have specific terms spelling out ownership of any new developments made using the confidential information while carrying out the work for the company.

Distribution of a new product will likely require advance notice of the product. If the company learns that its distributor will not sign a NDA, and to the contrary requires a specific waiver of confidentiality to accompany any offer or disclosure, the company should have any patent applications for its new product filed with the USPTO before any disclosure of any information to the distributor. This will protect any potential U.S. or foreign patent rights that the company may choose to pursue.

13.5 Assignment of Rights

An Assignment is a contract between two parties in which rights owned by one party are transferred to the other party. In the U.S., all rights to an invention are initially vested in the inventor(s). Accordingly, a formal Assignment is important to transfer rights from an inventor, or inventors, to the medical device company. For employees, the obligation to assign inventions made in the course of an employee's regular job duties can also be included in the employment contract. If the obligation to assign inventions is not included in an employee's employment agreement, a separate agreement can be entered into including the obligation to assign inventions to the company. This separate agreement should preferably be made and signed with the employee in connection with an annual review and raise so that there is no issue with respect to consideration for the agreement.

Example A sample Assignment contract from an individual inventor to a medical device company is set forth in Appendix C.

Many companies offer employee incentive programs that pay bonuses for creating inventions that help the company. This typically takes the form of a lump sum payment at the time a patent application is filed or a patent is granted. Some countries, such as Germany, have specific statutory requirements that define the amount that an employee must be compensated for any invention that is used by the company.

When working with third parties, it is common to include assignment terms in vendor contracts for inventions within the scope of the specific project. For patentable inventions, this can be critical if a breakthrough is made by the vendor in connection with the development of a company's product. If no agreement is reached before the contract work is undertaken, rights would belong to the vendor or the vendor's employee, creating the potential for being forced into a sole source of supply, or worse, having the product which the company paid the vendor to develop offered to third parties without any ability for the company to control it.

Example Assignments are critical for any intellectual property. For example, assume a startup medical technology company hires a contractor (writer, drafter, programmer, designer, etc.) whose work product would be eligible for copyright protection. In the absence of a written assignment for works made under contract, the copyrights would be owned by the contractor. It must be clear from the hiring contract that any IP rights created by the contractor are assigned to the technology company and that any work done by the contractor was done as a "work for hire," thereby assigning all copyrights to the medical device company.

13.6 Employee Education

Employee education is one of the cornerstones of a successful intellectual property program. Well-informed employees can be a company's best defense against loss of intellectual property rights. Inadvertent disclosure of confidential information or new inventions that are being developed can easily result in potential intellectual property rights being lost, or worse, landing directly in a competitor's hands. The only way to effectively address this is to train employees such that they understand the ramifications of their actions and the potential cost to the company in terms of lost intellectual property rights and lost profits on new products or developments that cannot be protected.

Many companies hold periodic intellectual property seminars that are presented by in-house or outside intellectual property counsel. These programs educate employees on the basic tenants of intellectual property law in a practical setting, and how they can handle particular situations. A typical program would include a review of what may constitute patentable subject matter given the company's technology field, (for instance, guidance would vary depending on whether the company focuses on a medical device that is largely mechanical in its operation, or a software-based medical device), the potential bars to patentability that an employee should be aware of, and a review of the medical device company's system for documenting inventions and subsequently handling those documents. Employees who regularly deal with third-party vendors should also receive special training related to risks associated with dealing with vendors.

13.7 Accurate Record Keeping

13.7.1 Patents

In the United States, patent applications filed on or after March 16, 2013 are subject to the America Invents Act "first to file" rule, meaning that the right to the grant of a patent lies with the first person to file a patent application, regardless of the date of actual invention. There is a public disclosure grace period, however, which allows the inventor to disclose the invention up to one year prior to filing. In the event that an inventor publicly discloses the invention before another inventor independently develops the same invention and files for a patent on the invention, the first inventor may be entitled to the patent protection due to the one year disclosure grace period. It is essential to maintain detailed records of when an invention disclosure is made.

Another document that the company should have is an employee invention submission form. This should include sections for a complete description of the invention, as well as any potentially critical events that occur, such as a disclosure to others. This form serves two purposes:

(1) It can serve as the vehicle for in-house review and a determination of whether patent protection is going to be pursued.
(2) If the company proceeds with a patent application, it can act as the vehicle for transmitting information on the invention to patent counsel for searching and/or the preparation of a patent application.

> ***Example*** A sample invention disclosure for completed for the CardioColander invention is included in Appendix A.

For each invention disclosure that is ultimately pursued as a patent application, a company should open a separate file as a place to store all information related to the invention, including copies of the relevant completed pages of the inventor's notebook, the invention disclosure form, any known prior art that might relate to the invention, and all correspondence and documents related to the preparation, filing, and prosecution of a patent application before the USPTO.

13.7.2 Trade Secrets

Trade secret protection depends on defining and following a strict set of policies for handling the information that is being protected as a trade secret. Trade secret policies should be clearly documented in a policy manual. All employees who have access to the trade secret information should receive regularly scheduled training on handling trade secret information and should be required to periodically review the company's policy regarding treatment of such information. Policies should outline procedures for identifying and marking trade secret information, including information or technology that is under development. Identification may be as broad as "all information related to project X" or as specific as a particular formula for a product, and can be established by management or counsel. Marking should be on both physical and electronic documents, using labels like "Confidential," "Secret," "Trade Secret" or "Proprietary Information" of the company.

Access to the trade secret information should be on a very strict "need to know" basis inside the company; and likewise access to the information should be on a "log in–log out" basis. Any electronically stored or transmitted information should be encrypted in accordance with a defined procedure. A policy should also define storage of the information when an employee is away from their work area, and may include a locked central or private storage area. The reason for this high level of security is that in a misappropriation lawsuit, a court's inquiry will not only focus on the bad acts of the accused party, but it will also examine and consider whether the company asserting its trade secret rights adequately maintained and safeguarded its trade secrets.

13.8 Patent and Trademark Searches

13.8.1 Patent Searches

There are a number of different patent searches that can be useful for a number of different purposes, including patentability, infringement clearance, validity and state of the art searches.

A patentability search is the most basic type of search, involving the examination of U.S. patents and patent publications, as well as potentially other foreign patent and non-patent literature, to determine whether an "invention" meets the USPTO requirements for patentability, based on the documents identified by the search. This is a useful tool for gauging the potential for patentability; however, it does not guarantee that a patent will ultimately be granted. The limitations on patentability searches are that they are generally not exhaustive, and other more pertinent references may ultimately be identified from areas not searched. Accordingly, while negative results can be relied upon, a positive search report merely leaves the possibility of patent protection open.

An FTO search is a search of U.S. patents that remain in force for potential infringement by a new product that is being developed. FTO searches should be performed prior to the product's launch. As the search requires a specific review of each independent claim of the relevant patents, it is more time-consuming and expensive. If done early enough in the design or development process, an FTO search allows the new product to be modified to avoid potential infringement.

A validity search is a prior art search of U.S. and foreign patent and non-patent documents that is compared to the claims of a specific patent. This can be used to determine whether the claims of a known or asserted competitor patent are valid. The results of the search can be used for negotiations, or can be used to invalidate the patent in a court or USPTO proceeding.

A state-of-the-art search looks at U.S. and possible foreign patent document collections for representative technology and developments in a particular field. This is used as a research tool for resolving a particular problem or for examining the type of work competitors have done in a given field.

13.8.2 Trademark Searches

Trademark searches determine whether a company can adopt a trademark for its product or service. A trademark search can be done in the Federal Trademark database, or can be done in one or more comprehensive databases. Trademark searches should be carried out before adopting a trademark to determine whether any third party has used the trademark for the same or similar types of goods, and may therefore have superior rights in the trademark. Adopting a third party's mark, whether knowingly or unknowingly, may result in a lawsuit for trademark infringement. A trademark search can help avoid this costly litigation.

13.9 Choosing the Best IP Protection for Your Medical Device Technology

The earlier a company decides what type of intellectual property to pursue, the lower the likelihood that rights will be inadvertently lost. For patents, it is important to observe specific timelines before the invention's first public disclosure, use or offer for sale. These dates will also determine certain statutory bars to patentability in the United States. For trade secrets, the earlier a decision is made to protect a new product or its method of manufacture as a trade secret, the easier it will be to ensure that some information is not inadvertently released.

13.9.1 Deciding Between Patent or Trade Secret Protection

Many medical device companies use a combination of patents and trade secrets to protect inventions. However, since these two forms of protection are mutually exclusive in a fundamental regard (i.e., whether or not to disclose an invention), companies must often choose between the two forms of protection. Each has advantages over the other that should be carefully considered when forming the appropriate intellectual property strategy.

13.9.1.1 Differences in Scope of Protection

A first consideration is the difference in subject matter between trade secrets and patents. Trade secret protection covers a wider range of possible innovations, and it has a potentially unlimited duration. The Uniform Trade Secret Act ("UTSA") defines trade secrets as any information, including a formula, pattern, compilation, program, device, method, technique, or process that derives independent economic value from being "secret" (more on trade secrets and the UTSA can be found in Chap. 8). While concepts, databases, and compilations are generally not patentable, the UTSA expressly protects them if they are valuable to the business and the business takes specific steps to maintain their secrecy. Further, other categories of information that may be subject to trade secret protection, where patent protection would not be allowable, include customer lists, product pricing, strategic planning, company policies, market analyses, etc.

13.9.1.2 Differences in Litigation Remedies

A second consideration is the different types of litigation remedies. A U.S. patent grants the owner the right to exclude others from making, using, or selling the invention throughout the United States for a limited time. In return for this right, the patentee must disclose to the public the information necessary to make and use the invention. Thus, with patent protection, even a competitor who independently, and without any knowledge of the patent, develops or reverse engineers an invention that is covered by the patent cannot practice the invention without infringing the patent.

In contrast, if a person independently discovers or develops a particular trade secret, independently of the "first" trade secret owner, the "first" trade secret owner has no recourse. Additionally, if a person who is privy to a trade secret unlawfully uses it, the owner can enforce the trade secret by filing a suit for misappropriation. However, once a trade secret is disclosed to the public, it is irretrievably lost. Trade secrets are litigated less frequently than patents because the owner may not want to disclose the trade secret as part of the litigation discovery process or in court proceedings.

13.9.1.3 Right Creation and Term of Protection

A third consideration concerns the process for creating rights and the term of those rights. Unlike patents, there are no formal application or registration requirements for trade secrets. Any valuable and confidential information used by a business is protected, as long as the business takes reasonable steps to maintain its secrecy. This generally means the initial cost of trade secrets is lower. However, sometimes the long-term costs of enforcing and updating internal procedures to keep information secret may be more than the cost to secure a patent.

The following chart summarizes these and other differences between patents and trade secrets (Table 13.1).

Patent protection is typically favored when:

- It is likely that a product can be reverse-engineered.
- The innovation might be discovered by others simultaneously.
- The technology is difficult or costly to keep secret.
- The technology must be disclosed to be of use.
- The subject matter is patentable.
- The commercial value of the innovation exceeds the registration and maintenance costs.

Trade secret protection is typically favored when:

- The subject matter may not be patentable.
- The subject matter is part of a relatively "crowded" art.
- Keeping the innovation a secret is realistic and would not place an undue burden on the corporation.
- The potential market is likely to last longer than 20 years.
- The technology is developing rapidly and the innovation is likely to be obsolete in a few years.

13.10 Developing and Managing an Intellectual Property Portfolio

It essential to have a well-organized and focused intellectual property management program in place to effectively develop and enforce a company's intellectual property rights. This plan should be an integral part of the company's business plan, and should be reviewed and updated as often as the business plan. Factors that should be considered when developing an intellectual property portfolio management program include:

(1) Strategic considerations in developing an IP portfolio.
(2) Administrative issues associated with managing the IP portfolio.
(3) Ongoing IP diligence protecting rights and pursuing others.

Table 13.1 Differences between patents and trade secret

	Patent	Trade secret
What is the protected subject matter?	Inventions (e.g., processes, machines, manufactures, compositions of matter, improvements of the foregoing, etc.)	Any information that gives the owner a competitive advantage over competitors and is maintained in secret (e.g., formulas, patterns, compilations, mailing lists, programs, devices, methods, techniques, processes, etc.)
What is the term?	20 years from patent application filing date	Indefinite, as long as information is kept secret and used in the business
How is it acquired?	Filing a patent application with the USPTO	Acquired upon creation. No formal application process
What are the requirements?	The invention must be patentable subject matter, useful, novel and non-obvious	The information must provide the owner with a competitive advantage and must be kept confidential
Anticipated costs?	Patent application filing fee, patent issue fee, post-allowance maintenance fees, and attorney time required to prepare and prosecute a patent application	No specific costs. However, costs are typically incurred in trying to maintain the subject matter as a secret (i.e., confidentiality and non-disclosure agreements, implementing internal policies for treatment of confidential information, etc.)
Can others use the invention/ information?	Others cannot practice the claimed invention without permission (i.e., a license). However, others may design around the invention	Trade secrets can lawfully be reverse-engineered by others. Additionally, others may use the information in accordance with a non-disclosure or confidential agreement
Can the rights be lost?	Patent rights can be terminated if the validity of the patent is challenged. In addition, rights to a patent can expire if the invention was in public use or sold or offered for sale more than one year prior to filing a patent application (typically through the inventor's own actions)	A trade secret is extinguished if it is disclosed to the public. For example, if an owner of a trade secret files a patent application for the invention, the trade secret is lost upon publication of the patent application or issuance of the patent
How is it enforced?	Assert rights against others for patent infringement	Assert rights against others for misappropriation of trade secret

13.11 Developing an Intellectual Property Portfolio Strategy that Fits the Medical Device Company's Business Goals

Developing a medical device company's IP portfolio strategy can be summarized in five (5) steps:

(1) Identifying the corporate strategy and determining how best to align the corporate strategy with the IP strategy.
(2) Identification of existing IP assets (by conducting an IP audit), and putting in place a process to identify and track new assets.
(3) Determining which of those assets are "core" assets (i.e., those assets that have a strategic importance to the company).
(4) Allocating resources to core and non-core assets as appropriate.
(5) Setting up a program to ensure that those assets are periodically reviewed and maintained to ensure that the IP portfolio is developed consistently with the corporate business plan.

13.11.1 Identifying Business Objectives and Goals and Aligning These with the Company's IP Strategy

The first step to developing a medical device company's IP strategy is to start by identifying the corporate business strategy. Once there is an in-depth understanding of what drives the company, it is easier to proceed through the next four steps for developing the company's IP portfolio strategy in a way that is well-coupled to the overall business strategy and objectives of the company.

13.11.2 Identification of IP Assets

To properly identify IP assets, it is crucial that personnel within the corporation who are knowledgeable about the company's IP or its technology plan conduct a thorough IP audit. This should also be performed in coordination with an experienced IP attorney. The IP audit should identify copyrights, trademarks, trade dress, trade secrets, confidential information, copyrights, mask works, domain names, industrial designs, patents, patentable inventions, and other IP assets owned by the company. Some registered assets (for example, patents, trademarks, and copyrights) require periodic maintenance in the form of payment of official fees and filing of official documents. If these fees are not paid or the official documents are not filed in a timely manner, these rights can expire. Consolidating all this information in a single location can help ensure that a company's entire intellectual property can be easily identified, tracked, and maintained.

Along with intangible assets, an intellectual property audit should identify any encumbrances on the company's IP assets. These encumbrances may include agreements (such as distribution agreements, licenses, software licenses, franchise agreements, assignments, covenants not to compete, employment agreements, third-party development agreements, and liens against the intangible assets) and any other liabilities that a company may have with third parties. The audit should identify whether or not these agreements affect the company's IP or other intangible assets.

Example ConglomoMed's patent portfolio contains over 100,000 issued patents and patent applications from around the world. The patent portfolio is broken down into four board technology segments, which represent the specific core areas of industry in which ConglomoMed competes. The breakdown is as follows:

(1) Common technologies and future business related IP—this segment includes technology related to laboratory equipment and corporate center technology.
(2) Electronic-based medical devices, including digital health technology.
(3) Ubiquitous healthcare product solutions.
(4) Technology solutions, including communications related technology and information processing technology for healthcare applications.

13.11.3 Periodic Review of IP Assets

Following the initial IP asset review, it is essential to periodically review those assets. These successive reviews ensure that existing IP assets are properly maintained, and an informed decision is made to retain, sell or dispose of non-used assets. Additionally, it is essential to establish a process for identifying and tracking new assets as they are created.

The frequency at which such reviews are conducted will vary greatly, depending on whether the company is a startup, an emerging entity or a long-established corporation. At the very least, a yearly audit should be conducted. Preferably, quarterly reviews of a current portfolio should be undertaken. The amount of resources to devote to the review will also depend upon the importance of the IP assets to the company. If the only core asset identified is the corporate name, which has been protected with a trademark registration, in-depth IP audits on a yearly basis may not be necessary.

Example ConglomoMed routinely reviews its IP for opportunities to enhance its portfolio. For instance, pending patent applications are periodically reviewed to determine if there is any content that can be protected via supplemental or divisional patent application procedures, or whether protection should be sought in additional countries. Non-core assets may also be dropped.

13.12 Administrative Issues for Long-Term IP Portfolio Management

After building an IP portfolio, administering it is a responsibility that requires integrating knowledge from various parts of an organization. A well-executed IP portfolio management program must be closely tied to a company's strategic plan, marketing initiative, legal department, and corporate research and development initiatives. Although tying all these aspects together in a comprehensive IP portfolio management program may appear daunting, there are several steps that make managing an IP portfolio a much easier endeavor, regardless of its size.

13.13 Ongoing IP Diligence: Protecting Rights and Pursuing Other

13.13.1 Defending Your IP

Aggressively pursuing competitors with litigation can be reckless if the company is vulnerable to an easy counterattack. Pursuing a solid defensive strategy means implementing the minimum measures necessary to protect core IP assets.

IP protection is, by its very nature, a defensive measure. Patents prohibit others from using the products or processes generated from research and development programs. For a technology-centric medical device or biotechnology company, this can represent the core of the business. Trademarks and service marks protect one of the most important aspects of a corporation's image; its name, logos, etc. Significant amounts of marketing and advertising budgets are typically allocated to using trademarks and service marks. Therefore, it is extremely important to protect these valuable assets. Copyrights protect a corporation's expression of ideas such as drawings, websites, marketing materials, documents, website designs, and software. Trade secrets, if handled properly, help protect customer lists, business plans and strategic plans. Since large amounts of a company's resources are typically allocated to these endeavors, it is essential to consider protection for all aspects of a company's IP portfolio.

Pursuing a vigorous defensive strategy has additional benefits. Such a strategy not only insures that the IP rights are protected, but can also strengthen those IP rights. For example, with effective trademark protection, if a trademark holder does not provide notice to competitors or the public at large from using a trademark to refer to a particular product (for example, using the term "Kleenex" as opposed to "facial tissue"), the trademark may ultimately become "genericized," whereby anyone can use the mark. Thus, rights to the mark will be lost and it will cease to become an asset for the company. This occurred with respect to the terms "aspirin" for the Bayer Corporation and the term "escalator" for the Otis Elevator Company.

A similar defensive strategy may be pursued using patent protection. If a corporation has a product that is protected by a patent, it can defend against another company trying to copy that product by asserting the patent against the infringer. If successfully implemented, such a strategy will protect the product and, as a result, the company's market share.

There are many different reasons why companies initiate intellectual property litigation. One of the primary defensive weapons that a company may have is to counter a claim of infringement with a counterattack using its own IP portfolio. When a medical device company that is aggressively pursuing patent infringement litigation realizes that the alleged infringer is not going to "cave in," the company may become more reasonable regarding its demands, or may back off from a litigation altogether. Without a robust defensive patent portfolio as a countermeasure, a company may become vulnerable to repeated attacks from different competitors. Often, entire industries are cross-licensed in this manner. However, such cross-licenses are only granted to those companies that have an IP portfolio to bring to the table.

Several measures that should be considered when putting together a strong defense include:

Are all core assets protected?

It is critically important to ensure that all identified core IP assets are properly protected. If they are not protected, a company may be vulnerable to a competitor entering the market with a similar competing product, thereby quickly eroding the company's market share or operating margins. Core IP assets being robustly protected is also attractive to potential investors (which is relevant for medical device or healthcare technology startups).

Having IP assets on fundamental technology will help insulate a company from attacks by competitors having related technology. Although such claims may be frivolous, defending against these attacks can drain smaller companies of desperately needed capital and divert management's attention from its primary focus.

Have clearance opinions been sought on all products to ensure that a company is not charged with infringement of competitors' IP portfolios—particularly patents or trademarks?

Just a critical as protecting a company's core assets, is ensuring that its products do not infringe competitors' IP rights. One way to determine this is to periodically review the industry for competitors' IP assets. If relevant IP assets are identified, it is recommended that a company seek a clearance opinion or a freedom-to-operate opinion from an experienced IP attorney. This can help ensure that the company's invention is free to be used or sold without charges of infringement by competitors.

Have vulnerabilities or gaps in competitors' product lines or IP assets been identified?

Once a company assesses its own vulnerabilities, it is critical to identify the vulnerabilities of its competitors. Although this does not require immediate or offensive action, such information will become invaluable should a competitor begin to make demands of the company.

If you have decided not to pursue patent protection for an invention, have you considered a defensive publication in order to keep a competitor from gaining rights to the idea?

As discussed above, a medical device company will need to make strategic decisions about which IP rights to expend resources for protection. If a decision is made to forego protection for a particular asset, the company should consider whether the asset could provide competitors with an advantage if it were independently developed. In order to keep competitors from gaining such a strategic advantage, publication of an article can dedicate information about the invention to the public in general and mitigate the harm resulting from a competitor's use of the invention.

13.13.2 Leveraging Your IP Rights

An IP portfolio is a significant strategic asset that can be leveraged to achieve favorable outcomes against competitors. A primary offensive weapon is the threat of an infringement suit. A company can also utilize licensing and the threat of (or actual) litigation in order to generate revenue from its IP assets. When deciding to aggressively pursue an offensive IP strategy, corporate management must be willing to back up any infringement allegations. There are several measures that should be considered when putting together a strong offense:

(1) Are your competitors litigation averse (i.e., do they have a history of settling litigations quickly)?
(2) Is your corporate management and Board of Directors willing to use the threat of litigation (or actual litigation) in order to enforce IP rights or to gain a strategic advantage over your competitors?
(3) Is the company willing to monetize its unused IP assets through franchising, licensing or sale?
(4) Is the company willing to develop IP assets solely for the purpose of monetizing them?
(5) Is the company willing to search for and purchase IP assets and other corporate assets from those that are available in the marketplace?

All of these measures should be considered carefully when crafting a winning offensive strategy. A company should carefully weigh the risks and benefits before embarking upon an IP strategy that incorporates an aggressive offense as a central part of its strategy. Such a strategy will only be successful if it has the full support of management and directors, as it will require significant allocation of corporate resources.

References

1. *Singer Mfg. Co. v. June Mfg. Co.*, 163 U.S. 169 (1896).
2. *Gilead Scis., Inc. v. Natco Pharma Ltd.*,753 F.3d 1208 (Fed. Cir. 2014).
3. *In re EchoStar Communs. Corp.*, 448 F.3d 1294 (Fed. Cir. 2006).
4. 35 U.S.C. § 102(a)(1).
5. *In re Hall*, 781 F.2d 897 (Fed. Cir. 1986).

Chapter 14
Tips for Avoiding and Preventing Intellectual Property Problems

14.1 Get Legal Advice Early

Many companies hesitate to hire an intellectual property lawyer. The company is reluctant to get a lawyer because they can be expensive, and the experience of working with a lawyer can sometimes be overwhelming, especially if it is the first time the company (in the case of a startup) has ever worked with an attorney. Additionally, lawyers are often overly concerned about the potential legal consequences of every action the company takes, which can be a daunting and discouraging attitude for a company that derives much of its energy and motivation from its ideas.

However, having an IP lawyer is going to help protect your company from potential pitfalls that might arise on your path to success. An IP lawyer may not be as expensive as you might initially guess, depending upon the specific needs of the company. Additionally, specific advice obtained at key points in the growth of the company may be invaluable. Isn't it better to avoid a problem in the first place than to stumble into the problem and then have to waste valuable time and resources to resolve it?

An IP lawyer will help a medical device company or manufacturer in a number of ways, including:

- Making sure that all of the intellectual property in the company, and the IP created by employees of the company, is properly assigned to the company. For example, it is essential that employees have an employment agreement that requires then to assign the rights of any inventions created within the scope of their employment to the company. Additionally, the company should require employees to promptly assign all rights to each invention to the company as they are discovered. This avoids the trouble of trying to track down and convince ex-employees to cooperate once they have left the company. The company needs to own all of the relevant intellectual property rights associated with its business, so that no one else can undermine the company's success by claiming a right to the company's underlying intellectual property. Fights over ownership rights of corporate IP assets have been the demise of many companies.

G. B. Halt Jr. et al., *FDA and Intellectual Property Strategies for Medical Device Technologies*, https://doi.org/10.1007/978-3-032-12649-8_14

- Making sure that the company complies with the strict deadlines associated with filing for certain intellectual property rights.
- Providing advice on the appropriate time to file for intellectual property protection, and what type of intellectual property protection is best suited for a particular piece of intellectual property.

14.2 Conduct Cost–Benefit Analysis on IP Protections

The cost of procuring intellectual property can be expensive, and medical device companies are often cost-conscious because their budgets must be carefully planned to fund research and development efforts, prototype testing and seek regulatory approval. As such, medical device companies need to carefully consider their options when it comes to intellectual property protection. While some means for protecting IP are very affordable, as is the case for trade secrets and copyrights, trademarks and patents can be substantially costlier to secure. Each form of intellectual property protection offers its own benefits and risks, and these factors should be carefully weighed by any medical device company when determining its IP strategy.

Medical device companies should conduct a cost–benefit analysis of their options when two forms of IP protection are available for a particular intellectual property aspect of a new medical device product. For instance, if the company has intellectual property that could be protected by keeping it secret (i.e., as a trade secret) or by means of patent protection, the company must make a choice between one form of IP protection over the other as patent protection and trade secret protection are mutually exclusive. The costs of each option need to be weighed against both the short-term and long-term potential risks associated with each option.

14.3 Do Things Correctly at the Outset

The heart and soul of a company are the intellectual property rights that protect the medical device company's product, services, and brand. Medical device companies and manufacturers can avoid many future headaches and potential legal troubles by researching and taking steps related to their intellectual property correctly from the outset. By having your company's intellectual property strategy planned out, the appropriate IP assignments in place, and IP applications completed correctly and filed in a timely manner, you can better protect the intellectual property rights of your company.

14.3.1 *Protecting the IP You Have*

Most new medical devices and procedures stem from an idea. The idea might be documented as design drawings, sketches, or prototypes, and these pieces of intellectual creation form the basis for the intellectual property rights that belong to the medical device company. It is important to seek the appropriate legal protection for intellectual property assets (such as a company name, an invention, or software code that makes up the core product or brand of the company), as soon as possible, regardless of whether the appropriate legal protection is patents, trademarks, copyrights or trade secret protection.

A medical device company should make sure all IP rights related to the company's products, services, or brand, are assigned to the company so that ownership of the IP assets is clear and properly recorded. The same should be done with any IP developed by employees, independent contractors, and other third parties that contribute to the company's intellectual property development.

14.3.1.1 Registering IP

One of the first things a company, should do is register the company name for trademark protection, and register the corresponding domain name. This is particularly important if the company is a startup. More mature companies also need to stay on top of registering IP assets on new products and services that are developed. Registration of IP creates an exclusive right in the intellectual property, and no one else is allowed to use the registered IP without the permission of the company.

Consider trademark protection, for example. While a company automatically gets trademark protection as a matter of common law simply by using its mark in commerce, protection without federal registration of the trademark with the United States Patent and Trademark Office (USPTO) provides only limited, regional protection. Federal registration will provide nationwide protection against others using your trademark without permission. Consumers, such as doctors, hospitals, and patients, will come to associate your trademarks with your company brand, and federal registration discourages and prevents others (i.e., competitors and copy-cats) from cashing in on your company's brand. Even if some other entity tries to use your company's trademark, federal registration allows your company to seek injunctions, damages, and other forms of recovery when others use your company's trademark without authorization.

14.3.1.2 Applying for Patent Protection

Thanks to the Leahy-Smith America Invents Act, the United States is a first-to-file patent system, which means that inventors are "in a race" to file their patent application with the USPTO in order to establish their claim to the invention. There are strict rules regarding disclosure of an invention to the public and seeking patent protection. If a medical device company makes a disclosure, presentation, or demonstration of their invention to the public, through a conference, lecture, or some other public demonstration (e.g., the company presents their invention at a technology conference), the company has one year to file a patent, or else the right to seek patent protection for that invention will be lost. Understanding the patent application filing timelines and procedures is crucial for a company seeking to maximize the value of its IP while minimizing the risk of losing the potential value that is contained in the intellectual property.

14.3.1.3 Use Contracts with Clear Clauses that Protect Your IP

Companies should use contracts to manage their business relationships with all parties involved in the development of the intellectual property. Employment contracts, independent contractor agreements, non-disclosure agreements, vendor and supplier contracts… every contract made with a person or entity that will have access to the company's intellectual property, or will contribute to the company's intellectual property, needs to include clear clauses addressing the confidentiality of the company's IP and the assignment of any IP to the company. The goal of these contract clauses is to ensure that there is no doubt regarding the ownership of all IP related to the company. It all belongs to the company.

14.3.2 Evaluate What IP Procurement Offers the Most Value

Since medical device companies often face strict budget constraints, they must carefully consider prioritizing IP procurement efforts. Which intellectual property rights should the company pursue first? A good strategy is to consider the protection options that are available to the company and determine which avenue of IP protection is going to provide the company with the most value. This is a strategic decision that will differ from company to company.

For instance, procuring trademark protection for the company name, registering a domain name, and securing patent protection for a medical device invention that is critical to the success of the company, are each IP procurements that carry a lot of value for the company. On the other hand, seeking trademark protection for the product name of every prospective product the company has planned might not make very much sense or carry much value for the company if those product lines aren't fully developed in the future. A product line could fall apart in the development stage, a vendor might go out of business, or consumers might not respond well to market testing, and the company might have to scrap the entire product line. Expending money and time on procuring IP rights that don't hold much value for the company is wasteful. Prioritize what IP rights to procure first, and save the procurement of less important IP rights for later.

14.4 Take IP Protection, Policing and Enforcement Seriously

14.4.1 IP is Important, but so is Getting Your Product to Market

Many medical device companies, and particularly startups, need to strike a balance between financing normal business operations and financing the procurement, maintenance, and enforcement of their IP rights. As discussed in earlier chapters of this book, IP procurement, maintenance, and enforcement can be costly in terms of resources, capital, and time. Yet, having secure and enforceable intellectual property rights could be a matter of long-term survival of the company. IP rights are the strongest line of defense that a company has against competitors, preventing competitors from stealing the company's products, ideas, and brand.

While IP is critically important to medical device companies, a company cannot neglect its normal business operations, or sacrifice operating capital that is earmarked for running the business, R&D, prototyping, regulatory review, marketing, etc. Long-term success stems from implementing a successful IP strategy while maintaining the company's overall health.

14.5 Don't Infringe the IP Rights of Others

Intellectual property litigation is one of the costliest things that can happen to a medical device company. Hundreds of thousands of dollars or more could be lost in mounting a defense to an infringement suit. In addition, when a company is involved in infringement litigation, it can scare away current or potential investors and even customers.

To avoid infringement litigation, companies should refrain from infringing the IP rights of others in the first place. It is good practice for companies to gather intelligence on competitors' intellectual property portfolios so that the company has a good understanding in terms of where its competitors stand, and the company can focus on differentiating itself from those competitors. It is also good practice to develop a thorough understanding of the intellectual property landscape for the particular technical area which the company is involved. Mapping the intellectual property of others can help a company develop a strategy for avoiding the inadvertent infringement of the intellectual property rights of others.

Additionally, if a company needs access to intellectual property rights held by another, it is important that the company identifies the owner of those rights and requests permission to use the intellectual property from the rightful owner. Deals can be struck and licenses can be obtained that permit the company to use the IP it needs, without trampling on the rights held by others. It may also be possible to purchase the necessary IP rights outright.

An ounce of prevention is better than a pound of cure when it comes to avoiding IP infringement. The company's focus needs to be on growing a strong, well-protected business that investors want to be a part of and that consumers need. Taking specific, strategic steps to avoid infringement litigation at the outset will save the company a lot of potential headaches and frustration in the long run.

14.6 Contamination of IP Rights

Contamination of IP rights is something that can occur when a company engages in collaboration with another party and that party contributes to the company's IP in a meaningful way, but no protections were in place to protect the company's claim of ownership to the resulting IP assets. For example, if no contract clauses were in place to grant the company all IP generated by the collaboration with the third party. Effectively, the third party contributed to the company's IP and thus staked an ownership claim for their own contribution. To think of this another way, the third party is contaminating the company's IP rights.

A medical device company can significantly reduce its risk for IP contamination by executing contracts with all third parties that specifically address the issue of intellectual property transfer to the company. By entering into express agreements with third parties concerning the ownership of any intellectual property that might be the byproduct of collaboration between the company and the third party, the company can protect its intellectual assets and avoid costly disputes over ownership of the intellectual property.

Chapter 15
Successful Implementation of a Medical Device Company's Regulatory Strategy

15.1 Talk with a Regulatory Lawyer About Regulatory Compliance Issues

When bringing a new medical device to market, it can be easy to become confused about regulatory compliance issues. The medical device can be subject to federal regulation through the FDA and state and/or local regulatory requirements. Further, if a company plans to sell the medical device internationally, there will be foreign regulatory requirements to consider. The best way to handle FDA regulatory compliance is to develop a regulatory strategy at the outset.

A company's regulatory strategy will encompass a collection of different plans, practices, and measures, that are often integral with one another, designed to ensure that the medical device company will be in compliance with the relevant laws and regulations. The first step in devising a regulatory strategy is knowing what laws and regulations apply to the company and its medical device.

The laws and regulations that will apply in any given situation will depend on a number of factors. Just a few examples of these factors include the classification of the medical device (e.g., Class I, Class II, Class III), the regulatory pathway through the FDA that is selected for the medical device, the state that the company is located in, whether the company wants to be in conformance with voluntary standards, and whether the company wants to sell the product abroad and in what jurisdictions. In addition to all of these considerations, it is important to emphasize that the regulatory law and the relevant guidances update and change frequently. Not only is it important to know the applicable laws, regulations and guidance, but it is also critical to be sure that they are the most up-to-date laws, regulations, and guidance.

With so many different aspects of compliance, ranging from choosing a regulatory pathway through the FDA, to labeling requirements, to demonstrating compliance with current Good Manufacturing Practices (cGMPs) during inspection, it is understandable that the regulatory process can seem overwhelming and daunting. Yet, a

G. B. Halt Jr. et al., *FDA and Intellectual Property Strategies for Medical Device Technologies*, https://doi.org/10.1007/978-3-032-12649-8_15

lack of specific knowledge of regulatory matters should not be discouraging. Comprehensive counseling from an experienced FDA regulatory attorney or consulting firm that specializes in regulatory matters will be immensely beneficial when identifying the relevant regulatory requirements applicable to a particular medical device and company, and in developing a company regulatory strategy. A knowledgeable professional can leverage their regulatory experience to provide guidance and thoughtful advice.

15.2 Develop a Sensible Regulatory Compliance Strategy and Prepare a Regulatory Plan

Regulatory compliance strategy focuses on evaluating the various pathways to market that are available for a medical device, and deciding how best to accomplish each one, weighing the benefits, risks, burdens, and costs associated with each potential option. The regulatory compliance strategy should also take into account the company's commercialization strategy, which is fairly straightforward in a single market, but gets increasingly more complex as the product is marketed in more countries. Careful planning can lead to early identification of potential problems and a regulatory compliance plan can be prepared that addresses, mitigates, or avoids those problems. Each medical device product will have its own regulatory strategy, which is likely to evolve as the medical device product transitions through development to market.

A well-planned and well-executed regulatory strategy that is in alignment with a company's business goals and commercialization plans can streamline the regulatory review process, in turn reducing the time to market. For instance, if a company sees an opportunity to capture a share of the market, it may be in the interest of the company to adopt a regulatory strategy that will get the product to market most quickly. Taking the time to anticipate and prevent problems from arising at the outset would further the company's goal of getting to market as quickly as possible. There can be a lot of value in taking the time to carefully plan out a regulatory strategy to prevent costly mistakes.

Regulatory strategy starts before an FDA submission is made, lasts well into commercialization of the medical device, and is a living process that must be revisited regularly and updated as needed. Many companies experience obstacles and hurdles during product development that require modifications be made to the regulatory plan or regulatory strategy. For example, regulatory bodies frequently issue new guidances on compliance with the law and regulations as well as new regulations and many companies find themselves needing to adapt their regulatory strategies accordingly.

Similarly, there may be reasons to adjust the regulatory strategy for a particular medical device from the business and product development standpoints. For instance, the company might make a business decision to scale back particular design features of a medical device or to drop a specific indication or two, in order to get the device cleared, granted, or approved sooner. Alternatively, circumstances might change for the company (e.g., cut ties with a vendor or component supplier) such that manufacturing the device to have certain features becomes prohibitively expensive.

15.2.1 Choose the FDA Regulatory Pathway for the Medical Device

One of the first steps involved in developing a sensible regulatory strategy is to choose the appropriate FDA regulatory pathway for your medical device. First, companies must identify the appropriate classification for the medical device (e.g., Class I, Class II, or Class III). How to make a determination as to the classification of your device is discussed in Chap. 2. Once the appropriate classification of the medical device is identified, the appropriate FDA premarket submission (e.g., 510(k), PMA, De Novo, etc.) pathway can be selected, and the corresponding submission prepared. Guidance on completing the submission process can be found in Chaps. 3–5 of this book.

In some situations, a medical device company may have some leeway in terms of which submission to make, and in such situations, the company needs to decide what strategy best comports with the company's business objectives. In a scenario where a company has a medical device product that is a Class III device, but the design of the device could easily be modified so that the device qualifies for classification as a Class II device (e.g., by scaling back a particular feature or use indication), the medical device company needs to think about how to proceed with its regulatory strategy. The company has a few options:

(1) The company could opt for the "quickest" path to market, which would involve modifying the device to be a Class II device and filing a 510(k), provided that there is an appropriate predicate device to demonstrate substantial equivalence. But in doing so, the company can only obtain FDA clearance (as opposed to FDA approval), which may impact the marketability of the product. Note: While the 510(k) pathway is usually quicker than the PMA pathway, it is not always a quicker process. Additionally, if there is not an appropriate predicate device, the alternative to a PMA would be a De Novo, which could end up being just as involved and lengthy of a process as submitting a PMA.

(2) The company could opt for the PMA route. The device would be a Class III device, and while the PMA process might take longer to complete, once the FDA approves the device, the product can be labeled as "FDA approved," which can enhance the marketability of the product. Choosing this option also has the added benefit of providing a competitive advantage over competitors who have modified their product in order to take the 510(k) pathway or who have yet to make a submission to the FDA.

15.2.2 Budget for the Regulatory Process

There are many different costs associated with the regulatory process. The FDA application itself can be quite costly (see Table 4.1 in Chap. 4), but there are so many other costs related to the regulatory process that precede the FDA application—and all of them need to be budgeted for. For instance, several critical components of regulatory compliance are design controls, establishing a Quality Management System (QMS) (discussed more in Sect. 15.3), and recording important documentation. It is important to realize that there are many costs associated with developing and implementing design controls (e.g., designers and engineers must be paid, software licensing fees for design tools must be purchased, overhead costs, etc.), building and maintaining a QMS (e.g., cost to purchase a QMS software solution, cost associated with a high-quality printer for a paper QMS, etc.), and managing a record system (e.g., costs associated with electronic storage space, such as electronic media or hard drives, server space, etc.). Further, funds must be allocated to build a prototype of the medical device (e.g., costs can include machining or manufacturing costs, cost of materials, etc.), to conduct non-clinical testing (e.g., in vitro and in vivo laboratory testing, toxicology testing, biocompatibility testing, mechanical and chemical property testing, software testing, etc.) and subsequent clinical trials (as appropriate) (e.g., costs associated with conducting human trials, exploratory trials, pivotal trials, etc.), and for international regulatory compliance (e.g., the costs associated with obtaining the required certifications, cost to secure licenses, cost to obtain marking such as the CE marking or the CCC marking, etc.)—again each of these activities will require a lot of funding.

Medical device companies can benefit from preparing a budget for the regulatory process early on, which can help reduce unexpected costs later down the line. Consider starting by preparing an extensive, itemized list of all of the regulatory compliance efforts that must be undertaken in order to obtain FDA regulatory approval, grant, or clearance. This list can be generated either from, or in conjunction with, the company's regulatory plan. Next, estimate what the costs will be associated with each regulatory compliance effort that has been identified on the budget. If working from previous experience going through the FDA regulatory review process, estimated costs can be drawn from previous regulatory compliance efforts for similar medical devices. However, if the medical device company is unfamiliar with the FDA regulatory process, or has no previous experience, it is highly advisable to research the anticipated costs associated with the planned regulatory compliance efforts.

15.3 The Importance of Developing Quality Management Systems

The FDA expects medical device companies to comply with current Good Manufacturing Practices (cGMPs) of the Quality System Regulation (soon to be the "Quality Management System Regulation") [1], and if a medical device company plans to market internationally, for instance, in Europe, then the company will also want to ensure that it is in conformance with international standard ISO 13485:2016. To aid in consistency and harmonization of device regulation, on January 31, 2024, FDA issued the Quality Management System Regulation (QMSR) Final Rule [2, 3]. The final rule amends the device cGMP requirements of 21 C.F.R. Part 820 primarily through incorporating by reference the device Quality Management System requirements of ISO 13485. While the scope of the Quality System Regulation will be retained, many of the provisions will be amended as well as the title of the regulation. Further, additional provisions and requirements will be created to clarify some of the concepts and expectations that are used in ISO 13485. The rule will be effective as of February 2, 2026. Until then, manufacturers are required to comply with the current Quality System Regulation. The FDA webpage, Quality Management System Regulation: Final Rule Amending the Quality System Regulation – Frequently Asked Questions, can be referenced for additional information on this upcoming change (https://www.fda.gov/medical-devices/quality-system-qs-regulationmedical-device-current-good-manufacturing-practices-cgmp/quality-management-system-regulation-final-rule-amending-quality-system-regulation-frequently-asked).

As such, quality systems and processes are critical to business success and need to be documented and recorded extensively for regulatory compliance. A thorough Quality Management System (QMS) is one of the most valuable tools that a medical device company has for establishing and maintaining compliance. A QMS also ensures that customers are getting a high-quality product.

A QMS is a formal system of quality controls, with emphasis on documenting the efforts that a company has taken to operate according to the company's quality policies using a number of quality control processes. It is a highly integrated collection of company policies, processes and procedures that are intended to improve quality, enhance controls, increase efficiency and productivity, reduce waste, address and reduce customer complaints, implement corrective and preventive action as needed and ensure that compliance is achieved and maintained. The standard operating procedures that define a QMS include design controls, production and process controls, specific document controls that are related to the medical device product, verification and validation for the design and process, supplier and purchasing controls, data analysis, risk management, customer complaint management, record-keeping policies, record retention, employee training, and reviews for design, quality, management, corrective and preventive action, etc.

A QMS also serves as a checks and balances system to ensure that the company is consistently doing what it takes to be and remain compliant. Not only does the

QMS include the policies, procedures, and processes that the company must follow in order to ensure good quality, but the QMS also provides reinforcing mechanisms and evaluation tools, such as audits and reviews, to verify that the policies, procedures, and processes of the QMS are being followed diligently.

Every company that develops a final finished medical device must develop a QMS (refer to 21 C.F.R. 820 for additional information; note that the class of the device [I, II, III] dictates which aspects of the regulation are required). Developing a QMS is something that yields as much as is put into it—if substantial thought and effort are put into the development of a QMS at the outset, the result will be a better, more efficient and effective QMS. Just as a company works hard to develop a top-notch medical device, the company must work equally hard to develop an efficient and effective QMS early in the medical device design and development process.

In terms of compliance there are four very important aspects of the QMS: design control, risk management, supplier controls, and corrective and preventive action.

Since medical devices that are marketed to consumers in the U.S. must be safe, design control is required for regulatory compliance [4, 5]. Design control is a framework of documentation and evidence that serves as proof that the device has been designed and manufactured to be safe and to meet the needs of consumers. Design control must:

- Address design and development planning.
- Identify design inputs (e.g., design aspects of the device that address user needs).
- Identify design outputs (e.g., what the device must do to adequately satisfy design inputs).
- Include design review processes and results.
- Address design verifications (e.g., affirm that the design inputs are adequately satisfy the design outputs).
- Address design validations (e.g., affirm that the device was properly designed to accomplish its purpose).
- Provide details concerning the design transfer process to manufacturing.
- Produce a design history file.

Risk management is another critically important part of a QMS. If a company is going to market a medical device to the public, the company is responsible for identifying hazards associated with the device or use (or misuse) of the device, conducting a risk analysis for the device, and developing and implementing risk control measures that can be monitored for effectiveness. The company must create risk controls to help reduce any risks associated with the product and its use to acceptable levels. In performing its risk analysis, the company must conduct a risk management review and produce a risk management report record. In many cases, risks can be managed through design control. Risk management procedures need to work in conjunction with design control, and vice versa.

Another variable that medical device companies need to account for in their QMS is Supplier Control. If the manufacture of the medical device depends on suppliers as a source of components or materials, then the medical device company must impose controls on suppliers to ensure that the supplier consistently provides the requisite level of quality. A bad batch of components or materials in the supply chain can have a disastrous impact on the quality of the final medical device product and could potentially lead to a regulatory compliance violation.

When something does go wrong and a process or procedure is discovered to be non-compliant, or could potentially place the company in non-compliance, the company must regain and maintain control over quality by taking Corrective and Preventive Action (CAPA). CAPA is more than just correcting a problem or non-compliance issue. Rather, CAPA is the plan that is enacted to eliminate the cause of the non-compliance issue. CAPA also involves implementing a change so that potential non-compliance issues are prevented from occurring in the first place. Any changes that are made to a QMS as a result of CAPA need to be recorded in a CAPA record and validated and verified to show that the CAPA has resolved the non-compliance issue and has not created any additional problems that need to be addressed.

Quality control is key to regulatory compliance, and a QMS is a useful tool for ensuring compliance. Each company must decide on how to implement their QMS. Some companies utilize a paper-based QMS. This involves generating and obtaining paper copies of all policies, processes, procedures, documents, records, reviews, audits, etc., which can become quite cumbersome. Paper-based QMS have a number of drawbacks. For instance, they are bulky, difficult to maintain and use, and hard to extract data from easily. Alternative systems and tools exist to make developing a QMS easier. For example, there are a number of commercially available software solutions that can be used to implement an electronic QMS. All policies, processes, procedures, documents, records, reviews, audits, forms, etc., can be stored electronically, which makes the system easier to use and maintain and data extraction more streamlined. Regardless of the method of implementation used, the most important function of the QMS is to serve as a record of how the company is striving for, achieving, and maintaining compliance.

15.3.1 Accurate Record Keeping is a Significant Portion of QMS and Regulatory Compliance

Records and other documentation are critically important for regulatory compliance. Keeping records well organized and easily accessible in a QMS can be helpful down the road, especially if the records need to be produced as evidence of compliance for an FDA inspection or if the company is ever involved in litigation. A number of regulatory frameworks require that document and record management systems be established for compliance.

A QMS is very helpful for achieving and maintaining compliance because effective QMSs are built around a robust document control management process. For instance, a medical device company's QMS must include a number of specific document controls related to the design and development and production history of the medical device. Some important document controls that are relevant to FDA regulatory compliance include a design history file, device master record, and device history record.

15.3.1.1 Building a Design History File

A design history file (DHF) is the collection of records and documentation related to the design and development of a medical device over time. The initial draft of the design, identification of user needs, design inputs and outputs, modifications or changes in the design, redesigned features, reports resulting from design reviews, design verification and validation procedures and reports, etc.—essentially everything related to the design and development of the device so that it is in compliance with the design control requirements of 21 C.F.R. § 820.30—should be recorded in the DHF.

If a particular record or reference is relevant to the design or development of the medical device, but would be better classified in some other portion of the record management system, is in a format that makes it difficult to include in the DHF, or is simply too large to include in the DHF (for example, a large data base or iterations of a computer program), then a reference to the record in question should be indexed in the DHF so that the record can be easily located if necessary. FDA inspectors often ask to see the company's DHF as part of a surveillance inspection (discussed more in Chap. 16), so it is important to have a DHF that is up-to-date, accurate and in compliance.

A good approach for building a DHF is to create a DHF at the outset and to compile the file as new design modifications and developments are made to the medical device. Since a lot of design and development of a medical device occurs from its ideation to when it is finally manufactured, the DHF can be quite large.

15.3.1.2 Compiling a Device Master Record

A necessary supplemental record to the DHF is a device master record (DMR). A DMR includes all the data and documentation related to the production process for building and testing of the medical device, i.e., all of the documentation related to how the medical device is actually made, tested, serviced, packaged, and labeled. A thorough DMR contains the necessary information required to instruct someone through exactly how to make the medical device step-by-step from the beginning to final product, i.e., how to make one individual medical device unit. Since the DMR is a how-to guide for making the medical device, the DMR necessarily must refer to a number of documents and records related to design and development of the device that are included in the DHF.

While there is a lot of carryover from the DHF to the DMR, the DMR includes additional information. Specifically, manufacturing specifications, manufacturing instructions, details on manufacturing processes (including equipment specifications, production environment specifications, production procedures, and production methods that are needed to make the device), engineering drawings, software specifications, component specifications, quality assurance and testing procedures, acceptance criteria for quality assurance equipment used with the medical device, quality assessments, installation and maintenance information, servicing protocols and methods, as well as packaging and labeling documentation for the medical device, are all included in the DMR [6].

Since the DMR requires reference to many documents and records contained in the DHF, medical device manufacturers preparing the DMR can include references to the DHF for any required documents or information needed to instruct the manufacturing process of the medical device contained in the DMR. In this way, records that are already included in the DHF do not need to be reproduced for inclusion in the DMR—a reference in the DMR to the appropriate location of the desired documentation contained in the DHF will usually suffice.

15.3.1.3 Maintaining a Device History Record

The device history record (DHR), on the other hand, is a history of the medical device that includes any records and documentation regarding the medical device products that are produced, i.e., the DHR is a production history that contains details about specific batches, lots, or units of finished medical devices. The purpose of the DHR is to log a history and to collect information demonstrating that the medical devices were produced in accordance with the DMR.

Information related to the manufacturing date of a batch of medical devices, the quantity produced in the batch, and the quantity that was released for distribution are included in the DHR. Additionally, the DHR includes acceptance records that indicate that each medical device was measured according to the DMR. In terms of labeling information, the DHR includes an entry for each medical device produced in the batch that includes the primary identification label and labeling used for each medical device that was produced, any unique device identifier (UDI), universal product code (UPC), or any other identification, including control numbers.

15.3.1.4 Records Related to Company Operations

Another collection of records that must be maintained accurately in a QMS is those related to company policies, processes, and procedures. The rules and guidelines by which the company, a department, or a division of the company operates are company policies; the activities that are involved to implement a company policy are the processes; and the specific detailed instructions on how to affect the process is the procedure.

In addition to developing written standard operating procedures that explain how the company is complying with FDA regulations, records will need to be carefully maintained that demonstrate that employees have been thoroughly trained on correct standard operating procedures and are following the standard operating procedures. Records must also keep track of when internal audits and assessments take place.

Another set of records that are very important to regulatory compliance is documentation related to the process by which the medical device company handles customer complaints [7, 8]. No matter how large or small a customer complaint might be, it needs to be taken seriously. Equal treatment must be given to complaints that originate from outside of the United States as well. Medical device companies must have a documented system in place for receiving complaints and a protocol for evaluating whether a complaint is directed to an event that needs to be reported to a regulatory agency, such as the FDA. The complaint handling system must also include a procedure for addressing the complaints. Addressing customer complaints largely involves conducting an investigation to identify the root cause of the problem, and taking a corrective action and/or a preventive action as appropriate so that the cause of the complaint does not recur. As such, complaints often need to be resolved through utilization of the company's CAPA process.

Records must be created for every complaint that is received and investigated and must specifically identify the medical device product involved (including relevant identification information, such as UDI and batch number), the date the complaint was received, the specifics of the complaint, the dates associated with when the complaint was investigated and a corrective action was taken, details about the corrective action that was taken to resolve the complaint, any reply made in response to the complaint, and contact information for the individual who filed the complaint. Complaints must be handled promptly.

Not only should records be created if action is taken in response to a complaint, but records should also be created to document when action is not taken. If upon review, it is determined that no corrective action is required to address the complaint, it is important to document the reasons why CAPA was not needed.

15.3.2 How to Handle Changes to a QMS

Change happens all the time in the medical device industry and many companies have to go through the process of modifying or adjusting their QMS to accommodate change. Companies need to have a process for managing change, and particularly for managing changes to the QMS.

Change management starts with having a plan. The plan must identify the changes, must address the purpose of the changes that need to be made, and must identify the potential consequences that could result from implementation of the change (including applicable regulatory submissions that may need to occur before the change(s) is implemented). The plan must also lay out details about how the integrity of the QMS will be maintained during the change, and once the change has been affected. It is also necessary that the plan identify the relevant resources that the change will require and what the availability of those resources is, i.e., will there be sufficient resources available to facilitate the change. Finally, the change management plan must examine the effect of the change on employees' responsibilities and/or authorities or must identify who will be accountable or responsible for the change. Once the company has established a change management plan, the next step is to implement the changes and verify that the changes have not created any new issues that need to be addressed.

Change needs to be documented and recorded to ensure quality. Documenting also helps improve traceability of the change across the various components of a QMS.

Example The DMR is one particular document control of the QMS that often requires updating throughout the manufacturing process. Whenever the manufacturing process associated with a medical device regulated by the FDA is changed, the changes must be accounted for in the documentation.

If a change is needed to the DMR, then a consequence of that change is that similar changes must also be made to the corresponding DHF and DHR because these documents are all integrally related. Since the changes to the DMR will have an impact on both the DHF and the DHR, the company must:

- Develop a plan.
- Identify the change that will be made to the DMR, and the consequences (i.e., similar changes must be made to the DHF and the DHR).
- Identify how the changes to all three documents will be made (*e.g.*, the changes require approval from the company's quality control department head).
- Indicate how the integrity of the QMS will be ensured (*e.g.*, including a dated and approved copy of the change management plan, or a reference thereto, in each of the DMR, DHF, and DHR).
- Identify what resources will be required to make the change and who will be responsible for the change (*e.g.*, the head of the quality department of the company must enter the changes into each of the DMR, DHF, and DHR).
- Implement the plan.
- Verify that the changes to the DMR, DHF, and DHR do not generate new issues that need to be addressed.

15.3.3 Conduct Regularly Scheduled Internal Quality Audits

Conduct regularly scheduled internal quality audits to ensure that the company's QMS is in good shape. An internal audit is an opportunity to identify problems, or potential problems, and to take steps to correct and improve the situation—a process which has long-term benefits for the company. For instance, think of internal quality audits as preparation for a FDA inspection.

Many medical device companies use internal quality reviews as a chance to get organized and to verify that the QMS functions as it should. For instance, a company might designate a team, comprising employees from various parts and levels of the company, to manage the internal audit. This results in a team with a diverse and varied experience and knowledge about the policies and procedures that the company follows in its day-to-day operations. A diverse team will be better able to assess whether compliance is occurring and to identify potential issues, since they have a more intimate understanding of the company's policies and procedures.

A common technique used by medical device companies to conduct internal audits is to utilize a checklist process that addresses the items that are anticipated as being part of a FDA inspection. Using a checklist ensure that each item is specifically reviewed and evaluated for compliance each time that an audit is conducted, while executing the audit as a process makes sure that auditors are involved and actively engaged in the process. Audit results can be compared over time to assess how well the company is adjusting its QMS to address problems and new challenges.

15.4 Simultaneous International Regulatory Compliance

It is not uncommon for medical device companies to want to participate in the global market. Just as medical devices are regulated by the FDA in the United States, other countries have their own counterpart authorities to the FDA. A few examples of FDA counterpart authorities include:

- Australia—Department of Health Therapeutic Goods Administration (TGA).
- Canada—Department of Health Canada (Health Canada).
- China—The China Food and Drug Administration (CFDA).
- France—French National Agency for Medicines and Health Safety Products (ANSM).
- Germany—The German Institute of Medical Documentation and Information (DIMDI).
- Japan—Pharmaceuticals and Medical Devices Agency (PMDA).
- United Kingdom—The Medicines and Healthcare products Regulatory Agency (MHRA).

In order to sell medical devices in foreign markets, medical device companies must comply with the regulatory requirements and/or standards for each foreign market. For companies looking to enter a foreign market, it is often a cost-effective approach to target two (or a few) initial markets in which to seek approval. There is added value in this approach if the regulatory requirements for the markets are similar, or if the first market has more stringent regulatory requirements that necessarily encompass all, or most, of the regulatory requirements of the second market (i.e., approval for the more stringent foreign market essentially translates to approval in the less stringent second foreign market). In this way, a medical device company's regulatory compliance efforts will serve double duty. Once the device is approved and being sold in the initial markets, and revenues permit, the company can then expand to additional countries.

Just to reiterate, the purpose of regulation is to ensure quality and safety of medical device products. Many medical device companies planning on entering foreign markets often structure their QMSs to be in conformance with ISO 13485:2016 and seek ISO 13485:2016 certification. While ISO 13485 is a voluntary standard in the United States, many foreign markets, such as the European Union and Canada, actually require ISO certification for regulatory compliance. Many medical device manufacturers seek ISO certification in parallel with FDA clearance, grant, or approval, since the quality system requirements for both are very similar. An added benefit of ISO certification is that it can enhance the marketability of the medical device product.

15.4.1 Example Markings Required in Other Countries

In order to participate in markets abroad, medical device manufacturers must be aware of the regulatory requirements for the foreign market, as well as the medical device marking requirements used in that specific foreign market. Markings on medical devices serve two main purposes. First, markings indicate that the medical device complies with a standard of safety, quality, and/or performance that is trusted in the market where the marking is required. Second, markings indicate that the sale of the product is legal in the country (or countries) where the marking is recognized.

15.4.1.1 CE Marking for Europe

One popular marking for medical devices is the CE marking. Medical device companies that are interested in entering the European market, for example, must obtain CE marking approval. Medical devices that comport with the essential requirements of the relevant European Union's Medical Device Regulation can be approved to be labeled with the CE marking, which stands for "Conformité Européene" or "European Conformity." Medical device products that bear CE marking can be legally placed on the market in the European Union (EU) and European Free Trade Association (EFTA) market.[1]

[1] CE marking is recognized in thirty (30) European countries: The twenty-seven (27) European Union (EU) member countries, including Austria, Belgium, Bulgaria, Croatia, Cyprus, Czech Republic, Denmark, Estonia, Finland, France, Germany, Greece, Hungary, Ireland, Italy, Latvia, Lithuania, Luxemburg, Malta, the Netherlands, Poland, Portugal, Romania, Slovakia, Slovenia, Spain, and Sweden; and three (3) European Economic Area member states, including Iceland, Liechtenstein and Norway.

15.4.1.2 CCC Certification for China

China has a certification for product quality standardization known as the China Compulsory Certification (CCC). Products that have been certified are labeled with a CCC marking according to the regulations provided by the Chinese Certification and Accreditation Administration. Medical devices that are sold in, or imported to, China usually require the CCC marking, but there are some classes of medical devices that are exempt from CCC certification requirements, and thus are also exempt from the CCC marking requirements. The classes of medical devices that do not require CCC include medical X-ray diagnostic equipment, hemodialysis equipment, artificial heart–lung machines, electrocardiograms, blood circuit piping in vitro for blood purification equipment, hollow fiber dialyzers, and latex condoms.

15.5 Have a Plan for Consumer Adoption and Payer Reimbursement

A company's regulatory strategy can sometimes be influenced by other areas of planning, such as consumer adoption planning and payer reimbursement. Who will use the product (e.g., consumers, patients, nurses, physicians, hospitals, etc.) and who will pay for the product (e.g., will health insurance providers pay for the device?) are important questions to answer and prepare for during the development of a medical device company's regulatory strategy. While consumer adoption and payer reimbursement are postmarket concerns, medical device companies need to plan for market during product development and regulatory strategy planning.

What follows is a high-level description regarding planning for consumer adoption and payer reimbursement in tandem with developing a regulatory strategy. It is important to note that payer reimbursement is very important to successful market entry but is also highly complex. Those who have limited or no experience with payer reimbursement should seek consultation with an experienced professional.

15.5.1 Plan for Consumer Adoption

For every medical device that is developed, there needs to be a consumer adoption plan. Having a clear and detailed path forward regarding consumer adoption in advance of going to market will allow a medical device company to situate itself in the best possible position for success once the device is ready for market. Inadequate planning can lead to delays in consumer adoption, and in the worst cases market failure.

Each consumer adoption plan is unique for each medical device. For starters, the consumer adoption strategy for a particular medical device will depend on the target consumer. The target consumer may not necessarily be the user of the medical device. Rather, the consumer may be the individual or entity that purchases the medical device.

When planning for consumer adoption, a good place to start is to identify all of the relevant consumers who will potentially purchase or use the medical device. Next, strategize about how to reach each consumer group most effectively. Finally, prepare a plan for implementing consumer adoption, including specific details about how implementation will be accomplished. In many cases, a consumer adoption plan will have multiple strategies for encouraging and promoting consumer adoption of the product.

Example If the medical device is a surgical tool that is used by surgeons to perform a specific procedure, the surgeon is the end user of the product. However, the hospital or clinic is the entity that will make the decision to purchase the medical device. In planning for consumer adoption, a medical device company would benefit from making efforts to get the product into the hands of surgeons, perhaps through demonstrations at surgical conventions, so that the surgeons can serve as champions of the product who advocate the benefits of the device to the hospital to encourage the hospital to purchase the device. However, the medical device company needs to also have a marketing plan focused on approaches for getting hospitals and clinics to adopt the product. For instance, obtaining a specific medical billing code designation for the device and the procedures for which it is to be used may help convince a hospital or clinic to adopt the device.

Example In another example, the consumer adoption strategy for an at-home biological test kit that is sold over-the-counter to consumers at local pharmacies will focus on raising consumer awareness about the product. The consumer adoption strategy might include providing brochures and posters to doctors' offices and pharmacies so that patients can learn about the benefits of the product, and where to get it.

Example In a third example, if the medical device is used in a clinical setting, then clinical adoption would require a strategy that involves getting the medical device into use in clinical settings. A medical device company with an investigational device exemption (IDE) might offer a research hospital the opportunity to partner together to conduct clinical studies. The company would provide the research hospital with a prototype version of the medical device free of charge to be used in the clinical studies and the research hospital would get the opportunity to conduct clinical research to assess the potential benefit that the device can have on improving patient care. There is a likelihood that numerous peer-reviewed publications will be written about the research, and these far-reaching publications will serve to spread the word about the medical device and its benefits. An additional benefit of a partnership of this nature is that the clinical data generated from the clinical studies conducted by the research hospital could be used in the FDA PMA or 510(k) submission for the device. This is an example of how consumer adoption planning and regulatory strategy can intersect.

15.5.2 Plan for Payer Reimbursement

Payer reimbursement plays a significant role in consumer adoption. If healthcare providers such as hospitals and physicians cannot get reimbursed by payers, such as large commercial insurers or Medicare, when a medical device is used or a procedure is performed, it is unlikely that providers will adopt the medical device product. In order to be commercially successful, medical device companies need to have a good understanding of how regulatory strategy can influence payer reimbursement.

Payer reimbursement involves three main components:

- **Payer Coverage**. Payer coverage refers to whether payers like Medicare or health insurance companies will pay for a medical device and/or the service or treatment for which the device is used. Since payers often offer a number of different coverage policies to consumers, there can be a lot of variance between what is covered by one payer versus what is covered by another payer.
- **Medical Billing Codes**. Coding is used in medical billing to help providers keep track of the procedures and services that are being provided to patients and to help payers assess reimbursement decisions. Medical, surgical, and diagnostic procedures or treatments are coded using Current Procedural Terminology (CPT) codes. Once coded, the provider submits a reimbursement claim to a payer and the payer evaluates the coding used in the claim to determine whether the claim is covered by the payer.
- **Payment**. Unlike coverage, which is focused on whether payers will pay for a medical device or procedure, payment refers to actually getting reimbursed, or paid. Sometimes a payer will cover a medical device, but will not pay the full amount, which can leave patients picking up the tab.

As noted in Chap. 4, reimbursement is often tied to specific labeling claims of the medical device. Off-label use of the device, i.e., any use outside of the indicated use(s) in the medical device label, would not be covered.

If it is a medical device company's goal to have as many payers cover its medical device as possible and to reduce cost to patients, the medical device company might include payer reimbursement considerations with its regulatory strategy in order to achieve the desired goals. For instance, it would be beneficial for the company to design its medical device such that the device qualifies for an existing CPT billing code that is known to be covered in-full by the top-ten insurance providers.

When a medical device is new, but is also clearly beneficial, it is sometimes possible to get the agencies responsible for creating new billing codes, such as Centers for Medicare & Medicaid Services (CMS), to add a new code or to modify one that already exists to accommodate the medical device. Similarly, sometimes insurance providers can be convinced to modify a coverage policy to accommodate a new medical device, or to add a new coverage policy in the right circumstances. Implementing either one of these approaches will take time and will require the cooperation of the coding agency and/or insurer. Since timing is everything in the rush to market, companies need to be strategic in terms of when to begin their efforts toward accomplishing this goal.

One final benefit to having a plan for payer reimbursement, which is particularly relevant to medical device startup companies, is that potential investors are attracted to startups that have payer reimbursement plans. Investors are drawn to opportunities that can de-risk the investment, and a payer reimbursement plan demonstrates forward thinking and that the company has an eye toward generating profits through reimbursement.

References

1. C.F.R. § 820.
2. FDA Webpage: Quality System (QS) Regulation/Medical Device Current Good Manufacturing Practice (CGMP). Available online at https://www.fda.gov/medical-devices/postmarket-requirements-devices/quality-system-qs-regulationmedical-device-current-good-manufacturing-practices-cgmp. Updated 31 January 2024. Accessed 20 August 2025.
3. Quality Management System Regulation (QMSR) Final Rule Amending the Quality System Regulation – Frequently Asked Questions. Available online at https://www.federalregister.gov/documents/2024/02/02/2024-01709/medical-devices-quality-system-regulation-amendments. Updated 21 November 2024. Accessed 20 August 2025.
4. C.F.R. § 820.30.
5. ISO 13485:2016.
6. C.F.R. § 820.181.
7. C.F.R. § 820.198
8. ISO 13485:2016, section 8.2.2.

Chapter 16
Tips for Avoiding and Preventing Regulatory Problems

16.1 Potential Pitfalls in Quality Management Systems

As discussed in Chap. 15, the Quality Management System (QMS) is a significant part of ensuring regulatory compliance. With that being the case, it is only natural that some of the biggest potential threats to the successful implementation of a regulatory strategy are pitfalls in the company's QMS. Problems in the QMS that are not identified and addressed immediately can harm quality, delay the regulatory review process, and lead to decreased profitability. There are a number of common QMS pitfalls that cause medical device companies a lot of grief. Being aware of these potential pitfalls, and addressing them before they get out of control, can help companies avoid and prevent regulatory problems down the line.

16.1.1 Disorganization and/or Waiting Too Long to Develop a QMS

There are a lot of companies, particularly startups, in the medical device space that focus intensely on designing and developing a new medical device and on the FDA submission process. However, by focusing solely on these aspects of bringing the device to market, a company runs the risk of neglecting its responsibility of building and maintaining a robust QMS. Companies without a QMS risk making mistakes, being inconsistent, creating confusion within the company, wasting time, and losing profitability.

A company that does not have a QMS and, more specifically a document control system, in place at the outset is at risk of producing a lot of records without a system to organize them. This is problematic for the company because it leaves the company's record-keeping system disorganized. It will take substantially more time, effort, and resources to implement a QMS after records have been created.

G. B. Halt Jr. et al., *FDA and Intellectual Property Strategies for Medical Device Technologies*, https://doi.org/10.1007/978-3-032-12649-8_16

Disorganization can be a problem when it comes to making an FDA submission and getting ready for an FDA inspection. The FDA requires evidence of quality controls from medical device manufacturers—and lots of it. So, when the company's documentary evidence of quality controls is not well organized and not gathered all in one system, the company can be left scrambling to pull all of the important and relevant documentation together in a hurry.

Furthermore, it is not uncommon for medical device companies that find themselves in a state of disorganization to discover that they are missing some critical step in the quality management process or some key record or documentation while trying to create a QMS so late in the game. The consequence of being disorganized, or not having a QMS in place early on, is that once deficiencies are discovered they must be addressed, which can delay regulatory review, production, and market launch.

The solution is that companies need to treat the development and implementation of a QMS as an important aspect of the company that is critical to regulatory compliance. In developing the company's QMS, particular attention should be paid to implementing sound document controls since documented evidence of compliance is so important to the FDA. There are many tools available to medical device companies to help them develop efficient and effective QMSs, but one thing that all good QMSs have in common is that they stem from a detailed QMS plan. Everything in business requires planning, and quality management should be no different.

16.1.2 QMS Shortcuts that End up Making More Work

Regularly auditing automated processes in a Quality Management System (QMS) is essential to ensuring they function as intended. While automation can greatly improve efficiency, and streamline workflows, it also carries the risk of amplifying mistakes. A QMS is a complex system with many moving parts, and if an automated process contains an error, that error can be replicated dozens or even hundreds of times before it is discovered.

A QMS is a complex system with a lot of moving parts. It only makes sense that one would look for ways to streamline the QMS processes to make it more efficient and easier to use. One shortcut that companies often implement in their QMS is automated processes. The trouble that can arise with automated processes is that if there is a mistake in the automated process, that mistake can be duplicated many times before it is discovered. Once a mistake due to an automated process in the QMS is discovered, not only must the automated process itself be updated to address the mistake, but also any output of the automated process needs to be reviewed and corrected to address the mistake—one mistake in an automated process can lead to a lot of extra work.

For instance, a company's QMS has an electronic record automated submission process for employees to use when archiving important design and development documentation. Through an automated software process, employees tick a box indicating what type of record they are submitting, and the software automatically transfers the record to the appropriate server location for storage, as per the company's document control procedures. However, if there is an error in the automated software process that designates the incorrect storage location for the type of record indicated for submission by the employee, important documentation related to the design and development of the medical device can end up misplaced or lost entirely. The error in the automated process might not be discovered until the company is being inspected by the FDA and the inspector requests to see particular design and development a file or record, but when the company goes to retrieve the file, the file is not where it is expected to be. Not only will the automated process need to be fixed, but the missing files will need to be identified, located, and transferred to their correct storage location.

Automated processes can be great for improving the efficiency and effectiveness of a company's QMS. However, care needs to be taken to ensure that the automated processes are functioning as intended. Just as companies should conduct internal audits to ensure they are operating in compliance with the law and federal regulations, companies that are utilizing automated processes in the QMSs should perform audits regularly to ensure that these processes are operating as intended.

16.2 Be Flexible, Changes to Your Plans Are Inevitable

In medical device development, change is inevitable. It is unrealistic to think that everything is going to go according to plan. Expect that circumstances will change, new obstacles will emerge, and unexpected issues will need to be addressed. Some changes will be deliberate, may originate from the company's own actions, or be brought on by company management, while other changes will be the result of having to adapt and adjust to new circumstances that are attributable to factors outside the company's control. It is important to remain flexible, with a mind toward solving problems and moving forward.

Companies should always plan for problems in order to mitigate the negative effects of those problems once they inevitably occur. But not all problems can be anticipated. Getting a medical device to market will generally take longer than expected because of unforeseeable events. Designs might have been based on metric units, but the supplier sent components in U.S. units, a batch of production materials might be discolored, an important digital design file might become unexpectedly corrupted, or the FDA review might take longer than the anticipated length of time. Not to mention that the governing regulations and laws could change, and the FDA guidances are updated frequently and sometimes regulations are as well. While all of these changes are outside of your control, your company must remain flexible in order to accommodate these changes, maintain quality, and ensure product safety.

16.3 Periodically Review and Update Your Strategies for Relevance

Developing a medical device and navigating the regulatory pathway is a living, fluid process. A medical device company must develop a number of different strategies to get a medical device product from the design stage to market, and many of these strategies must interplay with one another in order to reach commercial success. For instance, the company's regulatory strategy must be in alignment with its business and commercial strategy; the IP strategy for the medical device must reflect the outputs of the product development strategy; and the marketing strategy needs to be harmonized with the medical device labeling strategy.

For every strategy that the company has, there should be a corresponding documented plan. Plans are required for virtually every step of the regulatory process. A product design and development plan, a design transfer to manufacturing plan, a regulatory compliance plan, a quality control plan, a corrective and preventive action plan, are just a handful of the important plans that medical device companies must prepare in order to be compliant with FDA regulations.

However, the reality of the process of bringing a medical device to market is that circumstances will change over time. Designs will be updated, strategic partnerships might be made or lost, employees will come and go, and the various strategies used by the company may shift in terms of scope or focus. As an unavoidable consequence of change, the company's perfectly laid plans will need to be revised and updated to reflect the current thinking for its various strategies.

To help cut down on the risk of being non-compliant with federal laws and regulations, medical device companies should periodically review their strategies and make updates for relevance. Keeping all plans and company policies and procedures up-to-date through periodic review and revision will allow the company to stay in compliance.

16.4 Be Ready to Address Issues Immediately

In a regulatory environment, companies must always be prepared to address issues immediately. Preparation can be proactive in nature, such as being prepared for an FDA inspection or ISO audit, or reactionary in nature, such as being prepared to respond to FDA warning letters promptly. Even if the company is in a situation where it cannot address the problem immediately due to some obstacle, it is important that the company start taking steps toward addressing the issue, such as coming up with a corrective and preventive action (CAPA) plan for resolving the problem the minute it becomes possible to do so. Since it is a virtual certainty that issues will arise for any medical device manufacturer, companies need to be ready to handle problems as soon as possible.

16.4.1 Be Ready for FDA Inspections and ISO Audits

The FDA has the authority to conduct inspections and to obtain copies of records from medical device companies that sell products that are subject to FDA regulatory controls under sections 414 and 704 of the FD&C Act.

A site inspection conducted by a representative from the FDA is something that medical device manufacturers should always be prepared for. For those medical device makers that are sponsoring a PMA, the FDA will require that a site inspection be completed prior to approval, i.e., a PMA preapproval inspection. FDA may inspect relevant Sponsor facilities prior to granting or declining a De Novo (i.e., to inspect the Quality Management System, clinical and non-clinical data that support the submission, etc.). Similarly, even if the medical device is submitted via a 510(k), which does not *require* a site inspection before clearance, FDA suggests that manufacturers be prepared for an FDA inspection at any time, before or after clearance. Additionally, postmarket inspections can also be initiated by the FDA for the purpose of surveillance, for instance, after a company has been cited for issues during a previous inspection, or for-cause, meaning that there is a particular reason or event that prompts the inspection, such as a recall or receipt of consumer complaints.

For all class II and class III medical device makers, the FDA is required to perform an inspection every two years which evaluates the company's Quality Management System for compliance with the appropriate laws and federal regulations. Component manufacturers located in the United States may also be subject to periodic FDA inspection. During these inspections, the FDA is looking to see whether the company has defined, documented, and implemented, i.e., effectively established, and effectively maintained their Quality Management System.

The inspection can be unannounced, but usually the FDA will contact the company that will be inspected to schedule a date to begin the review. Each inspection is different, but inspections begin with the issuance of a FDA Form 482 Notice of Inspection, which is usually followed by a walk-through inspection of the facility and a number of requests that the company produces evidence, reports, and certain documentation that are required for compliance with regulations. FDA's inspection will be performed on the company's management controls, design controls, production and process controls, and corrective and preventive actions procedures. The techniques that are generally used by the FDA inspector are outlined in the FDA's August 1999 *Guide to Inspections of Quality Systems* Inspection Techniques guidance (https://www.fda.gov/files/Guide-to-Inspections-of-Quality-Systems.pdf, accessed 20 August 2025).

The inspection typically lasts for several days. Inspectors are looking for violations of current good manufacturing practices (CGMPs), examples of which include evidence of contaminated or poorly maintained manufacturing equipment, data integrity issues, inadequate process control, poorly trained employees, apparent failure to address issues, and drastic changes to production that do not comply with regulations. At the conclusion of the inspection, the FDA inspector will prepare an

FMD-86 Establishment Inspection Report, which identifies the conclusions and decisions made regarding the inspection. The report could call for the company to take voluntary action to correct minor issues, could recommend advisory action (such as the issuance of a warning letter to the company) or could recommend legal sanctions be placed on the company.

When issues with compliance are identified during an inspection, the company receives a FDA Form 483 Inspectional Observation, and/or a warning letter (in the event that violations have been identified during the inspection). Both a 483 observation and a warning letter require a response within fifteen (15) days of receipt. The FDA 483 Inspectional Observation is in essence a notification that issues exist that need to be addressed immediately. If not addressed promptly, you risk inevitably receiving a warning letter as well. A warning letter places mandatory requirements on the company, and a corrective action plan must be submitted to the FDA by the company, and continued correspondence with the FDA as corrective action is being taken is required.

Many potential issues that can be identified in an FDA inspection could be mitigated or eliminated in advance of an inspection. To say this another way, many of the issues that are commonly identified by FDA inspectors are addressable issues that the medical device company can resolve before the FDA steps in.

Preparation for FDA inspections is key. Medical device companies can avoid issues with FDA inspection by taking preventative measures to ensure compliance with the law and federal regulations. A few tips that can help medical device companies have a successful FDA inspection include:

- **Know What Laws and Regulations Apply to the Product and Company**. Ignorance about the laws and regulations that apply to a particular medical device or company is not a legal excuse to a violation. Companies must be proactive and know what laws and regulations apply to them and their products.
- **Comply with the Required Laws and Regulations**. Knowing the rules is just the first step, but medical device companies must follow up by complying with the necessary laws and regulations. Companies must actively develop policies, plans and procedures, and an effective Quality Management System that are compliant.
- **Maintain an Effective Quality System**. It is not enough merely to establish an effective quality system—companies must also regularly update and review their Quality Management Systems to ensure good controls are maintained to keep the company in compliance with the necessary laws and regulations.
- **Have Written Policies, Plans, and Procedures**. Written policies, plans, and procedures go a long way toward proving that company operations and controls are defined and documented.
- **Train Employees About Implementation of the Company's Policies, Plans, and Procedures**. When employees are well trained on company policies, plans and procedures, it shows. FDA inspectors may interview employees, and they recognize when employees know and understand acceptable CGMPs.
- **Ask Questions if Clarification Is Necessary to Determine What the Inspector Needs During the Inspection**. It's okay to ask for clarification from the FDA

inspector if there is uncertainty about what is being requested by the inspector. Further, make sure you provide the inspector with only what they ask for and avoid providing extraneous materials.

- **If Problems Are Noted During the Inspection, Express a Willingness to Correct the Problem.** Issues may be identified during the inspection, and if they are, it is important to demonstrate a willingness to address the issue. Even if you disagree with the assessment of a particular issue, being polite and respectful as you work with the FDA to resolve the matter is important for quick and effective resolution.

In addition to FDA compliance, many medical device companies also must conform with ISO standards (i.e., ISO 13485), and are subject to ISO audits, if the company wants to be ISO certified. ISO certification is required for medical devices that are to be marketed outside the United States, for instance in Canada or throughout Europe.

Companies can register for ISO conformance. ISO conformance results in ISO certification and is voluntary, whereas FDA regulatory compliance is required in the medical device space (for Class I, II, and III medical devices). ISO conformance and FDA compliance are two separate processes that evaluate different things and a satisfactory outcome of one process does not automatically mean that the other process will also be successful. For instance, ISO certification does not automatically equate to FDA regulatory compliance, and vice versa.

An ISO audit is similar to an FDA inspection in that a representative visits the site in order to conduct the audit. If the ISO auditor identifies a non-conformance, either major or minor, the company will have a set amount of time, based on the classification of the non-conformance issue, to submit a corrective action plan, or risk loss of the ISO certification. The company is given a set amount of time to address the non-conformance, and once corrected, the auditor will return to the site to follow-up.

16.4.2 Respond to Warning Letters

Medical device companies can be faced with a warning letter from the FDA if issues that are identified during the FDA inspection and noted in the FDA form 483 Inspectional Observation are determined by the FDA Office of Compliance to be sufficiently severe. While a FDA warning letter is informal and advisory in nature, it signifies that the medical device company is in violation of the law and regulations. It is something to take seriously, and medical device companies should not delay in responding to the warning letter and taking corrective and preventive action (CAPA) (discussed below in Sect. 16.4.2) voluntarily to achieve regulatory compliance.

First and foremost, it is critical to respond to communications from the FDA in a timely manner. Specifically, after a FDA inspection, medical device companies are required to respond to any FDA form 483 Inspectional Observations within fifteen

(15) days. Similarly, medical device companies are required to respond to any FDA warning letters within fifteen (15) days. The response must lay out, in clear detail, the steps that will be taken by the medical device company to correct the violation, along with the timeframe for completing the corrective action. The response should also explain the preventive action that the company is taking, or is going to take, to prevent recurrence of the violation in the future. If it has not been possible for the medical device company to take corrective action sooner, the response should include an explanation as to why there was a delay. The response should be accompanied by documented evidence of the completion of the corrective action, or if the corrective action is still to be performed, documented evidence of the corrective action plan.

Failure to respond to FDA warning letters and take appropriate CAPA can prompt the FDA to take enforcement action against the medical device company. Companies can be forced to pay penalties, have their products recalled, or the company could be shut down by an injunction. Since a medical device company's response to a FDA warning letter is such an important matter, and since the consequences for not responding thoroughly or timely are potentially so serious, it is a good idea to discuss a FDA warning letter with an experienced regulatory professional.

16.4.3 Take Corrective and Preventive Action as Needed

Businesses that are regulated by the FDA have a lot of responsibilities to juggle in order to be compliant with the applicable laws and regulations that govern their industry. Design and development compliance, supplier control, risk management, quality control, change management, manufacturing, packaging, labeling, and distribution are just a few of the responsibilities that a majority of medical device manufacturers must tackle to achieve and maintain compliance. With so much going on, it is always possible that a non-compliance issue will be discovered in one of the company's processes.

When non-compliance issues are discovered, medical device companies need to quickly take action to bring the process back in line with compliance. One of the biggest challenges medical device companies face is taking corrective and preventive action after an FDA inspection. In order to take corrective and preventive action (CAPA) as necessary, medical device companies need to have a CAPA process in place for this purpose.

For each non-compliance issue, the company needs to develop a CAPA plan. The CAPA plan involves analyzing the processes, controls, operations, records, complaints and any other relevant source of data to determine the cause of a non-conformance or the potential cause of a non-conformance (such as a non-conforming product). The analysis is meant to focus on the particular weakness in the QMS that may be the cause of the non-conformance problem. Once identified, an investigation must be conducted to understand why the non-conformance occurred. In view of this information, the company can prepare a plan that identifies the necessary actions that will be required to correct the non-conformance cause and the necessary actions required to prevent the non-conformance from recurring in the future. After implementing these changes in the QMS, the CAPA must be verified and validated to ensure that the problem has been addressed and that the CAPA does not have any adverse effects. The CAPA plan for addressing non-conformance issues, and any changes that are made to the QMS (such as updates to manufacturing procedures or processes that reflect the CAPA), need to be documented and recorded.

A common pitfall of CAPA is that companies develop CAPA plans that are reactionary (e.g., in direct response to receiving an FDA 483 citation or a customer complaint). While corrective action is a key component of CAPA, it is only one half of the solution. CAPA is also supposed to be preventative in nature, i.e., taking measures proactively to prevent non-conformance issues from occurring in the first place. Far too many medical device companies make the mistake of taking an "if it's not broken, don't fix it" approach to quality, and while the approach might be economical it is not always a good long-term approach. The preventative action component of CAPA can be immensely beneficial. If a quality review identifies a potential non-conformance issue that can be addressed with preventative action, a small amount of work upfront could translate to significant cost savings down the line.

While it is not always practical from a business standpoint to try and address all, or even most, potential non-conformance issues before they turn into an actual problem, it is prudent from a quality and regulatory compliance standpoint to strike a balance between reactionary corrective action and proactive preventative action in a CAPA strategy.

Chapter 17
Potential Combined USPTO and FDA Regulatory Pitfalls

17.1 Failing to Disclose FDA Information to the USPTO

Patent applicants have a duty to prosecute patents before the USPTO with candor and good faith, which includes a duty to disclose information known to the applicants to be material to patentability [1, p. 1128]. This duty extends to:

(1) Each inventor named in the application.
(2) Each attorney or agent who prepares or prosecutes the application.
(3) Every other person who is substantively involved in the preparation or prosecution of the application and who is associated with the inventor, the applicant, an assignee, or anyone to whom there is an obligation to assign the application [2].

A breach of this duty is a serious matter because it may constitute inequitable conduct. Inequitable conduct involves making (or failing to make) a disclosure of relevant, material information to the USPTO and involves an intent to mislead or deceive the USPTO. Inequitable conduct can arise in several ways, including making an affirmative misrepresentation of a material fact, failing to disclose material information, or submitting false or misleading material information, to the USPTO.

Inequitable conduct is very bad for patent holders because a finding of inequitable conduct as to any single patent claim can render the entire patent unenforceable. It could also lead to other related patents and applications in the same technology family being held unenforceable as well [3, p. 1352].

However, establishing inequitable conduct is a high bar. When a party seeks to prove inequitable conduct, it must show by clear and convincing evidence that the patent applicant made misrepresentations or omissions material to patentability, and that the patent applicant did so with the specific intent to mislead or deceive the USPTO. It must also be shown that the deceptive intent was the single most reasonable inference to be drawn from the evidence.

G. B. Halt Jr. et al., *FDA and Intellectual Property Strategies for Medical Device Technologies*, https://doi.org/10.1007/978-3-032-12649-8_17

An illustrative example of the dangers of making disclosures to one federal agency, such as the FDA, but not to another, such as the USPTO, can be found in the Federal Circuit case of *Bruno Indep. Living Aids, Inc. v. Acorn Mobility Servs.* In *Bruno*, the Federal Circuit affirmed a finding of inequitable conduct when the patent holder, Bruno, had failed to disclose to the USPTO information related to several invalidating prior art stairlifts that were submitted by Bruno to the FDA during the process of seeking approval from the FDA to sell a stairlift product that was covered by Bruno's patent [4, p. 1350–51]. In *Bruno*, after the inventor filed his patent application, he learned that stairlifts were regulated by the FDA as medical devices. The inventor then asserted to the FDA that the Cheney Wecolator and Stair-Glide stairlifts were "substantially equivalent" to his patented stairlift in handwritten annotations submitted to the FDA in his 510(k) submission in order to obtain FDA approval. However, the inventor failed to disclose this same information to the USPTO as prior art.

The Court held, that there was deceptive intent to withhold the disclosure of the prior art from the PTO because of the "fact that an official of Bruno, who was involved in both the FDA and PTO submissions, chose to disclose the Wecolator to the FDA but not to the PTO [4, p. 1354]." The Court further observed that in the absence of a credible, good-faith explanation from Bruno for disclosing certain information to the FDA but not disclosing the same information to the PTO, an inference could be drawn that the information had been withheld with deceptive intent [4, pp. 1350–1351]. The Court concluded that the inference of deceptive intent on the part of Bruno and/or his official was supported by damning evidence, including that the applicant had submitted material information to the FDA while simultaneously withholding the same information from the USPTO, "and attempting to explain the disparity of its FDA and PTO submissions in a way that 'strains credulity' [4, p. 1354]." Consequently, Bruno was found to have engaged in inequitable conduct by not disclosing the information contained in the 510(k) submission to the USPTO, and Bruno's patents were consequently rendered invalid.

In contrast, the case of *M. Eagles Tool Warehouse, Inc. v. Fisher Tooling Co.*, the Federal Circuit found that there was no inequitable conduct when a manufacturer failed to disclose that the claimed invention shared several parts with a device that the company had been selling for many years. The Court held that "a failure to disclose a prior art device to the PTO, where the only evidence of intent is a lack of a good faith explanation for the nondisclosure, cannot constitute clear and convincing evidence sufficient to support a determination of culpable intent" required to sustain a finding of inequitable conduct [5, p. 1345].

Instead, the Court noted that "[i]n an inequitable conduct determination based upon a nondisclosure, the applicant must know, or should have known, of the materiality of the reference for an inference of intent [5, p. 1341]." Therefore, the Court distinguished *M. Eagles Tool Warehouse* from *Burno* by highlighting that in *Bruno* disclosures of material prior art were made to one federal agency, i.e., the FDA, and simultaneously were withheld from the USPTO, whereas in *M. Eagles Tool Warehouse, Inc.*, there was no deliberate choice to disclose the information to one federal agency but not to another.

In the case of a 510(k) submission for a medical device, there is always a deliberate choice with regard to whether or not to disclose the information. Failing to disclose information contained in the 510(k) to the USPTO, like in *Bruno*, will be found to be a deliberate choice and likely tantamount to inequitable conduct. That is why it is critically important for manufacturers of medical devices to disclose all information contained in the 510(k) submission that is made to the FDA, including the predicate device, to the USPTO. Failure to do so not only jeopardizes the particular patent, but also all patents in the related technological family.

17.2 Failing to Disclose USPTO Information to the FDA

17.2.1 Be Truthful, Accurate, and Complete

Patent and regulatory teams often work in parallel, but for medical devices (including SaMD), those tracks converge at one critical legal point: nothing submitted to the Food and Drug Administration can be false or omit a material fact. Information that arises during prosecution at the USPTO, such as test results, admissions about device performance or limitations, acknowledgments about comparative technology, or statements about intended use, can become material to a 510(k)/De Novo/PMA if it relates to substantial equivalence (for a 510(k)), safety, effectiveness, or labeling. When such information exists, and is not disclosed to the FDA, the omission can create violations of multiple statutes and regulations, with consequences ranging from clearance/grant/approval delays to potential criminal liability. The guiding principle is straightforward: if the USPTO record makes a 510(k)/De Novo/PMA statement incomplete or misleading, the inconsistency must be addressed before the 510(k)/De Novo/PMA is signed and submitted.

Every 510(k)/De Novo/PMA must include a signed Truthful and Accurate Statement. By regulation, the submitter certifies, to the best of their knowledge, that all data and information in the premarket notification are truthful and accurate and that no material fact has been omitted. That certification is now codified at 21 C.F.R. § 807.87(l); the FDA also provides a template. In practice, this is where undisclosed USPTO information creates direct risk: if prosecution materials undercut the 510(k)/De Novo/PMA narrative on technology, performance, or risk, the "no material omission" clause is implicated.

Submitters must also choose either a public 510(k) Summary or a 510(k) Statement. The Statement is a commitment to make the 510(k) safety and effectiveness content available upon request once the FDA finds the device substantially equivalent. While the Statement text itself focuses on public availability, the companion confidentiality certification that many filers use to keep the existence of a not-yet-public submission confidential for a limited time expressly acknowledges that submitting false information to the government is prohibited by 18 U.S.C. § 1001 and by the FDCA (21 U.S.C. § 331(q)). Together, these provisions underline that a 510(k) premised on incomplete or contradicted facts is not just a paperwork problem; it carries legal consequences.

The federal false-statements statute fills in the criminal backdrop. Under 18 U.S.C. § 1001, it is a crime to knowingly and willfully make a materially false statement, conceal a material fact, or use a false writing in any matter within the jurisdiction of the federal government, which includes the FDA's 510(k)/De Novo/PMA processes. Material omissions or misstatements in a submission can therefore support prosecution under § 1001 independent of the FDCA.

The FDCA adds device-specific prohibitions and penalties. Section 301, codified at 21 U.S.C. § 331(q)(2), makes it a prohibited act for a device to submit any report required by or under the Act that is false or misleading in any material respect. Criminal penalties attach under 21 U.S.C. § 333, including misdemeanor liability without proof of intent and enhanced felony penalties for repeat violations or an intent to defraud or mislead. Read together, these provisions mean that a 510(k)/De Novo/PMA that omits or contradicts material information known from USPTO proceedings can trigger both false-statement and FDCA counts, alongside administrative remedies.

If the 510(k)/De Novo/PMA includes clinical data, there is a second layer of risk. FDA may disqualify clinical investigators who knowingly or deliberately submit false information in required reports, and Sponsors must certify compliance with human-subject protections, IRB requirements, and the IDE regulations (as applicable) when U.S. clinical investigations support a 510(k). Investigator disqualification threatens the credibility of the data set and can derail future studies that your portfolio depends on. The cross-reference to clinical-study compliance appears in 21 C.F.R. § 807.87(j), and the disqualification authority is at 21 C.F.R. § 812.119.

Post-submission marketing claims bring their own trap. Representing that a device is "FDA approved" based upon FDA's clearance of a 510(k) is misbranding. The regulation is explicit: any representation that creates an impression of official approval because the firm complied with the 510(k) regulations is misleading and constitutes misbranding. 21 C.F.R. § 807.97 discussed misbranding by reference to premarket notification (i.e., the 510(k)) and businesses should consult this regulation for guidance.

In practice, the compliance question is not abstract: it is about aligning what the company tells the USPTO and what it tells the FDA. Patent prosecution may emphasize novelty, unexpected results, or distinctions from prior art; 510(k)/De Novo/PMA practice often emphasizes similarity, unchanged risk profiles, and established technological characteristics. Those messages can both be true, but they must not contradict each other on matters material to substantial equivalence or safety and effectiveness. Examples that frequently matter include a bench or software test result submitted to the USPTO that reveals a failure mode material to the 510(k)/De Novo/PMA risk analysis; a claim-construction or prosecution argument that narrows a "controller" or "algorithm" in a way that makes the 510(k)/De Novo/PMA's technological-characteristics discussion incomplete; or inventor declarations that attribute a performance improvement to a parameter that the 510(k)/De Novo/

PMA treats as unchanged. The Truthful and Accurate Statement is the legal hook that makes those disconnects dangerous.

The fix is operational. Build a cross-functional disclosure review that runs before the initial 510(k)/De Novo/PMA filing and again before final responses to additional-information requests. Compare the 510(k)/De Novo/PMA device description, technological characteristics, performance testing, and indications for use to the pending claims, allowed claims, applicant arguments, and inventor declarations. If the USPTO record contains facts that are material to the FDA's review, incorporate them or address them directly so that the 510(k)/De Novo/PMA narrative remains truthful, accurate, and complete. When clinical data are involved, confirm and document compliance with Parts 50, 56, and 812, and verify that no investigator has submitted false information; if concerns arise, consult counsel about investigator disqualification risk and corrective steps.

The bottom line for device companies (including those developing SaMD) is straightforward. The 510(k)/De Novo/PMA process requires you to certify that nothing material has been omitted and to keep your statements accurate. If the USPTO file contains facts or analyses that matter to the FDA's substantial-equivalence or safety-and-effectiveness determinations, you must either incorporate them or correct the record. The legal consequences for failing to do so can include § 1001 exposure for false statements, FDCA prohibited-act liability and penalties under §§ 331 and 333, investigator disqualification where clinical data are involved, and misbranding if post-clearance communications claim "approval." Align the patent story with the regulatory story before filing the 510(k)/De Novo/PMA.

17.2.2 FDA's Application Integrity Policy

All applicants who submit a 510(k) submission must comply with the FDA's Application Integrity Policy (AIP). The AIP describes the FDA's approach for reviewing applications that raise significant questions regarding data reliability due to wrongful acts.

The AIP applies the term "applicant" broadly to include any person who submits to the FDA data or other information for the purpose of influencing or supporting a FDA decision regarding the grant of approval to market an FDA-regulated product and can include actions taken by an applicant's employees. Further, wrongful acts are defined by the AIP as "any act that may subvert the integrity of the review process [6, Sect. 1-1-3(9)]" such as "…making an untrue statement of material fact [6, Sect. 1-1-3(1)]." Under the AIP, an untrue statement of material fact is understood "as a false statement, misstatement, or omission of a fact [6, Sect. 1-1-3(9)]."

When an untrue statement of material fact is reported to the FDA, the statement is investigated in order to determine whether the statements raise significant questions regarding data integrity and reliability with respect to a regulated product [6, Sect. 1-1-3(9)]. The investigation involves searches conducted to reveal evidence of the wrongful acts. The searches can encompass formal and informal documents, such as correspondence, memoranda, or verbal communications that occurred in telephone conversations or one-on-one meetings [6, Sect. 1-1-3(9)]. During the AIP investigation and subsequent assessment, the FDA will delay conducting any substantive scientific review of the data in a pending application until the assessment is complete and questions regarding the data's reliability are resolved [7].

In the event that the investigation reveals that an untrue statement of material fact caused the data to be unreliable, the FDA may choose to refuse approval of a pending application or may choose to withdraw approval of a previously approved application [7]. To make matters more severe, the FDA has the power to seek recalls of marketed products and can potential take other actions, such as seizing medical devices, obtaining an injunction, imposing civil penalties, and seeking criminal prosecution as necessary and appropriate [7]. A Sponsor can avoid a lot of trouble and the threat of a recall by taking every precaution to not violate the AIP and should always make truthful statements to the FDA.

However, as it relates to patent prosecution, the AIP is potentially problematic. During patent prosecution, the patent examiner searches for prior art that the Examiner believes teaches the same or similar technical features of the claimed invention. The examiner submits these prior art findings to the patent applicant through an Office Action. The applicant then responds to the Office Action by highlighting the technical differences between the cited prior art and the claimed invention. In this process, a problem arises when the prior art is related to the "predicate device."

A "predicate device" must have the same technological characteristics as the predicate device or different technological characteristics that do not raise different questions of safety and effectiveness than the predicate device [8]. The FDA considers the technological characteristics to include features such as the materials used in making the device, the design of the device, the energy source used by the device or other technical features of the device [9].

If, during the patent prosecution process, the applicant argues that the claimed invention is patently different in terms of materials, design, energy source, or other features of the prior art predicate device, the applicant/Sponsor may accidentally be raising a significant question regarding the integrity of the statements made in the 510(k) submission made to the FDA. Specifically, in the 510(k) filing, the Sponsor is required to make a statement that the claimed invention has the same technological characteristics.

Although the terminology "the same technological characteristics" in a patent context is quite different from in an FDA context, if this apparent contradiction is brought to the attention of the FDA it may be sufficient grounds to warrant an FDA AIP investigation be opened. At a minimum, the AIP investigation will delay regulatory approval, and at worst may result in the application being denied or a recall of already sold products.

It is critically important that a medical device manufacturer cautiously distinguish the novelty under 35 U.S.C. §102 and non-obviousness under 35 U.S.C. §103 of a claimed invention over the cited prior art when the cited prior art is the predicate device. Failing to heed this caution could potentially result in delays with the 510(k) due to an AIP investigation.

17.3 A Cautionary Tale: The *Belcher v. Hospira* Case

The 2021 case *Belcher Pharmaceuticals, LLC v. Hospira, Inc.*, exemplifies this risk of portraying a product characteristic as known or unremarkable to the FDA (to facilitate approval), but later touting the same feature as novel or critical to the USPTO (to secure a patent) [10]. In *Belcher*, a pharmaceutical patent was invalidated due to contradictory representations made to the FDA and the USPTO, providing a cautionary tale for all innovators.

Belcher Pharmaceuticals obtained FDA approval for an injectable L-epinephrine formulation via a 505(b)(2) New Drug Application (NDA). In seeking approval, Belcher's Chief Science Officer relied on data from a *reference drug product* and prior literature. Notably, Belcher informed the FDA that its formulation's pH range (2.8–3.3) was essentially the same as that of an existing product and characterized a manufacturing change to reach that pH as a "very minor change." In other words, for regulatory purposes, the company downplayed any innovation, implying the formulation's characteristics were old and well-known. This strategy of leveraging known safety and efficacy data can expedite FDA approval, but it planted the seeds for later legal trouble.

Around the same time, Belcher was prosecuting a patent on the same epinephrine formulation (U.S. Patent No. 9,283,197). During patent examination, the USPTO initially rejected Belcher's claims as obvious in view of prior art that disclosed an epinephrine solution with a broader pH range of 2.2–5.0. To overcome this, Belcher argued that the narrower 2.8–3.3 pH range was "unexpectedly found to be critical" for reducing racemization (i.e., a degradation process). This argument for a critical, novel feature persuaded the examiner—the USPTO allowed the claims specifically because Belcher "demonstrat[ed] the criticality" of the 2.8–3.3 pH range in achieving stability. The patent issued in 2016, and Belcher later sued Hospira for infringement.

In the ensuing litigation, Hospira uncovered Belcher's FDA submissions and prior art references. It became apparent that Belcher had withheld key information from the USPTO—namely, the *same reference product and literature* that Belcher cited to the FDA, showing the 2.8–3.3 pH range was already known to stabilize epinephrine. The contradiction was stark: Belcher told the USPTO that the pH range was a novel, critical invention, while telling the FDA it was an established, minor tweak drawn from prior art.

The district court found that Belcher's Chief Science Officer, who oversaw both the NDA and patent prosecution, intentionally withheld material prior art and made inconsistent disclosures, amounting to inequitable conduct. The withheld references (including prior epinephrine products and scientific articles) revealed the allegedly "critical" pH range and impurity levels, meaning that Belcher's "improvement" was already publicly known. Had the patent examiner known of this information, the patent would not have been granted.

The Federal Circuit affirmed that Belcher's patent was unenforceable for inequitable conduct, citing the clear materiality of the withheld information and the intent to deceive the USPTO by hiding that information. In Judge Reyna's words,

"Belcher's alleged critical improvement over the prior art was already within the public domain, just not before the examiner." Because Belcher was aware of this (from the FDA process) and chose not to inform the USPTO, the courts concluded that the only reasonable inference was a specific intent to mislead the Patent Office.

The outcome for Belcher was severe. Its patent was rendered entirely unenforceable, extinguishing the company's ability to assert it against competitors. This occurred even though the patent had been formally issued and would have otherwise been valid. This is a dramatic demonstration of how failing to reconcile statements between these two important federal agencies, the FDA and the USPTO, can nullify years of research and development and patent prosecution effort. The *Belcher* case stands as a warning that one cannot play fast and loose with the duty of candor: any short-term gains from inconsistent arguments can lead to long-term catastrophe when a patent is stricken for inequitable conduct.

In the wake of *Belcher*, the USPTO and lawmakers have shown heightened concern about inconsistent statements. The USPTO issued a notice in 2022 entitled *Duties of Disclosure and Reasonable Inquiry During Examination, Reexamination, and Reissue, and for Proceedings Before the Patent Trial and Appeal Board*, [11]. This notice emphasizes that applicants must not take positions in patent prosecution that are inconsistent with statements submitted to other government agencies, including the FDA. The notice reinforced that the duty of candor and good faith [12] covers such scenarios, and even imposed a *duty of reasonable inquiry*—suggesting that patent practitioners should affirmatively check for statements made to other agencies that might be material. Moreover, the PTO warned that deliberately insulating the patent team from the regulatory team (to claim lack of knowledge) would not excuse compliance; any "established practices that prevent those with a duty of candor from learning of such documents" violate the duty [11]. In short, regulators expect internal coordination: siloed teams are no defense. Congress has similarly called for the PTO to ensure that drug manufacturers cannot unfairly use the patent system by making conflicting statements to different agencies. All these signals point to a future in which FDA and patent submissions will be cross-checked more rigorously for alignment.

For medical device companies (including those developing SaMD), this means that consistency is not just ethical best practice, but likely to be subject to regulatory scrutiny. Companies should assume that anything submitted to the FDA might eventually be seen by the patent office (and vice versa). Therefore, harmony between the technical narratives in both domains is essential. When the same people are involved in both regulatory and patent processes (as was the case in *Belcher*), they carry an even greater responsibility to ensure they aren't saying one thing in one forum and the opposite in another.

17.4 Infringing on IP of Predicate Device

The FDA requirement of "substantial equivalence" for obtaining 510(k) approval for a medical device, as well as the USPTO requirements of novelty under 35 U.S.C. §102 and non-obviousness 35 U.S.C. §103, must both be satisfied in order to obtain a patent protection for a medical device that is able to enter the market using the expedited 510(k) approval process. While these requirements from the two federal agencies may seem similar the surface, it is essential to remember that they are distinct and often diametrically opposed.

For example, in order to obtain 510(k) approval, a manufacturer of a medical device must demonstrate to the FDA that the device is "substantially equivalent" to a "predicate device" that is already available on the market. For the purposes of 510(k) approval, "substantial equivalence" of the "predicate device" means that the device:

(i) Has the same technological characteristics as the predicate device.
(ii) (I) has different technological characteristics and the information submitted that the device is substantially equivalent to the predicate device contains information, including appropriate clinical or scientific data if deemed necessary … that demonstrates that the device is as safe and effective as a legally marketed device, and (II) does not raise different questions of safety and effectiveness than the predicate device [8].

However, in order to obtain a patent, the manufacture of the medical device must demonstrate that the device is novel under 35 U.S.C. §102 and non-obviousness under 35 U.S.C. §103. Therefore, the manufacturer will be unable to obtain a patent if:

i. The invention is described in a printed publication, or in public use, on sale, or otherwise available to the public before the effective filing date of the claimed invention [13].
ii. The differences between the claimed invention and the prior art are such that the claimed invention as a whole would have been obvious before the effective filing date of the claimed invention to a person having ordinary skill in the art to which the claimed invention pertains [14].

Accordingly, the requirement for "substantial equivalence" for a 510(k) approval and novelty and non-obviousness for patent procurement appear to be contradictory. On the one hand, the Sponsor needs to demonstrate to the FDA that the device has the same technological characteristics as a predicate device that has been commercially available prior to the 510(k) submission. On the other hand, patent applicants must demonstrate to the USPTO that the device is novel and inventive in light of the entire corpus of human knowledge available as prior art.

On this issue, the Federal Circuit has stated that "FDA equivalence is irrelevant to patent law because it involves fundamentally different inquiries [15, p. 1349 n. 3]." Specifically, the Federal Circuit has held that in contrast to FDA equivalence, "equivalency for purposes of patent infringement requires an element-by-element comparison of the patent claim and the accused product, requiring not only equivalent function but also equivalent way and result [16, p. 1298]." Different attributes of a given medical device product may be relevant to bioequivalency but are not equivalent for the purposes of infringement, and vice versa [16, p. 1298]. The courts have repeatedly refused to allow FDA 510(k) notifications of substantial equivalence to be used as an admission of infringement in patent infringement cases [17, p. 1382]. As such, it is possible for a medical device to not infringe on the predicate device.

For instance, in *Innovative Therapies, Inc. v. Kinetic Concepts, Inc.*, the Federal Circuit held that patents owned by Kinetic Concepts, Inc. (KCI) for a negative pressure wound therapy were not infringed by Innovative Therapies, Inc.'s 510(k) filing that listed KCI's device as its predicate device. The Court reasoned that even though Innovative Therapies, Inc. had obtained expedited premarketing approval of its medical device based on representations made to the FDA by Innovative Therapies, Inc. that its device had the "same technological characteristics" as the predicate device, i.e., KCI's previously approved wound therapy device, Innovative Therapies, Inc.'s mere indication that the KCI device was a predicate for its own medical device was insufficient standing alone to constitute an act of infringement [17, p. 1379–1380].

Likewise, in *Abbott Point of Care, Inc. v. Epocal, Inc.*, the Alabama Northern District Court found Epocal's "EPOC Blood Analysis System" did not infringe the patents covering the predicate device, i.e., Abbott's i-Stat Analyzer, identified by Epocal in its 510(k) submissions to the FDA [18, p. 1]. In fact, the Alabama Northern District Court excluded all evidence regarding Epocal's 510(k) submissions to the FDA from the patent infringement trial, including any descriptive statements about the EPOC device, and any statements related to how the EPOC Blood Analysis System had substantial equivalence to the Abbott i-Stat predicate device because the Court was concerned that the jury could be misled into believing that the statements made by Epocal to the FDA were supportive of a finding of patent infringement [18, p. 11].

Similarly, in *Am. Med. Sys. v. Laser Peripherals, LLC*, the District Court for the District of Minnesota rejected Am. Med. Sys.'s (AMS) contention that representations made by Laser Peripherals to the FDA that the accused products were "equivalent" to AMS's AddStat products, was an admission of infringement of the patents held by AMS [16]. The Court reiterated that Courts have repeatedly refused to allow FDA 510(k) notification of substantial equivalence to be taken as admission of patent infringement because in the FDA context, the term "equivalence" has a different meaning than when the term is used in the patent infringement context [19].

However, in *Bard Peripheral Vascular, Inc. v. W.L. Gore & Associates, Inc.*, the District Court for the District of Arizona held, and the Federal Circuit (i.e., the appeal court) affirmed, that a jury was properly able to determine that Bard's patents were infringed in a situation where an expert relied on specific statements within the 510(k) submission to provide testimony during the infringement lawsuit that parts of Gore's device were covered by Bard's patent claims. Statements regarding specific limitations of the patent claim were addressed in Gore's 510(k) submission for its Propaten Vascular Grafts product. In the 510(k) submission, Gore told the FDA that its product, Propaten Vascular Grafts, had the same physical structure and tissue ingrowth characteristics as several predicate Gore-Tex Vascular Grafts that were already available on the market [20, p. 1092]. Based on these two statements, the expert was able to draw conclusions that Gore's Propaten Vascular Graft product infringed on patent claims held by Bard.

While 510(k) bioequivalency statements made to the FDA comparing a new medical device to a predicate device that is patented are potentially relevant to patent infringement analysis, a bioequivalency statement is not sufficient to establish infringement [16, p. 1298]. Instead, the courts will often exclude the 510(k) from evidence in a patent infringement lawsuit because bioequivalency statements pose a likelihood that jurors will become confused by the difference between the meaning of the term "equivalent," since this term has a different meaning under each of the USPTO and FDA regulatory regimes. Sponsors generally do not need to be concerned that they are making an admission of patent infringement when indicating a competitor's device as a "predicate device" on a 510(k).

17.5 Statements About Safety and Efficacy in the Patent Specification

In the United States patent system, a patent specification must teach a person of ordinary skill in the art how to make and use the claimed invention. In other jurisdictions, such as a Europe and Japan, the specification must also explain how the claimed invention is an improvement over the prior art. However, statements that the invention is an improvement over the prior art can potentially be problematic from both a patent and regulatory perspective.

For example, a claim was made as to the clinical effectiveness of a specific dosage range of a pain management drug in the written description of U.S. Patent No. 5,549,912 in *Purdue Pharma L.P. v. Endo Pharms., Inc.* Perdue's patent specification described the invention as a novel controlled-release oxycodone formulation method that substantially reduced the range of daily dosages required to manage pain in most patients. Perdue's patent specification further noted the clinical significance of its inventive formulation by making comparisons to other opioid analgesics, such as morphine, that require the patient to take twice the dosage range of the present invention, and noted that the inventive formulation and dosing range leads to a more efficient titration process for pain management (i.e., the process associated with adjusting a patient's pain medication dosage in order to provide acceptable pain relief with the least amount of side effects).

At the time of filing the patent application, Purdue did not have any clinical evidence to support the statements about clinical significance and efficacy in the patent specification. Instead, Purdue relied on the inventor's insight that led to the discovery of the reduced dosage range and envisioned clinical benefit associated with it. Yet, Purdue failed to indicate to the USPTO during patent prosecution that these statements were made based on the inventor's insight and lacked scientific proof—an omission that could be construed as inequitable conduct [1]. In determining whether Purdue had committed inequitable conduct, the Federal Circuit held that the omission that the clinical significance statements lacked scientific proof was material to patentability and could potentially be used to establish inequitable conduct, which in turn would invalidate the patent [1, p. 1133].

Statements in the specification of a patent application regarding the safety and efficacy of a medical device or drug can also be problematic from a regulatory perspective. The 510(k) submission requires that the new device not raise different questions of safety and effectiveness than the predicate device [8]. If the patent specification states that the medical device invention has improved safety and efficiency over what is known in the art, this likely presents a different question of safety and effectiveness to the FDA. Once a different question of safety and effectiveness is raised during the regulatory review process, the medical device is no longer eligible for expedited review under a 510(k) and must be handled under a PMA pathway. Patent specifications must be drafted with care in order to identify the improvements over the prior art as required in many countries, while minimizing statements regarding the safety and efficacy of the inventive medical device.

References

1. *Purdue Pharma L.P. v. Endo Pharms., Inc.*, 438 F.3d 1123 (Fed. Cir. 2005).
2. C.F.R. § 1.56(a).
3. *United States Water Servs. v. Novozymes A/S*, 843 F.3d 1345 (Fed. Cir. 2016).
4. *Bruno Indep. Living Aids, Inc. v. Acorn Mobility Servs.*, 394 F.3d 1348 (Fed. Cir. 2005).
5. *M. Eagles Tool Warehouse, Inc. v. Fisher Tooling Co.*, 439 F.3d 1335 (Fed. Cir. 2006).
6. *FDA Application Integrity Policy*, Section 1–1–3(1). Available online at https://www.fda.gov/downloads/ICECI/EnforcementActions/ApplicationIntegrityPolicy/ucm072631.pdf. Accessed 20 August 2025.
7. *FDA Compliance Policy Guidance Manual*, Section 120.100. Available online at https://www.fda.gov/ICECI/ComplianceManuals/CompliancePolicyGuidanceManual/ucm073837.htm. Accessed 20 August 2025.
8. U.S.C. § 360c(i)(1)(A).
9. C.F.R. § 807.100.
10. *Belcher Pharmaceuticals, LLC v. Hospira, Inc.*, 11 F.4th 1345 (Fed. Cir. 2021).
11. *Duties of Disclosure and Reasonable Inquiry During Examination, Reexamination, and Reissue, and for Proceedings Before the Patent Trial and Appeal Board*, 87 Fed. Reg. 45,764 (July 29, 2022).
12. C.F.R. § 1.56.
13. U.S.C. §102(a).
14. U.S.C. §103.
15. *Johns Hopkins Univ. v. Datascope Corp.*, 543 F.3d 1342 (Fed. Cir. 2008).

16. *Abbott Labs. v. Sandoz, Inc.*, 566 F.3d 1282 (Fed. Cir. 2009).
17. *Innovative Therapies, Inc. v. Kinetic Concepts, Inc.*, 599 F.3d 1377 (Fed. Cir. 2010).
18. *Abbott Point of Care, Inc. v. Epocal, Inc.*, No. 5:2008cv00543, Doc. No. 358, (N.D. Ala. Nov. 5, 2012).
19. *Am. Med. Sys. v. Laser Peripherals, LLC*, 712 F. Supp. 2d 885 (D. Minn. 2010).
20. *Bard Peripheral Vascular, Inc. v. W.L. Gore & Assocs.*, 586 F. Supp. 2d 1083 (D. Ariz. 2008).

Appendix A
Sample Preliminary Invention Disclosure

Preliminary Invention Disclosure

It is important to have complete, accurate information about the inventor(s) and the invention. If you have any doubts about any information, include it. Every applicant for a patent is under a duty to advise the U.S. Patent and Trademark Office about all known prior art. Use additional sheets of paper if necessary. If you have any questions, please get in contact with us. Mail and/or fax this form to:

Volpe Koenig
Counselors at Law
United Plaza, Suite 1600
30 South 17th Street
Philadelphia, PA 19103

Telephone: (215) 568-6400 Fax: (215) 568-6499

(1) Title of Invention:

CardioColander LDL Cholesterol Blood Filter Device

(2) Description of the Invention:

The CardioColander LDL Cholesterol Blood Filter Device is an implantable filter for insertion into the vasculature of a patient. Insertion of the CardioColander device is accomplished through a percutaneous port implanted in the patient's skin. The percutaneous port has an attached access catheter or tube through which the CardioColander is inserted into the patient and subsequently removed once it has performed its blood filtering function.

The CardioColander is formed from densely coiled stainless steel fiber. The coils are tightly formed and linked together to form a mesh filter element having a front and rear surface as well as an edge. Additionally, the filter element has extending from its edge in both the forward and backward direction, cylinder shaped sleeves. The sleeves and edge of the filter support the filter once it is inserted into the patient's vasculature by supporting the filter against the vessel walls. The sleeves provide support for the filter, and also serve to funnel blood flow to the filter element.

G. B. Halt Jr. et al., *FDA and Intellectual Property Strategies for Medical Device Technologies*, https://doi.org/10.1007/978-3-032-12649-8

The CardioColander is formed from stainless steel fiber, but can also be made from other biocompatible materials, such as polymers and other metals. Additionally, the CardioColander can be coated in a polymer, which may or may not include additional therapeutic elements, such as drugs, growth factors, etc.

While the CardioColander is envisioned by the inventor for use in the vasculature of a patient to filter the blood of LDL cholesterol, the overall design and function of the CardioColander could be adapted for other medical procedures in other vessels in the body.

(3) First written description of the invention was made by: Date: Today's Date

This invention disclosure form. However, the broad concept of an implantable LDL cholesterol filter was written about by the inventor 2 months ago.

(4) First drawing of the invention was made by:
Drawing No. 3 Date: Today

Please find attached to this invention disclosure three drawings of the CardioColander invention (labeled as Figs. 5-7).

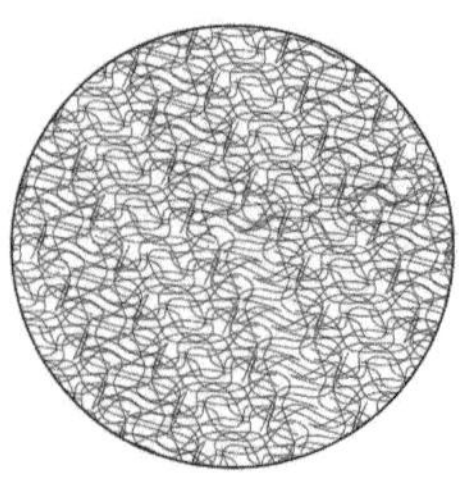

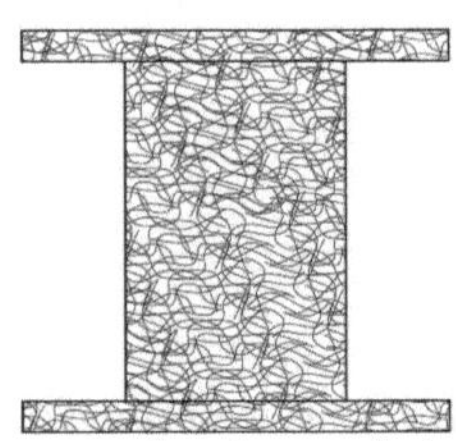

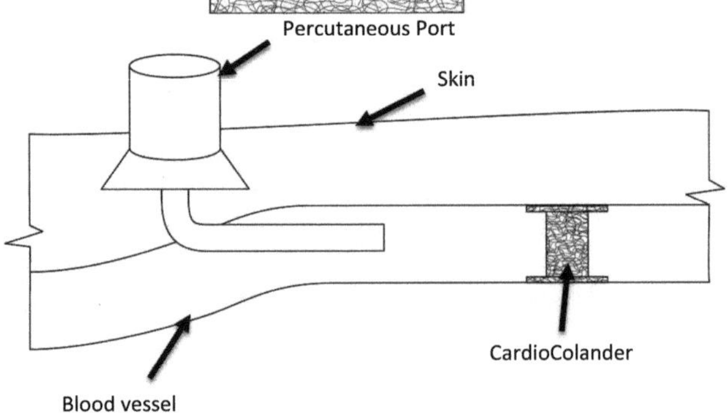

(5) First model was made by: **Date: 4 months ago**

Medicus Societas 3D printed a mock prototype of the CardioColander 4 months ago. Medicus Societas obtained a working prototype, created by Fiberfacture, 2 months ago.

(6) Invention was first disclosed to: **Date: 6 months ago**

The inventor approached Medicus Societas 6 months ago with the idea for the CardioColander LDL cholesterol blood filter.

(7) First effort to sell the invention: **Date: N/A**

Medicus Societas has not attempted to sell the CardioColander at this time. FDA approval for the CardioColander product is still pending.

(8) First use of invention: **Date: 1 month ago**

Medicus Societas has been approved for testing of the CardioColander product. Approval for initial testing was obtained 1 month ago.

(9) Description of circumstances(s) and date(s) of any disclosure(s) that are not described above:

Medicus Societas disclosed the CardioColander invention in our De Novo/Class III FDA PMA application, which was filed 4 months ago. Also, one of our employees gave a presentation in which the CardioColander device was disclosed. The presentation was to a group of college students, and took place about 3 months ago.

(10) If you know of other products, processes or machines like yours, describe them and how your invention differs from them.

Medicus Societas is quite sure that the CardioColander is a De Novo medical device for FDA regulatory purposes. We researched the invention, but found no similar medical devices to serve as a predicate device in our FDA application. The closest device we could find is a cardiovascular stent having a mesh "catch." The catch is meant to collect plaque and other debris that is generated when the stent is used to scrap plaque from blood vessel walls.

(11) Identify any known patents and/or other publications which relates to your invention. (Attach copies, if available.)

Medicus Societas has not identified any relevant patents. However, we are interested in obtaining a prior art search as part of the legal services you provide to us.

(12) In the case of a process invention has a product ever been made by the new process? Yes__No __.
If yes, has the product ever been offered for sale, sold, or used? Yes__ No_. If yes, please describe the circumstances with dates.

The CardioColander is made using a patented fiber manufacturing technique. Medicus Societas has licensed the patents related to the manufacturing process from Fiberfacture. The construction of the CardioColander is unique (it was not a product made by Fiberfacture prior to our interactions with them, nor does Fiberfacture make anything like the CardioColander).

(13) If anyone other than a named inventor had anything to do with this invention, state the person(s) full name, company affiliation, address and title and describe that involvement.

As discussed above, Fiberfacture is the source of the underlying manufacturing technique. Specifically, Fiberfacture owns the process patents for manufacturing fibers in the tightly coiled fashion that we use in our CardioColander product. However, the design of the CardioColander, and its use, is an invention made by Medicus Societas. We can provide the contact information for Fiberfacture upon request.

(14) If this invention is assigned or is to be assigned, identify the assignee.

The inventor of the CardioColander works at Medicus Societas. As part of his employment agreement, he has assigned the CardioColander invention to Medicus Societas.

Inventor:__________________	Inventor:__________________
(Print full name)	(Print full name)
Citizenship:_____________	Citizenship:_____________
Home Address:___________	Home Address:__________
_______________	________________
_______________	________________
Telephone No.:( ) -	Telephone No.: ( ) -
Social Security No.:_______	Social Security No.:_______
Signature______________	Signature________________
Date______________	Date______________

Attachments to this form: Three drawings of the CardioColander invention.

Appendix B
Sample Draft Licensing Agreement

DRAFT LICENSING AGREEMENT

This Agreement made this _________ day of ________________, 20__ by and between Medicus Societas, Inc., a Pennsylvania corporation, having a place of business at ____________________________________ (hereinafter "MEDICUS SOCIETAS") and CongloMed, Inc., a Delaware corporation, having a place of business at ____________________________________(hereinafter "CONGLOMOMED").

WHEREAS, MEDICUS SOCIETAS has developed a product (hereinafter "the MEDICUS SOCIETAS Product") intended for distribution and sale in a nationwide market and has expertise in producing such product; and

WHEREAS, CONGLOMOMED has a nationwide distribution network and is a manufacturer and distributor of similar types of products and is desirous of obtaining the MEDICUS SOCIETAS product and know-how as well as the right to manufacture, distribute and sell the MEDICUS SOCIETAS Product;

NOW, THEREFORE, in consideration of the above and of the mutual covenants and obligations contained herein, and intending to be legally bound, the parties hereto agree as follows:

1. DEFINITIONS

1.1 "Licensed Goods" shall refer to MEDICUS SOCIETAS' CardioColander product as described in MEDICUS SOCIETAS U.S. Patent Application No. XX/XXX,XXX.

1.2 "Licensed Territory" shall mean North and South America.

1.3 "Affiliate," with respect to either party, shall mean any person or entity who controls, is controlled by, or is under common control with a party to this agreement, and includes, but is not limited to, parent corporations, subsidiaries, and sister corporations.

1.4 "Net Selling Price" is the gross selling price (*i.e.*, the dollar amount) actually paid to and collected by CONGLOMOMED in a bona fide arms-length transaction consummated or intended to be consummated by transferring title in a Licensed Good, less:

(a) returns actually credited;

(b) actual losses experienced by CONGLOMOMED as a result of credits issued for such things as expired shelf life; and

(c) shipping charges separately charged to transferee;

however, no deduction shall be made for any other costs incurred, such as, but not limited to, costs of manufacture, sales, distribution, or exploitation of the Licensed Goods.

2. LICENSE

2.1 MEDICUS SOCIETAS hereby grants CONGLOMOMED and Affiliates a sole license for the manufacture, distribution and sale of the Licensed Goods in the Territory.

G. B. Halt Jr. et al., *FDA and Intellectual Property Strategies for Medical Device Technologies*, https://doi.org/10.1007/978-3-032-12649-8

MEDICUS SOCIETAS will provide technological support and know-how to manufacture the Licensed Goods to CONGLOMOMED.

3. ROYALTY

3.1. CONGLOMOMED agrees to pay a royalty to MEDICUS SOCIETAS of ____ percent (_%)on the Net Selling Price of the Licensed Goods. Royalties shall be payable on sales made during the period starting on the date of this agreement and ending on January 1, 2026, and for each annual period following therefrom, during the term or extended term of this agreement.

3.2. Royalty payments made under this agreement shall be made within sixty (60) days after the end of each quarterly period during the term of this agreement.

3.3. CONGLOMOMED agrees to submit to MEDICUS SOCIETAS within sixty (60) days after the end of each quarterly period a written royalty report setting forth the amount of royalties due for the preceding quarterly period and the manner in which CONGLOMOMED calculated said royalties. CONGLOMOMED agrees to keep complete records covering all royalty-bearing activities specified in this agreement in sufficient detail under its current accounting to enable the royalties payable hereunder to be determined and verified.

4. AUDITS

4.1. The parties hereby agree that MEDICUS SOCIETAS shall be permitted, at MEDICUS SOCIETAS' expense, to have a mutually agreed upon independent certified public accountant audit each royalty report submitted by CONGLOMOMED to MEDICUS SOCIETAS, within six (6) months from the date it is received by MEDICUS SOCIETAS. CONGLOMOMED shall make its records available to said accountant and cooperate by providing all available records essential to the verification of the report being audited and said accountant shall maintain confidential all information learned in the course of examining CONGLOMOMED's records, with the exception of a report to MEDICUS SOCIETAS with the accountant's findings as directly related to CONGLOMOMED's obligations to make royalty reports and payments. In the event of a finding by such accountant of a material variance with the report issued by CONGLOMOMED, then CONGLOMOMED shall reimburse to MEDICUS SOCIETAS the audit costs paid to the accountant and pay to MEDICUS SOCIETAS any additional royalties determined to be due.

5. PATENTS

5.1. Should CONGLOMOMED obtain information that any patents owned by MEDICUS SOCIETAS are or may be infringed, it shall provide such information to MEDICUS SOCIETAS, but shall have no further responsibility or obligation. Any patents obtained by CONGLOMOMED relating to the Licensed Goods or improvements to the Licensed Goods shall be the property of MEDICUS SOCIETAS. CONGLOMOMED shall promptly review any papers and execute, acknowledge and deliver all such papers as may be necessary or desirable, in the sole discretion of MEDICUS SOCIETAS, to obtain or maintain patent protection for the Licensed Goods and to confirm the ownership of any such patent by MEDICUS SOCIETAS.

6. TRADEMARKS

6.1. CONGLOMOMED shall use the trademark “CARDIOCOLANDER,” in such form as specified in writing by MEDICUS SOCIETAS, on the Licensed Goods in the Licensed Territory. CONGLOMOMED shall also use in connection with the trademark a “™” or, where U.S. Federal Trademark Registration has been obtained, an “®”. Once approved, CONGLOMOMED shall not depart from the approved form of the “CARDIOCOLANDER” mark on any materials requiring approval without the approval of MEDICUS SOCIETAS in accordance with paragraph 6.4 of this Agreement.

6.2. In order to assure the development, manufacture, appearance, quality and distribution of the Licensed Goods is consonant with the quality of the trademark, MEDICUS SOCIETAS retains the right to review the Licensed Goods.

6.3. CONGLOMOMED shall submit to MEDICUS SOCIETAS for approval samples of all Licensed Goods prior to any distribution or sale thereof by CONGLOMOMED.

6.4. Any such submission of the Licensed Goods for approval which is not disapproved within fifteen (15) days shall be deemed approved. Any disapproval by MEDICUS SOCIETAS shall be submitted to CONGLOMOMED in writing within the aforesaid fifteen (15) days together with remedial changes, which would remedy such disapproval.

7. ENFORCEMENT

7.1. If MEDICUS SOCIETAS obtains patent protection for the Licensed Goods, and any such patent is infringed by a third party, CONGLOMOMED and MEDICUS SOCIETAS may take appropriate action to suppress such infringement. As patent owner, MEDICUS SOCIETAS shall have the first right, but not the obligation to take action. In the event that MEDICUS SOCIETAS takes action against an alleged infringer, MEDICUS SOCIETAS shall be entitled to the entire recovery. If CONGLOMOMED requests MEDICUS SOCIETAS in writing to suppress any infringement, identifying in the request the infringer and the circumstances of the infringement, and MEDICUS SOCIETAS fails to file suit against the identified infringer or to otherwise take action to cause the identified infringement to cease within sixty (60) days of CONGLOMOMED’s request, CONGLOMOMED shall have the right to file suit against and to negotiate and enter into a settlement with the identified infringer. If CONGLOMOMED files suit, MEDICUS SOCIETAS is under no obligation to bear any cost of the suit. MEDICUS SOCIETAS, at CONGLOMOMED’s expense, shall join in the suit and render assistance and sign all papers, as may be reasonably required in connection with such enforcement. MEDICUS SOCIETAS shall be entitled to 20% of any court awarded or lump sum recovery, less costs, as a result of any enforcement of such patents by CONGLOMOMED.

7.2 If CONGLOMOMED becomes aware of a third-party infringement of the “CARDIOCOLANDER” mark in connection with electronics products in the Licensed Territory, CONGLOMOMED shall provide notice and the details of such infringement to MEDICUS SOCIETAS, and MEDICUS SOCIETAS shall take appropriate action to suppress such infringement.

8. TECHNOLOGICAL SUPPORT, QUALITY CONTROL AND PERFORMANCE ASSURANCE TESTING

8.1 MEDICUS SOCIETAS shall provide instructions and know-how for the production of the Licensed Goods.

8.2 MEDICUS SOCIETAS and CONGLOMOMED agree to jointly develop a Quality Assurance Plan to meet all applicable agency regulations and certifications for the Licensed Goods. CONGLOMOMED shall test each production batch of the Licensed Goods and shall maintain test records in accordance with the Quality Assurance Plan.

9. PRODUCT LIABILITY AND WARRANTY CLAIMS

9.1. CONGLOMOMED shall assume all liability for all claims of any nature with respect to the Licensed Goods distributed or sold by CONGLOMOMED. CONGLOMOMED hereby agrees to indemnify, defend and hold MEDICUS SOCIETAS harmless, from and against any loss, liability, damages and expenses (including reasonable attorney's fees and expenses) which may be incurred or for which MEDICUS SOCIETAS may be obligated to pay or for which MEDICUS SOCIETAS may become liable or be compelled to pay in any action, claim or proceeding against CONGLOMOMED and/or MEDICUS SOCIETAS for or by reason of any acts, whether of omission or commission, that may be claimed to be or are actually committed or suffered by CONGLOMOMED in connection with CONGLOMOMED's performance of this agreement. The provisions of this paragraph and the obligations under the same shall survive the expiration of this agreement. CONGLOMOMED shall maintain and procure at CONGLOMOMED's expense a comprehensive general liability policy including, but not limited to, contractual advertising and products liability coverage's with a policy limit of not less than $______ million per occurrence. Such policy shall be in full force during the entire term of this agreement and shall be placed with a responsible insurance carrier and shall name MEDICUS SOCIETAS as an additional insured and provide for at least thirty (30) days prior written notice to MEDICUS SOCIETAS of the cancelation or modification of such policy.

9.2. In the event that CONGLOMOMED does not obtain and maintain the aforesaid policy continuously in effect, upon prior written notification to CONGLOMOMED, MEDICUS SOCIETAS may obtain such insurance policy on behalf of CONGLOMOMED. All premiums for such insurance policy shall be deducted from royalties due under paragraph 2 of this agreement.

10. TERM AND TERMINATION

10.1. This agreement will have an initial term of seven (7) years from the execution date.

10.2. This agreement shall automatically be renewed for subsequent five (5) year terms, subject to the right of either party to terminate upon written notice one (1) year prior to the expiration of the initial term or any subsequent renewal term.

10.3 Notwithstanding the aforesaid, this agreement shall be subject to the rights of earlier termination by the party indicated as hereinafter set forth:

(a) By MEDICUS SOCIETAS in the event that CONGLOMOMED fails to make royalty payments following ten (10) days prior written notice and demand to cure from MEDICUS SOCIETAS; provided, however, if CONGLOMOMED cures such default within the ten (10) day period then such notice shall be of no force and effect; or

(b) By either party in the event that the other party breaches any other material obligation imposed upon it under this Agreement and fails to cure such breach within a period of thirty (30) days after notice and demand for cure from the party not in breach; provided, however, if the defaulting party cures its breach within the thirty (30) day period, then such notice shall be of no force and effect; or

(c) By either party immediately upon the other party becoming bankrupt, insolvent, making an assignment for the benefit of creditors, applying for or consenting to the appointment of a trustee or receiver or if bankruptcy proceedings are instituted against CONGLOMOMED.

11. RESOLUTION OF DISPUTES BETWEEN THE PARTIES

11.1. This agreement shall be deemed entered into the Commonwealth of Pennsylvania and shall be construed and governed solely by the laws of Pennsylvania.

11.2 In the event of any dispute, difference or question arising between the parties in connection with this Agreement or any clause or the construction thereof, then and in every such case, unless the parties concur to the appointment of a single arbitrator, the difference shall be referred to three (3) arbitrators; one to be appointed by each party, and the third being nominated by the two so selected by the parties, or if they cannot agree on a third, by the American Arbitration Association. In the event that either party within one month of any notification made to it of a demand for arbitration by the other party, shall not have appointed its arbitrator, such arbitrator shall be nominated by the American Arbitration Association. The arbitration shall take place in Philadelphia, Pennsylvania. The arbitrators must base their decision with respect to the difference before them on the contents of this Agreement and the attachments thereto, and the decision of any two of the three arbitrators shall be binding on both parties. The arbitrators shall apply the law of the Commonwealth of Pennsylvania.

12. PUBLIC STATEMENTS. Any public statements or publicity concerning the existence or contents of this Agreement shall be subject to review and approval by the other party. MEDICUS SOCIETAS and CONGLOMOMED will consult with each other concerning the means by which customers and potential customers shall be informed of this Agreement.

13. GOVERNMENTAL APPROVALS. CONGLOMOMED will at its own expense apply for and obtain the approvals of such governmental and regulatory entities as necessary to the marketing and sale of the Licensed Goods in the Territory. MEDICUS SOCIETAS will furnish necessary technical support to assist in obtaining any such approvals.

14. GENERAL PROVISIONS

14.1 This agreement sets forth the entire agreement and understanding between the parties hereto relating to the subject matter hereof, and supersedes any prior or contemporaneous oral or written representations, inducements or promises not contained herein.

<u>14.2</u> No amendment or modification of this agreement shall be valid or binding unless the same shall be made in writing and signed on behalf of each party by a duly authorized representative.

<u>14.3</u> This Agreement and all rights and obligations herein shall be binding upon and inure to the benefit of and be enforceable against the parties and their successors or assigns.

MEDICUS SOCIETAS and CONGLOMOMED shall make no assignment, pledge or hypothecation of this agreement or its performance thereunder without the express written permission of MEDICUS SOCIETAS and CONGLOMOMED.

14.4 The failure to enforce any of the terms and conditions of this agreement by either of the parties hereto shall not be deemed a waiver of any other right or privilege under this agreement or waiver of the right thereafter to claim damages for any deficiencies resulting from any misrepresentation, breach of warranty, non-fulfillment of any obligation of any other party hereto.

14.5 If any term or provision of this agreement is held to be invalid or unenforceable by reason of any rule of law or a public policy, this agreement shall be deemed amended to delete the term or provision so held to be invalid or unenforceable therefrom and all other remaining terms and provisions of this agreement shall remain in full force and effect. Provided, however, if the invalid or unenforceable provision contains a material term or condition of this Agreement then either party shall have the right to terminate upon five (5) days prior written notice following the determination of such invalidity or unenforceability. If any provision is inapplicable to any circumstance, it shall nevertheless remain applicable to all other circumstances.

15. NOTICE

15.1 Any notice or statement by any party shall be deemed to be sufficiently given when sent by receipted facsimile with a copy by prepaid, trackable overnight delivery, to the notified party at its address set forth above and to its counsel. These addresses shall remain in effect unless another address is substituted by written notice.

[Add Names and Addresses]

In witness whereof, the parties hereto have caused this agreement to be signed, sealed and delivered on the date indicated above.

ConglomoMed, Inc.

____________________	__________________________
Date	BY: Name: Title:

Medicus Societas, Inc.

____________________	__________________________
Date	BY: Name:

Appendix C
Sample Assignment

ASSIGNMENT

[LAST NAME, FIRST NAME], residing at ______________________________, a citizen of the United States (hereafter the undersigned), is the inventor of ______________________________ for which the undersigned executed an application for United States Letters Patent, U.S. Patent Application No. ________________, filed __________________, 20__.

The undersigned hereby authorizes assignee or assignee's representative to insert the Application Number and the filing date of this application if they are unknown at the time of execution of this assignment.

Medicus Societas, Inc., a Pennsylvania Corporation, having a place of business at ____, (hereafter referred to as the assignee), is desirous of acquiring the entire right, title and interest in said invention, all applications for and all letters patent issued on said invention.

For good and valuable consideration, the receipt and sufficiency of which is acknowledged, the undersigned, intending to be legally bound, does hereby sell, assign and transfer to the assignee and assignee's successors, assigns and legal representatives the entire right, title and interest in said invention and all patent applications thereon, including, but not limited to, the application for United States Letters Patent entitled as above, and all divisions and continuations thereof, and in all letters patent, including all reissues and reexaminations thereof, throughout the world, including the right to claim priority under the Paris Convention or other treaty.

It is agreed that the undersigned shall be legally bound, upon request of the assignee, to supply all information and evidence relating to the making and practice of said invention, to testify in any legal proceeding relating thereto, to execute all instruments proper to patent the invention throughout the world for the benefit of the assignee, and to execute all instruments proper to carry out the intent of this instrument.

The undersigned warrants that the rights and property herein conveyed are free and clear of any encumbrance.

EXECUTED under seal on this ___ day of __________________________, 20__ at

__________________________.

(Place)

Witness:

__________________________ ______________(L.S.)

First and Last Name

G. B. Halt Jr. et al., *FDA and Intellectual Property Strategies for Medical Device Technologies*, https://doi.org/10.1007/978-3-032-12649-8

State of

ss.

County of

On this _____ day of ______________, 20__ before me personally appeared [NAME], to me known to be the person described herein and who executed the foregoing instrument, and acknowledged that he executed the same knowingly and willingly and for the purposes therein contained.

Witness my hand and Notarial seal the day and year immediately above written.

____

Notary Public

My Commission Expires:

Appendix D
Sample Confidential Disclosure Agreement

CONFIDENTIAL DISCLOSURE AGREEMENT

This Agreement made by and between MEDICUS SOCIETAS, having an address of ________________, and CONGLOMOMED, having an address of______________________.

WITNESSETH THAT:

WHEREAS, MEDICUS SOCIETAS has developed a new semiconductor product, (hereinafter referred to as "the invention"), and is in possession of certain related confidential and proprietary information (hereinafter referred to as "proprietary information");

WHEREAS, MEDICUS SOCIETAS is interested in disclosing the invention and proprietary information to CONGLOMOMED, in confidence, for further evaluation for product production and distribution rights; and

WHEREAS, CONGLOMOMED is interested in receiving such information, in confidence, for conducting such further evaluation.

NOW, THEREFORE, for and in consideration of the foregoing premises, and of the mutual promises set forth below and for other good and valuable consideration, the receipt and sufficiency of which is hereby acknowledged, the parties, intending to be legally bound, hereby agree as follow:

1. MEDICUS SOCIETAS agrees to disclose to CONGLOMOMED in confidence proprietary information relating to the invention.

2. CONGLOMOMED agrees to accept and hold in confidence any and all proprietary information disclosed by MEDICUS SOCIETAS under this Agreement, except:

(a) information which at the time of disclosure can be shown to have been in the general public knowledge;

(b) information which, after disclosure, becomes part of the public knowledge by publication or otherwise, except through breach of this Agreement by CONGLOMOMED;

(c) information which CONGLOMOMED can establish by competent proof was in its possession at the time of disclosure by MEDICUS SOCIETAS and was not acquired, directly or indirectly, from MEDICUS SOCIETAS; and

(d) information which CONGLOMOMED receives without restriction from a third party, provided that such information was not obtained by said third party, directly or indirectly from MEDICUS SOCIETAS.

3. CONGLOMOMED agrees that the proprietary information received from MEDICUS SOCIETAS shall not be used by CONGLOMOMED, other than for evaluation and consideration for use as noted above and as otherwise agreed between MEDICUS SOCIETAS and CONGLOMOMED.

G. B. Halt Jr. et al., *FDA and Intellectual Property Strategies for Medical Device Technologies*, https://doi.org/10.1007/978-3-032-12649-8

4. This Agreement shall not be construed as granting any license or any other rights to CONGLOMOMED.

5. CONGLOMOMED agrees to restrict access to the proprietary information to those employees, agents and representatives who are engaged in that actual evaluation of the invention on a need-to-know basis and will require all such employees, agents and representatives to agree to maintain the proprietary information in confidence.

6. CONGLOMOMED agrees that any improvements or modifications developed by CONGLOMOMED in connection with the invention shall belong to MEDICUS SOCIETAS, and CONGLOMOMED shall assign all rights in any such improvements or modifications to MEDICUS SOCIETAS.

7. Upon completion of its evaluation, CONGLOMOMED agrees to return to MEDICUS SOCIETAS all information concerning the invention, including all photographs, diagrams, drawings, descriptions, prototypes, and notes, and any copies thereof.

This Agreement shall be binding upon and shall inure to the benefit of and be enforceable by and against the respective heirs, legal representatives, successors, assigns, subsidiaries, and affiliated or controlled companies of the parties hereto.

This Agreement shall be construed, interpreted and applied in accordance with the law of the State of _____________. With respect to the subject matter of this Confidential Disclosure Agreement, the foregoing constitutes the entire and only understanding between the parties, and this Confidential Disclosure Agreement supersedes any prior or collateral agreements or understandings between the parties with respect to confidentiality.

IN WITNESS WHEREOF, the parties hereto have caused this Agreement to be executed as of the last date written below.

CONGLOMOMED

Date:_____________ By: __________________

Name:

Title:

MEDICUS SOCIETAS

Date:_____________ By: __________________

Name:

Title:

Index

G. B. Halt Jr. et al., *FDA and Intellectual Property Strategies for Medical Device Technologies*, https://doi.org/10.1007/978-3-032-12649-8

MIX
Papier aus verantwortungsvollen Quellen
Paper from responsible sources
FSC® C105338

If you have any concerns about our products,
you can contact us on
ProductSafety@springernature.com

In case Publisher is established outside the EU,
the EU authorized representative is:
Springer Nature Customer Service Center GmbH
Europaplatz 3, 69115 Heidelberg, Germany

Printed by Libri Plureos GmbH
in Hamburg, Germany